Nondestructive
Evaluation
of Materials

SAGAMORE ARMY MATERIALS
RESEARCH CONFERENCE PROCEEDINGS

Available from Plenum Press

9th: **Fundamentals of Deformation Processing**
Edited by Walter A. Backofen, John J. Burke, Louis F. Coffin, Jr., Norman L. Reed, and Volker Weiss

10th: **Fatigue: An Interdisciplinary Approach**
Edited by John J. Burke, Norman L. Reed, and Volker Weiss

12th: **Strengthening Mechanisms: Metals and Ceramics**
Edited by John J. Burke, Norman L. Reed, and Volker Weiss

13th: **Surfaces and Interfaces I: Chemical and Physical Characteristics**
Edited by John J. Burke, Norman L. Reed, and Volker Weiss

14th: **Surfaces and Interfaces II: Physical and Mechanical Properties**
Edited by John J. Burke, Norman L. Reed, and Volker Weiss

15th: **Ultrafine-Grain Ceramics**
Edited by John J. Burke, Norman L. Reed, and Volker Weiss

16th: **Ultrafine-Grain Metals**
Edited by John J. Burke and Volker Weiss

17th: **Shock Waves**
Edited by John J. Burke and Volker Weiss

18th: **Powder Metallurgy for High-Performance Applications**
Edited by John J. Burke and Volker Weiss

19th: **Block and Graft Copolymers**
Edited by John J. Burke and Volker Weiss

20th: **Characterization of Materials in Research: Ceramics and Polymers**
Edited by John J. Burke and Volker Weiss

21st: **Advances in Deformation Processing**
Edited by John J. Burke and Volker Weiss

22nd: **Application of Fracture Mechanics to Design**
Edited by John J. Burke and Volker Weiss

23rd: **Nondestructive Evaluation of Materials**
Edited by John J. Burke and Volker Weiss

Forthcoming Volume

24th: **Risk and Failure Analysis for Improved Performance and Reliability**
Edited by John J. Burke and Volker Weiss

Nondestructive Evaluation of Materials

Edited by

John J. Burke

Army Materials and Mechanics Research Center
Watertown, Massachusetts

and

Volker Weiss

Syracuse University
Syracuse, New York

PLENUM PRESS · NEW YORK AND LONDON

Library of Congress Cataloging in Publication Data

Sagamore Army Materials Research Conference, 23d, Raquette Lake, N.Y., 1976.
  Nondestructive evaluation of materials.

  (Sagamore Army Materials Research Conference Proceedings: 23.)
  Includes index.
  1. Non-destructive testing—Congresses. 2. Munitions—Materials—Testing—Congresses. I. Burke, John J. II. Weiss, Volker, 1930-   III. Title. IV. Series: Sagamore Army Materials Research Conference. Proceedings; 23.
  UF526.3.S3 no. 23 [TA417.2] 623'.028s [620.1'127]
  ISBN 0-306-40185-1   79-12538

Proceedings of the Twenty-third Sagamore Army Materials Research Conference
on the Nondestructive Characterization of Materials, held at the Sagamore
Conference Center, Raquette Lake, New York, August, 1976

© 1979 Plenum Press, New York
A Division of Plenum Publishing Corporation
227 West 17th Street, New York, N.Y. 10011

SAGAMORE CONFERENCE COMMITTEE

Chairman
JOHN J. BURKE
Army Materials and Mechanics Research Center

Program Director
VOLKER WEISS
Syracuse University

Secretary
ARAM TARPINIAN
Army Materials and Mechanics Research Center

Conference Coordinator
JOSEPH A. BERNIER
Army Materials and Mechanics Research Center

PROGRAM COMMITTEE

JOHN J. BURKE
Army Materials and Mechanics Research Center

M. J. BUCKLEY
Air Force Materials Laboratory

G. DARCY
Army Materials and Mechanics Research Center

H. HERGLOTZ
E. I. du Pont de Nemours & Company

GEORGE MAYER
Army Research Office

R. B. THOMPSON
Rockwell International Science Center

VOLKER WEISS
Syracuse University

S. WEISSMANN
Rutgers University

Arrangements at Sagamore Conference Center
Helen B. DeMascio
Syracuse University

Preface

The Army Materials and Mechanics Research Center of Watertown, Massachusetts in cooperation with the Materials Science
Group of the Department of Chemical Engineering and Materials
Science of Syracuse University has conducted the Sagamore Army
Materials Research Conference since 1954. The main purpose of
these conferences has been to gather together over 150 scientists
and engineers from academic institutions, industry and government
who are uniquely qualified to explore in depth a subject of
importance to the Department of Defense, the Army and the scientific
community.

This volume NONDESTRUCTIVE EVALUATION OF MATERIALS,
addresses the areas of x-ray, ultrasonics and other methods of
nondestructive testing.

We wish to acknowledge the dedicated assistance of Joseph
M. Bernier of the Army Materials and Mechanics Research Center
and Helen Brown DeMascio of Syracuse University throughout the
stages of the conference planning and finally the publication
of this book. Their help is deeply appreciated.

Syracuse University
Syracuse, New York   The Editors

Contents

SESSION I
X–RAY
S. Weissman, Moderator
H. K. Herglotz, Moderator

1. Overview of X-Ray Diffraction Methods for
   Nondestructive Testing 1
       L. V. Azároff

2. Detection of Fatigue Damage by X-Rays 21
       S. Taira and K. Kamachi

3. A Historical Example of Fatigue Damage 55
       H. K. Herglotz

4. The Application of X-Ray Topography to Materials
   Science 69
       S. Weissman

5. Residual Stress Measurement by X-Rays: Errors,
   Limitations and Applications 101
       C. O. Ruud and G. D. Farmer

6. Flash X-Ray 117
       Q. Johnson and D. Pellinen

7. Neutron Radiography for Nondestructive Testing . . . 151
       J. John

8. Nondestructive Compositional Analysis 183
       L. S. Birks

9. Review of European Advances in Nondestructive
   XRD Testing 195
       E. Macherauch

10. Polychromatic X-Ray Stress Analysis and Its
    Application (Nondestructive Distribution
    Measurement of Sub-Surface Physical
    Quantities)   221
        M. Nagao and S. Kusumoto

                        SESSION II
                          SONIC
                    T. Moran, Moderator
                 R. B. Thompson, Moderator

11. Overview of Ultrasonic NDE Research   257
        R. B. Thompson

12. Electromagnetic Ultrasonic Transducers   283
        T. J. Moran

13. Ultrasonic Spectroscopy   299
        O. R. Gericke

14. Acoustic Interactions with Internal Stresses in
    Metals   321
        O. Buck and R. B. Thompson

                        SESSION III
                     OPTICAL METHODS
                    G. Darcy, Moderator

15. Optical Probing of Acoustic Emission Waves   347
        C. H. Palmer and R. E. Green, Jr.

16. Recognition and Analysis of Distributed Image
    Information   379
        M. G. Dreyfus

17. Moire Method - Its Potential Application to
    NDT   385
        A. S. Kuo and H. W. Liu

18. Image Processing in Nondestructive Testing   409
        D. H. Janney

19. Automated Detection of Cavities Present in the
    High Explosive Filler of Artillery Shells   421
        R. P. Kruger, D. H. Janney and
        J. R. Breedlove, Jr.

SESSION IV
OTHER METHODS
G. Mayer, Moderator

20. Advanced Quantitative Magnetic Nondestructive
    Evaluation Methods - Theory and Experiment   435
        J. R. Barton, F. N. Kusenberger,
        R. E. Beissner and G. A. Matzkanin

21. Positrons as a Nondestructive Probe of Damage
    in Structural Materials   479
        J. Wilkenfeld and J. John

22. Quantitative Methods in Penetrant Inspection   505
        P. F. Packman, J. K. Malpani and
        G. Hardy

Index   525

Chapter 1

OVERVIEW OF X-RAY DIFFRACTION METHODS FOR NONDESTRUCTIVE TESTING

L. V. Azároff

University of Connecticut
Storrs, Connecticut

ABSTRACT

Following a brief historical review, applications of x-ray diffraction to the nondestructive examination of materials are considered. Recent instrumental developments, including x-ray sources, monochromators, and detectors, and their utilization in experimental procedures are discussed. Applications of "routine" x-ray diffraction procedures as well as specialized techniques like residual stress analysis, small-angle scattering, and x-ray topography are illustrated by examples, including biological, metallic, and polymeric materials. The conclusion reached is that the future of x-ray diffraction in the study of materials has been made brighter than ever by the introduction of more intense x-ray sources and continuously improving instrumentation.

HISTORICAL INTRODUCTION

The discovery that crystals can serve as three-dimensional gratings for the diffraction of x-rays was made by Max Laue in 1912. The importance of this discovery was solemnized within two years by the award of a Nobel Prize to Laue in 1914. That very same year, two papers were published describing, respectively, what has become known as the kinematical theory [1] and the dynamical theory [2] of x-ray diffraction. So comprehensive were these analyses that their author, C. G. Darwin, was able to show eight years later [3] that an abnormal absorption effect that was troubling W. H. Bragg and his co-workers was explicable in terms of previously derived dynamical interactions of incident and scattered x-rays -- an effect Darwin dubbed extinction.

1

Working independently, in a country that was soon to be separated from England by the trenches of World War I, a young German physicist was developing for his doctoral dissertation a rather daring treatment of the interaction of electromagnetic radiation with a three-dimensionally periodic array of electric dipoles. When he learned of Laue's discovery, P. P. Ewald realized that his analyses could be extended to include x radiation [4] and his formalistically more elegant version of the dynamical theory was published in 1917. Thus, between them, the two theoretical physicists living on opposite shores of the English Channel laid the theoretical foundation for x-ray diffraction that has endured to the present day. The kinematical and dynamical theories have been rederived, reformulated, extended, and specialized, but neither their comprehensiveness nor their rigor has required significant amendment.

The consequence of these early successes has been that the field of x-ray crystallography that grew out of them has been concerned primarily with applications of x-ray diffraction theory rather than with fundamental discoveries in x-ray physics. This is not said to minimize the enormous contributions that x-ray diffraction has made directly to the advancement of the sciences of biology, chemistry, metallurgy, physics, etc., or indirectly to the creation of whole new disciplies like materials science, or new technologies like solid-state electronics, high-speed computers, laser telecommunication, and many, many more. The applied nature of x-ray diffraction is emphasized in order to help place in the proper perspective the concerns of current practitioners [5] about an apparent decline of academic and governmental support for research in x-ray crystallography. It also sets the stage for what follows; namely, a citation of some applications of x-ray diffraction to urgent problems of our day with primary emphasis on what might be properly called the nondestructive testing of materials.

KINEMATICAL THEORY

Present at the University of Numich during those fecund early years also was Peter Debye, who began to examine how x-rays interact with noncrystalline, i.e., aperiodic matter. By 1915, Debye had demonstrated that scattering by any collection of atoms, even in a gas, gave rise to discernible interference effects. He suggested the terms internal interference effect for that produced within cohesive atomic groupings, e.g., molecules in a gas or liquid, and external interference effect to describe possible interactions between x-rays scattered by such groupings. In gases, the external interference effect is dominant and causes additional scattering intensity at small angles close to the direct beam. In liquids and amorphous or glassy materials, internal interference gains

importance in direct proportion to the short-range order present.
Once such ordering has reached crystalline proportions, a polycrystalline aggregate results and, as is well known, this was
recognized before 1917 by Debye and Scherrer [6].

The presence of three-dimensional periodicity in crystals
limits the constructuve external interference effect to discrete
angles determined by the dimensions and shape of the lattice.   The
internal interference effects due to the specific atomic array
within each unit cell determine the intensities diffracted at these
angles.   The formulation of x-ray intensities in terms of this
crystal structure is that given by Darwin in 1914 and it includes
a complex quantity called the "phase" of the diffracted beam that
is "lost" when the real intensity is measured experimentally.
Although some fifty years had to pass before effective direct
methods for deducing the missing phases were developed, increasingly
more complicated crystal structures were being determined and they
have had an inordinate impact on biological and physical sciences
generally.   The sophistication of experimental equipment available
for measuring accurate x-ray intensities and the application of high-
speed computers to crystallographic problems enable electron density
distributions to be determined nowadays to within a tenth of one
electron per $Å^3$.   The importance of this to establishing chemical
valences, dipole moments, and other quantities critical to quantum
chemistry is no less than the early structure determinations were
to the development of modern-day crystal chemistry and mineralogy.
Similarly, precise measurement of bond distances and angles is
critical for many other basic studies.   It is noteworthy that, from
time to time, Nobel Prizes are still being awarded to scientists
basing their discoveries on x-ray crystallographic methods.

But precise electron densities are qually as important in the
development of new materials for solid-state electronics or new
pharmaceuticals for combating disease.   In fact, one can find
numerous examples of the application of perfectly "routine" x-ray
diffraction methods that have far-reaching consequences.   Starting
in 1935, G. L. Clark and collaborators used quantitative x-ray
diffraction analysis to demonstrate that crystalline quartz rather
than other forms of SiO_2 were responsible for silicosis contracted
by miners, stone cutters and others [7].   They also were able to
demonstrate that the addition of as little as one percent of powdered
aluminum was effective in neutralizing the pathogenic effect of
quartz.   Some forty years later, environmental quality has become
a national issue, both medically and politically.   It is essential,
in this connection, to establish crystallogrphically rather than
just chemically which toxic constituents are present in the environment.   In a series of studies only recently addressing themselves
to matters of specificity, it has been demonstrated, for example,
that Ni_3S_2 but not NiS or pure nickel powder is highly carcinogenic
[8].   Since nickel and sulfur are ubiquitously present in many

industrial processes as well as in domestic and automobile fuels,
the timeliness of systematic research in this area is self-evident.

DYNAMICAL THEORY

The burgeoning successes of the kinematical theory in determining crystal structures and in dealing with most other studies of
the crystalline state rapidly caused it to eclipse in importance the
dynamical theories developed at the same time. It took two totally
unrelated events to bring the dynamical theory back from virtual
obscurity. One was the discovery of the transistor and the subsequent high degree of interest in relatively perfect crystals. The
other was a chance discovery in 1951 by H. N. Campbell [9] that
large single crystals may exhibit an anomalous increase in the
transmitted (direct) beam intensity when set at the Bragg angle for
a set of transverse planes. The same effect actually had been discovered by Borrmann ten years earlier in Germany [10] but, again, a
war had served to sever communications within the scientific
community.

The anomalous transmission effect attracted a great deal of
theoretical as well as experimental interest at once. Theorists
quickly discovered that the rudiments were already included in the
original dynamical theories [11]. Experimentalists were attracted
by the fact that the effect manifests itself primarily in very
perfect crystals. This opened up extensive possibilities for
studying crystal perfection generally as well as the different
kinds of imperfections that crystals can contain [12-16]. In the
Laue or transmission arrangement shown in Figure 1, when a parallel
monochromatic beam (S_0) is incident on a set of transverse planes

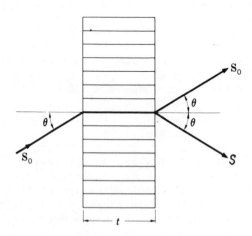

Fig. 1.   None

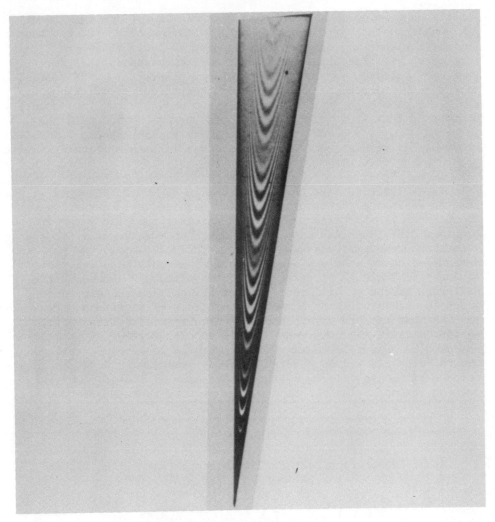

Figure 2.   Section topograph of a perfect silicon crystal showing
            variations in diffracted intensity (recorded by
            S. Kajimura).

at the appropriate diffraction angle θ, it actually propagates along
the planes and emerges at the opposite side as two beams: the
forward-diffracted beam (S_0) and the normally-diffracted beam (S).
Actually, four separate standing fields are set up within the crystal
under such conditions and the relative intensities of the normally
and forward-diffracted beams depend on the crystal thickness, $\underline{t}$,
its shape, and its perfection. If the crystal is wedge-shaped, for
example, the oscillating energy flow between these four subfields
produces a set of alternately high- and low-intensity bands called

pendellösung fringes, as can be seen in Figure 2. The details of how such topographs are prepared and how they can be used to observe any defects that may be present have been reviewed elsewhere [12-16]. Some of the most recent applications are discussed further by Professor Weissman in the present volume (Chapter 4). The combination of experimental measurements and theoretical calculations has reached such a high degree that it is now possible to draw intensity contours directly on teletype printouts for comparison with the recorded x-ray topographs [17]. Such comparisons make it possible, for example, to distinguish extrinsic from intrinsic stacking faults present in a crystal.

X-ray fringes, like those recorded in Figure 2, also can be used to determine the structure-factor magnitudes (diffraction amplitudes) of the reflecting planes directly from the inter-fringe spacings [18]. By combining such measurements with x-ray interferometer measurements, discussed below, structure factors can be determined on an absolute scale with a remarkably high accuracy [19]. As shown in Table 1, structure-factor magnitudes measured by two different groups on different silicon crystals agreed with each other to within 0.1%, or more than one order of magnitude better than the bes measurements possible with conventional (rocking-curve) methods. Quite recently, the "forbidden" 222 reflection of silicon was used to produce a section topograph of a wedge-shaped silicon crystal with the aid of a rotating-anode tube [20]. From the inter-ference-fringe spackings, a structure-factor of $1.64_5 \pm 0.03_4$ was deduced, exemplifying the kind of sensitivity that is afforded by

TABLE 1

Atomic Scattering Factors Determined by the Fringe Method
for Silicon at 20°C   [11]

h k l	T.K.*	A.H.**
111	10.66_4	10.66_5
220	8.46_3	8.43_6
3̇33	5.84_3	5.83_0
440	5.40_8	5.38_8
444	4.17_2	4.17_7

* Measured by S. Tanemura and N. Kato

** Measured by E. Aldred and M. Hart

this method. Regrettably, at present it is limited to crystals that can be grown to an appropriate size without loss of intrinsic perfection.

When the possible beam paths of a perfect crystal set to transmit x-rays at an appropriate diffraction angle, Figure 1, are examined, it quickly becomes clear that a second crystal, similarly placed to intercept the forward diffracted and the normally-diffracted beams will diffract each of these backward so that they must intersect each other. Placing a third crystal at this intersection then completes an x-ray interferometer as shown in Figure 3. The first to build an operating instrument, Bonse and Hart had the genius to construct it from a single crystal of silicon by cutting out the undesired parts of the crystal, thereby assuring correct alignment and positioning of all three functioning components [21]. Such a monolithic interferometer operates just like its optical analog so that it is possible to produce Moiré fringes in the exit beams due to the overlap of the standing wave patterns set up in the intersecting beams just ahead of the analyzer crystal. Deformation of the analyzing crystal then changes the Moiré pattern in predictable ways thus enabling the direct imaging of any defects present. Of even greater interest is the possibility of establishing a length scale for submultiples of the wavelength of light, on an absolute scale, and without recourse to characteristic x-ray emission spectra. Experiments to this end currently are in progress in the U.S., U.K., and Germany. Preliminary results at the National Bureau of Standards indicate error limits of 0.25 ppm.   The details

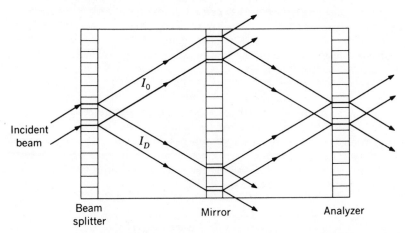

Figure 3. X-ray interferometer first proposed by Bonse and Hart [21] The forward-diffracted beam, $\underline{I}_0$, and the normally-diffracted beam, $\underline{I}_D$, are again effectingly "reflected" back by the central crystal (mirror) and made to intersect with each other prior to entering the analyzer crystal.

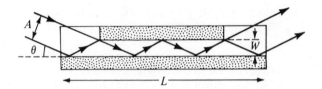

Figure 4. Bonse-Hart multiple reflection monochromator.

of this work as well as some provocative suggestions for future
applications of x-ray interferometers are contained in a recent
review of this subject [22].

APPLICATIONS OF X-RAY DIFFRACTION

X-Ray Monochromators

The revival of interest in the physics of x-ray diffraction
in the second half of the twentieth century coincided with the
development of highly perfect single crystals, as already noted
above. One of its practical outgrowths was the realization by
Bonse and Hart in 1965 that several successive reflections by a
perfect crystal could be used to produce a highly parallel monochromatic x-ray beam without significant loss in x-ray intensity
[23]. Such a monochromator can be manufactured, for example, by
cutting a parallel-sided channel in a single crystal of silicon
or germanium (Figure 4). If the intensity following a single
reflection at an angle $\theta+\Delta\theta$ is reduced by a factor $\underline{R}$ relative to
the intensity at θ, then $\underline{n}$ such reflections will reduce the reflec-
tion at $\theta+\Delta\theta$ by $\underline{R}^{\underline{n}}$. For $\underline{Cu}$ $\underline{K\alpha}$ radiation, for example, $\underline{R}$ at θ is
about 95% so that, for $\underline{n}$ = 5, $\underline{R}^{\underline{n}}$ is reduced by only 23% at θ,
whereas only 4 seconds of arc away, the tails of the reflection
are reduced by a factor of 10^8. As illustrated in Figure 5, the
halfwidth of the reflection curve remains approximately the same,
whereas the great enhancement of the peak-to-background ratio
makes this kind of monochromator extremely useful in applications
requiring measurements or x-rays scattered close to the direct
beam. Several modifications of this monochromator design have
been proposed subsequently and a full discussion of them can be
found in Chapter 7 of Reference [11].

Small-Angle Scattering

It is self-evident that the multiple reflection monochromators
are particularly well suited for small-angle scattering experiments.
Realizing that for certain applications the x-ray beam divergence
(cross fire) should be limited in more than just one plane, Bonse

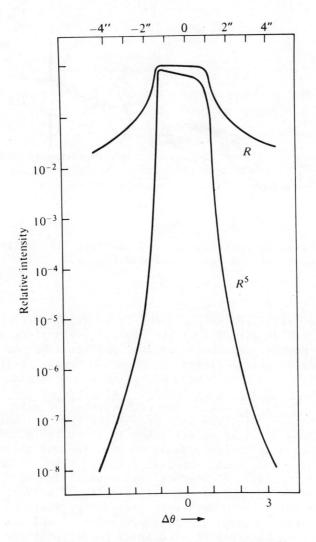

Figure 5.   Transmitted intensities of a singly-reflected and fivefold-
             reflected x-ray beam.   [23]

and Hart suggested the arrangement shown in Figure 6. A third
single crystal set to reflect from planes that are perpendicular to
those in the two normally positioned grooved crystals, has the effect
of reducing the "vertical slit height" effectively to zero; all this
without a very significant decrease in the transmitted intensity.

As is well known, small-angle scattering effects can be observed

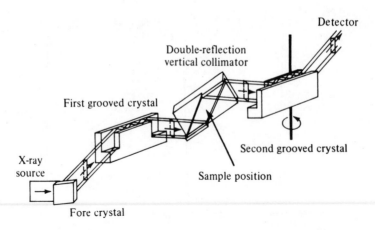

Figure 6.   Possible three-crystal arrangement for limiting both
            vertical and horizontal divergence.   [24]

whenever the irradiated sample consists of small particles or regions
having electron densities that are distinctively different from the
surrounding matrix. As the individual particles become progressively
smaller, they cause all diffracted beams to become progressively
broader. Since the direct-beam direction also contains the zero-
order diffracted beam, it is similarly affected by particle-size
broadening. In fact, because the pathlengths for different parts
of the scattering units are very nearly the same at small deviations
from $\theta = 0°$, any internal interference is virtually eliminated so
that the broadening effect depends on the size and shape of the
particles and not on their internal detailed electron density
(crystal structures). This means that crystalline and noncrystalline
materials can be studied with equal facility. In fact, because
double diffraction by fortuitously parallel crystals can mask or,
at the very least, hinder the interpretation of small-angle effects,
some of the more interesting studies of polycrystalline materials
have fallen out of favor.

     Thus the application of small-angle scattering has shifted
from metallurgical problems [25] to the study of surfaces [26],
polymers [27], and other macromolecules [28]. The method is
undergoing a renaissance of interest because it brings unique
measuring abilities to otherwise intractable problems. For example,
gas adsorption methods can be used to measure specific areas of
colloidal-size aggregates unless they are so densely packed that
the gas does not permeate the sample uniformly. X-rays, of course,
encounter no such penetration resistance. Unlike electron microscopes, which require the sample to be in a vacuum, x-ray scattering

can be conducted in a totally nondestructive manner, even on proteins
retained in original solutions. Moreover, it can provide information
on their size, shape, and degree of swelling, or on how these parameters are affected by changes in temperature, pH, or concentration.

The large repeat spacings in macromolecules also cause ordinary
coherent diffraction of x-rays to occur at very small angles.
Although these diffraction angles can be increased by using longer
wavelength radiation, the loss in x-ray intensity sets practical
limits. Here, the ready availability of high-intensity, rotating-
anode generators has been influential in increasing the popularlity
of small-angle studies. The considerably more intense x-ray beams
emitted by synchrotrons hold out even more promise for the future.
An interesting innovation for conventional x-ray sources was described by H. K. Herglotz at the 20th Sagamore Army Materials Research
Conference [29]. X-rays having two different wavelengths, produced
in the same bimetallic anode, are incident on a polymer. If their
wavelengths are suitably chosen, e.g., Cu $K\alpha$ (λ=1.54Å) and Al $K\alpha$
(λ=8.34Å), the shorter wavelength x-rays produce a "large-angle"
pattern while the longer wavelength rays contribute to small-angle
scattering only.

X-Ray Detectors

Advances in solid-state electronics, made possible by the
fruits of x-ray diffraction studies, have had a reciprocating
effect by increasing the sensitivity and accuracy of x-ray detection.
With the advent of Li-drifted germanium (or silicon) detectors, it
has become possible to substitute energy (wavelength) discrimination
for angular discrimination in x-ray diffractometry [30]. Not only
is the required equipment mechanically simpler (the detector, however,
must be maintained at all times at cryogenic temperatures), but it
is considerably faster, since the time-consuming angular scanning is
replaced by electronic scanning of the spectrum diffracted at a
single setting of the detector. Intensity measurement requires a
prior analysis of the continuous spectrum emitted by the x-ray
source but this is straightforward and makes use of the same multi-
channel analyzer. Positional measurement of reflections, however,
can be made rapidly and precisely; accuracies of 0.01% have been
reported for routine interplanar spacing measurements [31].

Energy-scanning diffractometers offer many experimental advantages, not the least of which is the minimization of instrumental
instabilities during the shortened recording periods. Thus, small-
angle scattering from polyethylene can be recorded in about 100 sec
with good correlation to conventional and much longer scanning
methods [32]. In the case of proteins that are radiation damaged
or in studies at cryogen temperatures, or other extreme conditions,
this added speed becomes even more important. The elimination of

angular scanning is particularly desirable in the study of scattering
from liquids. Geometrical correction of measured x-ray intensities
is greatly simplified because all data are recorded at a single
setting of the detector and the reduced measuring times enable
higher statistical precision in intensity measurement. In a recent
study of the structure of mercury, the chief difficulty was interference from characteristic emissions from the x-ray tube and from
the sample studied [33]. Nevertheless, the pair-correlation function (liquid structure) obtained compared very well with one
conventionally determined.

In concluding this discussion, it should be noted that combinations of angle- and energy-scanning in diffractometry are also
possible [34] and offer intriguing possibilities. In this connection
it is interesting to consider reviving the original Laue arrangement
which has been relegated to symmetry (orientation) recognition only
because all orders of an hkl reflection are diffracted at the same
angle. Energy-scanning detectors would enable measurement of each
reflection intensity separately, thereby removing the degeneracy.
This, in turn, would enable space-group determinations and,
possibly, crystal-structure analyses to be carried out in the Laue
arrangement, i.e., by using the entire x-ray spectrum rather than
just one characteristic line.

Residual Stress Analysis

X-ray diffraction is an important tool in residual stress
analysis because it provides a uniquely nondestructive method that
is equally applicable to metals, ceramics, and polymers. Its
importance in the nondestructive characterization of materials can
be gauged also from the number of papers devoted to this topic at
the present conference. In fact, international symposia have been
regularly scheduled to examine this topic and 40 papers were presented at the last one convened [35]. But as in other applications
of x-ray diffraction, progress is being made not in the underlying
theoretical underpinnings but in developing better techniques of
execution and in finding novel applications to new areas - and the
progress being made is real. Not only are better calibration
procedures being discovered [36] and more reliable methods being
standardized [37], but significant advances have been made in the
instrumentation so that fully-automated portable diffractometers
are available for in-situ stress measurements with great ease and
reliability [38,39]. In fact, things have become so routine that
the computer locates the peaks, calculates the peak angles,
calculates the residual stress from the measured angles, and types
a report - all within about 10 minutes!

The reliability and versatility of x-ray diffraction methods
also continue receiving testimonials. Maso and Castex determined

the residual stresses responsible for buckling structural steel
elements using in-situ equipment. Comparison of preliminary
diffractometer measurements with results of strain gauges, hole-
drilling methods, and sectioning methods gave excellent agreements
[40]. Ferritic steels containing about 0.1% carbon, 1.6% Mn, and
less than 0.1% Cr, Ni, and Ti, have been shown to display an
inverse relationship between the presence of microstrains and their
resistance to shock loading [41]. When the thermal stress behavior
of tungsten-copper composites was studied by x-ray diffraction, it
could be shown that residual stresses found present at room temperature were produced by the thermal coefficient mismatch of the
componenets [42]. Moreover, the thermal stress behavior was found
to be nonlinear and to exhibit a hysterises loop.

Even the very small piezelectric strains ($\sim 10^{-5}$ to 10^{-7}) produced in a single-domain ferroelectric crystal, when an electric
field is applied, have been measured using a simple method of x-ray
differential photometry capable of strain resolution of 3×10^{-7}
[43]. Similarly, the small strains induced by lattice mismatches in
epitaxially grown crystals have been measured by Hart, who employed
a two-crystal diffractometer arrangement to improve the sensitivity
[44]. Finally, microbeam techniques have been employed by Taira,
Tanaka, and Ryn to study stress concentrations around the tip of a
sharp notch in plain carbon steels [45]. Since cleavage crack
propagation is a key to understanding the effect of fatigue crack
growth under external loading, such studies provide still another
example of how x-ray diffraction methods find application in a
variety of areas of materials science. Professor Taira will have
more to say on this application in the next chapter, but it should
be noted here that x-ray methods have not been limited to studying
fatigue failures in metals, but have also been applied to studying
the deformation of high-impact polystyrene prior to and during
the occurrence of crazing and fracture [46]. Comparisons with ESR
signals obtained from the samples established the number of chains
actually broken during fracture and promises to become a very
potent method for the study of fatigue failures in polymers.

CONCLUSION

After looking back on the first 30 years of small-angle x-ray
scattering in 1969, its chief protagonist, André Guinier, remarked:
"No major change can be foreseen unless the power of x-ray sources
is increased by a factor of 10 or 100, which is rather unlikely"
[47]. In the intervening seven years, improved rotating-anode
designs have provided tenfold intensity increases and the intensities from storage-ring synchrotrons exceed conventional sources
by many orders of magnitude [48]. As discussed further by
Dr. Johnson in Chapter 6, very high x-ray intensities also can be
generated during short bursts from plasmas. For example, a

Figure 7. 110 x-ray topographs of a BaTiO₃ crystal prepared with synchrotron radiation. The crystal in (a) was prepared at 25°C while the crystal in (b) was heated to 140°C, above the phase transition. Changes visible at intermediate temperatures and other interesting details are given in the original paper [52].

special camera has been designed to record the x-rays emitted by
a laser-induced iron plasma having a duration of about 50 psec [49].
It should not be forgotten, however, that the "effective" intensity
in x-ray diffraction can be increased by improving the sensitivity
of the detector as well as by beefing up the generator so that the
combined use of more intense sources and Li-drifted solid-state
detectors promises many exciting developments in the future.

Already Chikawa has found it possible to display an x-ray
section topograph on a video tube by using a rotating Mo target
operating at 60kV and 0.5 amp [50]. By moving both the crystal and
the TV camera while transmitting the signals to an image-storage
tube, transmission topographs examining approximately 13 x 9 mm of
the crystal were produced in 10 sec. Applications so far have
included observation of the melting and growing of silicon crystals
and the study of thermally produced slip bands attendant thereto.

Intensity gains up to 10^6 can be expected once dedicated
storage-ring synchrotrons become accessible. Meanwhile, Hart has
found that section topographs of silicon and LiF crystals could be
prepared with the same high resolution as those from the best conventional sources, but the use of synchrotron radiation with nuclear
emulsions required exposure times measured in seconds! The possibilities that such rapid examination by x-ray diffraction opens up are
mind boggling. X-ray topography could be used to follow phase
transformations, Figure 7, observe the migration of defects under
the action of thermal gradients, mechanical stresses, or other
agents. Small-angle scattering could be used to observe nucleation
and growth phenomena in alloys, or to follow the kinetics of poly-
merization and related phenomena under a variety of external
conditions. Even the two-wavelength technique proposed by Herglotz
takes on a new importance because synchrotrons emit a very broad
spectrum of radiation concurrently. In fact, the high intensity
and very high degree of polarization and collimation of synchrotron
radiation calls for a whole new approach to its optimal utilization.
Instead of seemingly running out of steam, I believe that the
utilization of x-ray diffraction is entering an era of exciting new
developments and challenging new applications.

REFERENCES

1. Darwin, C.G., "The Theory of X-Ray Reflection", Phil. Mag., 27 (1914) 315-33.

2. Darwin, C.G., "The Theory of X-Ray Reflection, Part II", Phil. Mag., 27 (1914) 675-90.

3. Darwin, C.G., "The Reflection of X-Rays from Imperfect Crystals", Phil. Mag., 43 (1922) 800-29.

4. Ewald, P.P., "Zur Begründung der Kristalloptik III. Röntgen-strahlen", Ann. Physik, 54 (1917) 519-97.

5. Status and Future Potential of Crystallography, Report of Conference sponsored by U.S.A. National Committee for Crystallography (National Research Council, Washington, 1976).

6. Debye, P. and P. Scherrer, "Interferenzen an regellos orientierten Teilchen im Röntgenlicht", Physik Z., 28 (1917) 290-301.

7. Clark, G.L., Applied X-Rays, 4th Ed., McGraw-Hill Book Co., Inc., New York, 1955, pp. 536-38.

8. Sunderman, F.W., Jr. and R.M. Maenza, "Comparisons of Carcino-genicities of Nickel Compounds in Rats", Res. Commun. Chem. Pathol. Pharm., 14 (1976) 319-30.

9. Campbell, H.N., "X-Ray Absorption in a Crystal Set at the Bragg Angle", J. Appl. Phys., 22 (1951) 1139-42.

10. Borrmann, G., "Über Extinktion der Röntgenstrahlen von Quartz", Physik Z., 42 (1941) 157-62.

11. Additional references and a complete discussion of the dynamical theory is presented in Chapters 3, 4, and 5 of X-Ray Diffraction by L.V. Azároff, R. Kaplow, N. Kato, R. J. Weiss, A.J.C. Wilson, and R. A. Young, McGraw-Hill Book Co., Inc., New York, 1974.

12. Azároff, L.V., "X-Ray Diffraction Studies of Crystal Perfection", Progr. Solid State Chem., 1 (1964) 347-79.

13. Bonse, U., M. Hart and J.B. Newkirk, "X-Ray Diffraction Topo-graphy", Advan. X-Ray Anal., 10 (1967) 1-8.

14. Austerman, S.B. and J.B. Newkirk, "Experimental Procedures in X-Ray Diffraction Topography", Advan. X-Ray Anal., 10 (1967) 134-52.

15. Amelincks (Ed.), Modern Diffraction and Imaging Techniques in Materials Science, North Holland Publ. Co., Amsterdam, 1970.

16. Isherwood, B.J. and CA.A. Wallace, "X-Ray Diffraction Studies of Defects in Crystals", Phys. Technol., 5 (1974) 244-58.

17. Authier, A., "X-Ray Topographic Determination of the Intrinsic
     and Extrinsic Nature of Stacking Faults", Phys. Status Solidi,
     A27 (1975) 213-22.

18. Kato, N. and A.R. Lang, "A Study of Pendellösung Fringes in
     X-Ray Diffraction", Acta. Crystl., 12 (1959) 787-94.

19. Tanemura, S. and N. Kato, "Absolute Measurement of Structure
     Factors of Si by Using Pendellösung and Interferometry Fringes",
     Acta Crystl., A28 (1972) 69-80.

20. Fehlman, M. and I. Fujimoto, "Pendellösung Measurement of the
     222 Reflection in Silicon", J. Phys. Soc. Japan, 38 (1975)
     208-15.

21. Bonse, U. and M. Hart, "An X-Ray Interferometer", Appl. Phys.
     Letters, 6 (1965) 155-56.

22. Hart, M., "Ten Years of X-Ray Interferometry", Proc. Roy. Soc.
     London, A346 (1975) 1-22.

23. Bonse, U. and M. Hart, "Tailless X-Ray Single-Crystal Reflection
     Curves Obtained by Multiple Reflection", Appl. Phys. Letters,
     7 (1965) 238-40.

24. Bonse, U. and M. Hart, "A New Tool for Small-Angle X-Ray
     Scattering and X-Ray Spectroscopy: The Multiple Reflection
     Diffractometer", in H. Brumberger (Ed.), Proc. Conf. X-Ray
     Small-Angle Scatt., Gordon and Breach, Inc., New York, 1965.

25. Guinier, A. and G. Fournet, Small-Angle Scattering of X-Rays,
     John Wiley and Sons, Inc., New York, 1955.

26. Brumberger, H., "Determination of Specific Surfaces by Small-
     Angle X-Ray Scattering Methods", in H. von Olpen and W. Parrish
     (Eds.), X-Ray and Electron Methods of Analysis, Plenum Press,
     New York, 1968, pp. 76-85.

27. Alexander, L.E., X-Ray Diffraction Methods in Polymer Science,
     John Wiley and Sons, New York, 1969.

28. Pilz, I., "Small-Angle X-Ray Scattering", in S.L. Leach (Ed.)
     Physical Principles and Techniques of Protein Chemistry,
     Academic Press, 1973, pp. 141-243.

29. Herglotz, H.K., "Characterization of Polymers by Unconventional
     X-Ray Techniques", in J.J. Burke and V. Weiss (Eds.),
     Characterization of Materials in Research, Syracuse U. Press,
     Syracuse, NY, 1975, pp. 103-36.

30. Giessen, B.C. and G.E. Gordon, "X-Ray Diffraction: New High
     Speed Technique Based on X-Ray Spectroscopy", Science, 159
     (1968) 973-75.

31. Fukamachi, T., S. Hoyosa and O. Terasaki, "The Precision of
     the Interplanar Distances Measured by an Energy-Dispersive
     Diffractometer", J. Appl. Cryst., 6 (1973) 117-22.

32. Schultz, J.M. and T.C. Long, "Energy-Scanning Small-Angle X-Ray Scattering, Polyethylene", J. Mater. Sci., 10 (1975) 567-70.

33. Prober, J.M. and J.M. Schultz, "Liquid-Structure Analysis by Energy-Scanning X-Ray Diffraction: Mercury", J. Appl. Cryst., 8 (1975) 405-14.

34. Laine, E., I. Lachteenmaki and I. Hamalainen, "Si(Li) Semi-conductor Detector in Angle and Energy Dispersive X-ray Diffractometry", J. Phys., E7 (1974) 951-54.

35. Proceedings of 3rd International Symposium on X-ray Analytical Methods, Nice, France, Sept. 16-20, 1974.

36. Louzon, T.J. and T.H. Spencer, "X-Ray Diffraction Stress Measurements of Thin Films", Solid State Technol., 18 (1975) 25-28.

37. Shiraiwa, T., "Recent Progress in X-Ray Stress Measurement", in 1973 Symposium on Mechanical Behavior of Materials, Kyoto, Japan, Aug. 21-23, 1973. Soc. Mater. Sci., Kyoto, 1974, pp. 1-11.

38. Kawabe, Y., H. Okashita, R. Shimizu, K. Yasui, H. Sekiguchi and K. Morishita, "Shimadzu X-Ray Diffraction Stress Analyzer SMX-50", Shimadzu Reviews (Japan), 31 (1974) 45-56.

39. Barbier, R., "In Situ Measurement of Stresses by X-Ray Diffraction", Ref. [35], pp. 97-103.

40. Maso, J.C. and L. Castex, "X-Ray Diffraction Study of Residual Stresses in Mild Steel Construction Elements", Ref. [35], pp. 92-96.

41. Lysak, L.I. and L.O. Andruschik, "Phase Composition and Fine Structure in Shock Resistant Steel Used in Shipbuilding", Metallofizika (U.S.S.R.), 55 (1974) 31-35. (in Russian)

42. Takeda, H. and M. Morita, "X-Ray Measurement of Thermal Stress in Composites", J. Soc. Mater. Sci. Japan, 24 (1975) 35-40.

43. Lissalde, F., "X-Ray Determination of Piezoelectric Coefficients in Ferroelectric Crystals", Ref. [35], pp. 52-57.

44. Hart, M., "Measurement of Strain and Lattice Parameter in Epitaxial Layers", J. Appl. Cryst., 8 (1975) 42-44.

45. Taira, S., K. Tanaka and J.G. Ryn, "X-Ray Diffraction Approach to the Mechanics of Fatigue and Fracture in Metals", Mech. Res. Communic., 1 (1974) 161-66.

46. Nielsen, L.E., D.J. Dahm, P.A. Berger, V.S. Murty and J.L. Kardos, "Fracture Processes in Sytrene Polymers: SAXS and ESR Studies", J. Polym. Sci. Phys. Ed., 12 (1974) 1239-49.

47. Guinier, A., "30 Years of Small-Angle X-Ray Scattering", Phys. Today, 22 (1969) 25-30.

48. Madden, R.P., "Synchrotron Radiation and Applications", in L.V. Azároff (Ed.) X-Ray Spectroscopy, McGraw-Hill Book Co., Inc., New York, 1974, pp. 338-78.

49. McConaghy, C.F. and L.W. Coleman, "Picosecond X-Ray Streak Camera", Appl. Phys. Lett., 25 (1974) 268-70.

50. Chikawa, J., "Technique for the Video Display of X-Ray Topographic Images and Its Application to the Study of Crystal Growth", J. Cryst. Growth, 24-25 (1974) 61-68.

51. Hart, M., "Synchrotron Radiation - Its Application to High-Speed, High-Resolution X-Ray Diffraction Topography", J. Appl. Cryst., 8 (1975) 436-44.

52. Bordas, J., A.M. Glazer and H. Hauser, "The Use of Synchrotron Radiation for X-Ray Topography of Phase Transitions", Philos. Mag., 32 (1975) 471-89.

Chapter 2

DETECTION OF FATIGUE DAMAGE BY X-RAYS

S. Taira and K. Kamachi

Kyoto University and Yamaguchi University, Japan

ABSTRACT

X-ray diffraction is one of the means for investigating
the microscopic structure of crystalline materials. X-ray
diffraction is advantageous when it is applied to metal materials;
it responds very sensitively to changes in the metal's crystalline
structure. Another characteristic advantage of the x-ray diffraction approach is its nondestructive nature in the measurement of
crystalline material parameters, enabling us to observe the process
of mechanical phenomena of metals, such as fatigue.

The x-ray diffraction patterns obtained on a fatigued material
include a great deal of information covering the macroscopic and
microscopic characters consistent with the nature of the material.
Residual stresses measured by x-ray and x-ray diffraction line
broadening are the parameters easily obtained by conventional
x-ray diffraction techniques and are taken as the macroscopic and
submacroscopic material parameters whose changes are taken as the
conventional measure of fatigue damage. The microscopic material
parameters that are extracted mainly by means of the x-ray microbeam
technique are another measure of damage in fatigue. The basic
concept of defining macroscopic, submacroscopic and microscopic
material parameters that are related with nondestructive detection
of fatigue damage is discussed on the basis of engineering applications. Some examples of practical detection of fatigue damage are
presented.

INTRODUCTION

X-ray diffraction technique is known as one of the means to detect the microscopic structure of crystalline materials. The characteristic advantage of the x-ray diffraction approach for investigating the mechanical behavior of crystalline materials is the nondestructive nature of its experimental procedure. The change in microscopic structure of materials during mechanical phenomena such as fatigue can be detected throughout the process from the initial to the final stages of the phenomenon. The evaluation of crystalline material parameters supplies various information for interpreting the mechanisms of such a macroscopic mechanical behavior of materials [1-3].

X-ray diffraction has long been used for many purposes mainly in the field of fundamental studies, such as physics and chemistry, for investigating the crystallographic structure of materials. The x-ray diffraction techniques have been developed in order to observe the nature of materials by emphasizing a particular site in the crystallographic structure. A tremendous number of works concerned with the application of x-ray diffraction have been reported so far in the field of materials research mainly from a microscopic viewpoint. Nowadays we have abundant information on and knowledge of the microscopic structure of various sorts of crystalline materials discovered through the use of x-ray diffraction techniques.

Looking at the current status of materials research, we can see that the application of x-ray diffraction to the investigation of the mechanical behavior of materials such as fatigue is the tool to bridge works from the macroscopic viewpoint and microscopic studies, in that the x-ray diffraction process interprets the macroscopic mechanical properties in terms of the microscopic nature of materials. In the present paper, the philosophy of the application of x-ray diffraction techniques to studies of mechanical behavior of materials is briefly discussed, citing the case of the application to studies of fatigue of metallic materials.

In Japan, a group working on the x-ray approach was organized twenty years ago and has been conducting cooperative research work in the name of the X-ray Committee on the Study of Deformation and Fracture of Solids (simply called the X-ray Committee) of the Society for Materials Science, Japan [4]. Also, it should be mentioned that the efforts of the manufacturers of x-ray diffraction instruments in improving and developing new equipment to be used for engineering purposes in cooperation with the activities of the Committee has greatly enabled the wide distribution of and therefore a high potential for the x-ray approach in universities and various industries in Japan. It should be noted that the content

of the present chapter is very much associated with the results of
the collaborative works of the Committee.

X-RAY DIFFRACTION TECHNIQUES AND INSTRUMENTS
FOR USE IN THE STUDY OF THE MECHANICAL BEHAVIOR OF MATERIALS

There are several sorts of x-ray diffraction techniques which
can be applied to studies of the mechanical behavior of materials.
Each technique supplies information on the status of the crystal
lattice structure characteristic of that technique. The x-ray
diffraction technqiues that have been adopted as tools for the
present study are shown in Table i, exhibiting relevant material
parameters. The material parameters are classified into three
stages according to the range of domain that is related to the
parameter.

By x-ray stress measurement the mean value of crystal lattice
distortion existing in an area of irradiating x-rays of 2 to 10 mm^2
on the surface of materials may be obtained, by the precise
measurement of peak shifts in an x-ray diffraction profile. Thus,
the residual stress is counted as a macroscopic parameter of the
material.

TABLE 1.

Parameter	X-ray Technique	Technical Term
Macroscopic Parameter	—— X-ray Stress Measurement	———— Macroscopic Residual Stress
Submacroscopic Parameter	— ⌈Diffraction Profile for X-ray Stress Measurement … ⌊Diffraction Profile by Diffractometer	—— { Integral Breadth / Half-Value Breadth }—— of Diffraction Profile
Microscopic Parameter	— ⌈X-ray-microbeam Technique Diffraction Profile Analysis ⌊⌈Fourier Analysis / Integral-breadth Analysis⌋	—— { Subgrain Size / Microlattice Strain / Misorientation }——— Excess Dislocation Density — { Particle Size / Microstrain }— Dislocation Density

It is known that the diffusiveness in the diffraction profile corresponds to irregularities in the microstructure and the measurement of the integral breadth B is utilized as a submacroscopic parameter that estimates the approximate degree of these irregularities.

Wehn we use a very fine beam of x-rays 50 to 150μ in diameter (the x-ray microbeam technique), the resolving power is increased notably and is available for detecting changes in microscopic parameters such as sub-grain size t, microlattice strain $\Delta d/d$ and misorieintation β with respect to indivisual crystals in polycrystalline material. It is also useful for detecting the change of cyrstalline structure in a local area like the tip of propagating fatigue crack, as will be discribed later.

An ordinary type x-ray diffractometer is also used for microscopic study. It has strict high fidelity to the geometry of the x-ray diffraction and supplies an accurate profile of the object to be analyzed. The analysis of the profile gives information on other series of microscopic parameters such as particle size and microstrain at the area of irradiation. These parameters are derived from the diffraction theory. Although there is still room for further discussion of their physical interpretation, in this research they are important as material parameters which represent the average microscopic structure over an area from 1 to 10 mm^2.

As described above, the macroscopic, submacroscopic, and microscopic parameters are obtained mainly with these x-ray diffraction techniques as are shown in Table 1, and by appropriately combining these techniques in the study of the mechanical behavior of materials, the parameters that control the phenomena may be extracted to facilitate discussions on their mechanisms.

X-Ray Stress Measurement

The strain is determined by precisely measuring the shift of the peak of the x-ray diffraction profile. The stress is calculated by simple elastic formula using the values of measured strain. The diffraction profile is given by either the film-densitometer system or the counter system. Recent improvements in instrumentation have enabled notable advances in the accuracy of measuring stress by use of the counter system, and the new development of the x-ray stress analyzer enables the scanning of one profile of diffraction to finish in a few minutes. Thus the field of application of the x-ray stress measurement has been greatly enlarged in evaluating residual stresses in machine and structure components in industries.

Presently, the speciality in the application of the x-ray diffraction method for measuring macroscopic residual stress in a

wide variety of fields in universities and industries in Japan is
the adoption of the parallel beam method instead of the so-called
parafocusing method. The details of the advantage and disadvantage
of parallel beam geometry in x-ray stress measurement for engineering
purposes has been reported elsewhere. So far, the application of
x-ray stress measurement has been appreciated in the case of components of materials hardened by heat treatment and heavy work-
hardening in evaluating residual stress produced during the manufacturing process, since in the case of hardened materials we can
easily obtain a regular and continuous diffraction profile that
enables us to obtain a reliable and reproducible value of measured
stresses.

Traditional Wagon Type x-ray stress analysers are shown in
Figure 1 and examples of recent development of Portable Type x-ray
stress analysers are shown in Figure 2. For both modes of A and
B, the goniometer for parallel beam diffraction geometry is
installed in the measuring head, which is easily fixed at a part
of large scale structures. Oscillating beam mechanism facilitates
a reliable measurement of residual stress existing in coarse
grained materials such as welded part of large constructions. The
parallel beam principle is rather simple in diffraction geometry
as compared with the parafocusing principle and is not so sensitive
to the misfit of specimen surface. A diffraction profile can be
obtained that is sufficient for use in stress measurement even
though precision must be sacrificed.

Of course, photographic method is also widely used for the
measurement of local stress in the area usually of 1 mm in diameter.
Observation of the diffraction patterns on a film give us some
advantage to obtain important information of character of crystallographic structure on the surface such as the grain size, the
approximate extent of deformation of crystals, the quality of heat
treatment and also the extent of preferred orientation, besides
of use for the stress measurement. Direct observation of diffraction
patterms is one of the important methods of nondestructive detection
of fatigue damage for a person with experience in this field of
study.

Submacroscopic Parameter in Relation to Fatigue Process

The broadness of the line profile obtained in residual stress
measurement is related not to the macroscopic stress but to the
microscopic structure. The quantitative expression of the integral
breadth can be obtained when it is known to undergo regular changes
with the number of cycles during the fatigue process of the material.
Figure 3(a) gives the change in the integral breadth during the
fatigue process of an annealed 0.76%C steel [5], showing its fundamental composition in three successive increasing stages, and

OVERALL VIEW

MODEL A

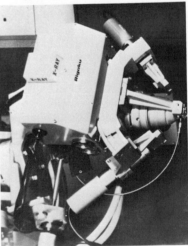

MODEL B

MEASURING HEAD

Figure 1. Wagon Type X-ray Stress Analyser

OVERALL VIEW AND OPERATING STATUS

MODEL A

MEASURING HEAD

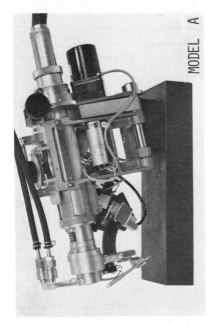

MODEL B

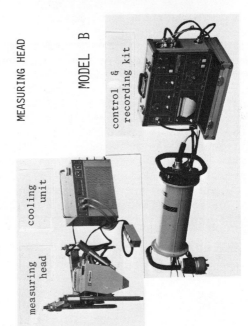

control &
recording kit

cooling
unit

measuring
head

Figure 2. Portable Type X-ray Stress Analyser

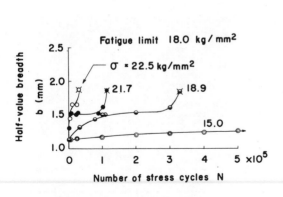 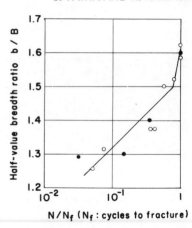

(a) Half-value breadth versus   (b) Half-value breadth
     number of stress cycles   ratio versus cycle
     relation   ratio relation

Figure 3.   Change in half-value breadth during fatigue of annealed
            0.76%C steel.

Figure 3(b) is the nondimensional diagram of the breadth ratio b/B
with respect to the initial breadth B and the number ratio N/N_f
with respect to the cycle number at fracture N_f. It is found that
the fatigue process is given by a single broken line independent
of the magnitude of the applied stress. Figure 4 gives similar
plots of the fatigue process of a hardened 0.79%C steel [6]. The
initial state is broad but becomes narrower with the stress cycles,
and this process is composed of three regular, successive, decreasing
stages. This behavior of the breadth of the x-ray diffraction profile during the fatigue process of metal can be interpreted in
terms of changes in microscopic parameters as will be described
later. It has been chosen as a subject of our joint research. It
is utilized in many industries for design improvement, rough
evaluation of fatigue damage, and life prediction.

CHANGE IN MACROSCOPIC AND SUBMACROSCOPIC MATERIALS
PARAMETERS DURING FATIGUE IN ENGINEERING APPLICATION

     The interest of the cooperative studies on fatigue damage
detection has been directed to collect the data on the change in
the material parameters, mainly in the change of diffraction line
broadening and that of residual stresses during fatigue, for a
great variety of metals and alloys that are used for engineering
structures.

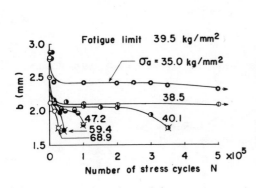

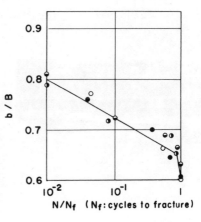

(a) Half-value breadth versus
     number of stress cycles
     relation

(b) Half-value breadth
     ratio versus cycle
     ratio relation

Figure 4.   Change in half-value breadth during fatigue of cold-
            worked 0.76%C steel.

As far as the x-ray diffraction techniques are the experimental
tools to observe the properties of surface layer of bulk material,
the nature of the surface has to be the point of primary interest.
The surface of a material, whatever it may be, is produced by a
wide variety of processes generally accompanying worked layer in
the vicinity of the surface. The property of the surface and the
depth of worked layer would depend upon the mechanism of surface
formation; for instance, by mechanical forming such as sutting,
grinding, honing, lapping and polishing, or recent material removal
techniques such as electrochemical cutting and grinding, spark
discharge machining, electron or ion beam, laser and plasma arc
material removal techniques. Also, additional surface processes
such as burnishing or shot peening are applied for improving property
of surface. These surfaces are inevitably provided with residual
stresses of macroscopic and microscopic order, and these are compre-
hensively expressed as irregularity of structures.

On the other hand, there are many components that are subjected
to stress relieving process by annealing. Macroscopic as well as
microscopic stresses are removed more or less, depending on the
annealing temperatures.

X-ray observation of fatigue damage has been practiced on the
change in material parameters, mainly that of residual stresses and
diffraction line broadening, occurring on the surface of specimens
subjected to fatigue stress cycling.

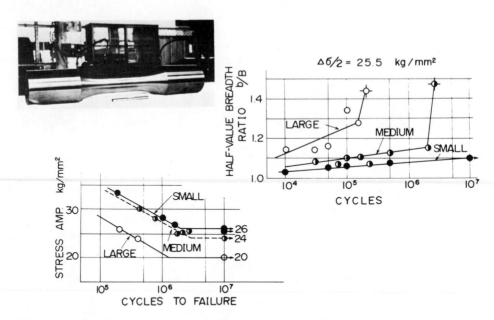

Figure 5.   Change in half-value breadth during fatigue of steel,
            indicating influence of size effect on fatigue life.

        Ohuchida, Nishioka and Nagao [7] have developed a series of
studies on size effect in fatigue on railroad shaft steel. The
fatigue life of metal is accompanied by a size effect, and in
tests on the same material with fatigue stresses of the same
nominal value, the life of large specimens is extremely short as
compared with that of small specimens. This is shown in Figure 5
by the S-N curves for small (10 mm dia.), medium (20 mm dia.) and
large (100 mm dia.) specimens and is a serious and as yet unsolved
problem in designing. However, as the x-ray diffraction technique
is applied to investigate line broadening, it becomes clear that
the development of damage in a large specimen is extremely rapid
in comparison to that in a small specimen from the half-value
breadth ratio versus number of cycles diagram shown in the figure.
This problem, which has heretofore been treated only by fracture
life comparison, can now be viewed in a different light in the
examination of the development of fatigue damage.

        Let us show the results of another series of work by Ohychida,
Nagao and Kuboki on the size effects of smooth and notched bars
of a normalized medium carbon steel (0.22%C) for railroad wheel

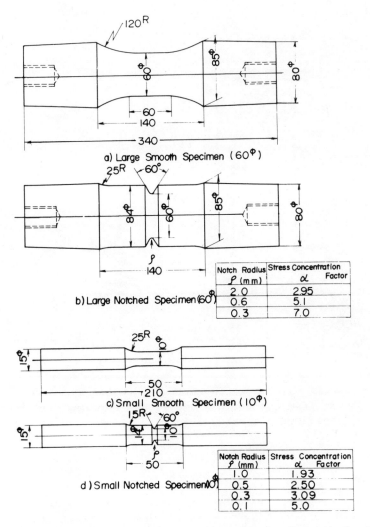

Figure 6. Specimens for studying size effect of notched and smooth specimens in fatigue life.

shaft [8]. The specimens of different size and shape; large size smooth, large size notched (60 mmϕ at test section), specimens and small size smooth, small size notched (10 mmϕ at test section), specimens as are shown in Figure 6 were prepared. These specimens were annealed in vacuum and tested in rotating bending fatigue and x-ray diffraction profile were taken on individual specimen surface at several numbers of stress cycling. The test conditions and a

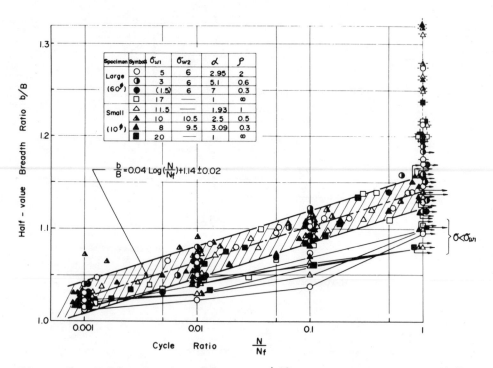

Figure 7.   Half-value breadth ratio b/B versus cycle ratio (N/N_f)
diagram for notched and smooth specimens of different
specimen dimensions.

part of test results are shown in Table 2. The results are presented at the b/B vs N/N_f diagram as shown in Figure 7. As is seen
in the figure, most of the plots are involved in a shaded range
showing regular increase in broadness with increase of number of
stress cycling. The points for stresses less than fatigue limit
are connected with full line and it is seen that these b/B values
are apparently less than the level of lower limit of shaded band.
As far as the change in half-value breadth ratio during fatigue is
concerned, the mode of its change in relation with number ratio
showed no difference between at the surface of smooth specimens
and at the surface of notch root in spite of the existing difference
of stress concentration factor.

Thus, it would be said that the x-ray diffraction line

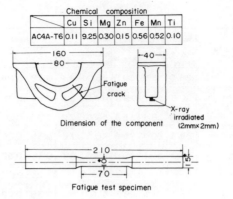

(a) Bearing cap of aluminum alloy casting and fatigue
     test specimen of same material

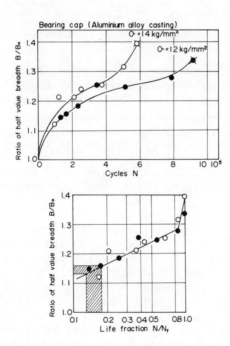

(b) Half-value breadth ratio versus number of cycle
     relation and master chart for fatigue damage
     detection of the material

Figure 8. X-ray damage detection for bearing cup of cast aluminum
           alloy.

TABLE 2

Specimen	Number of Cycles [*] $Nf = \times 10^6$	Phenomena	Mechanical Properties (kg/mm^2, %)				Impact Value $kg\text{-}m/cm^2$	Chemical Compositions (%)					
			Yield Stress	Tensile Strength	Elongation	Reduction of Area		C	Si	Mn	P	S	Cu
17871-12	0.6126	Failure	36.0	55.9	30.0	48.0	9.4	0.32	0.26	0.69	0.008	0.010	0.09
68549-12	0.6233	"	39.6	58.5	29.0	44.8	8.6	0.34	0.28	0.69	0.005	0.008	0.13
17407-30	1.2227	"	36.0	58.1	31.6	54.0	7.2	0.33	0.28	0.66	0.012	0.011	0.11
18147-8	1.4188	"	39.3	60.4	25.0	40.5	7.8	0.35	0.26	0.73	0.012	0.009	0.09
68359-12	2.6230	"	38.0	59.1	30.0	49.0	9.7	0.36	0.32	0.73	0.008	0.009	0.13
73076-10	3.6422	"	—	—	—	—	—	—	—	—	—	—	—
74856-9	9.9310	"	36.7	61.4	28.8	51.0	5.7	0.36	0.28	0.67	0.016	0.009	0.12
17678-22	10.0072	No Failure	37.0	57.5	30.0	51.0	8.4	0.33	0.31	0.67	0.007	0.008	0.10
17688-30	10.0230	"	38.3	60.7	27.0	42.7	6.3	0.34	0.35	0.68	0.007	0.009	0.10
17931-28	10.1047	"	39.6	63.3	25.6	39.1	6.0	0.37	0.34	0.73	0.021	0.008	0.12
17801-4	10.2068	"	39.0	59.8	28.6	45.9	8.6	0.34	0.30	0.71	0.009	0.009	0.12

[*] Rotating Bending Stress: 24 kg/mm^2

broadening detected at the specimen surface, i.e., the averaged surface micro-yielding, is proved to be the controlling factor in fatigue damage for high cycle fatigue independent of the size or shape of the specimens.

Figure 8 is another example of the application of x-ray diffraction technique to the fatigue damage detection. A crack was found, during in-service inspection, in a bearing cap made of an aluminum alloy casting of which chemical composition is shown in the figure.

In order to check the safe use of the same components that are used in other machineries, fatigue damage evaluation has been practiced by producing a master chart for fatigue damage on the basis of diffraction line broadening of this material. The dimension of fatigue specimen for obtaining a master chart of fatigue damage of the material is shown in the figure. The change in the half-value breadth ratio is plotted against number of stress cycles and the half-value breadth ratio versus number of cycles ratio is presented as is shown in Figure 8(b). It is worthy to note that the master chart for fatigue damage detection is very similar to the one that is shown in Figure 3. The shaded range in the figure shows the b/B value measured on the same components in use.

The residual stress induced in the components during manufacturing process is another sort of material parameter as described above. It has been reported that the residual stresses are, in many cases,

varied during fatigue mainly in the mode of fading in different
rate according to the magnitude of initial value, to the amplitude
of applied cyclic stress or strain and also to the property of
materials [9].

   A great number of test results have been reported on the change
of residual stresses during fatigue and the contribution of residual
stresses to the resistance of materials against fatigue life is
discussed [10]. It would be generally recognized that residual
stresses stored in rather soft materials by cold-working such as
surface machining, drawing or shot peening, tend to fade during
fatigue. Residual stresses stored in very high hardness materials
such as martensite structure, indiced by quench-temper processing,
are hard to fade [11]. The mode of fading of residual stresses is
more or less related to the variation of profile line broadening
and therefore it is availed for nondestructive detection of fatigue
damage in combination with the change of profile line broadening.

   Ohuchida, Nagao and Kuboki studied the fading of residual
stress due to fatigue stressing in large wheel shafts (100 mm in
dia.) of the same dimension in relation with their fatigue life [12].
Among 35 tested specimens, 26 specimens failed before 10^7 cycles
and 9 specimens endured 10^7 cycles, while specimens were made of
nominally same material as is shown in Table 3 and nondestructive
detection of defects by magnetic method did not show any recognizable
defect from them.

   Figure 9 shows the change in surface residual stress during
fatigue of lathe finished large wheel shaft specimen. All specimens
were tested at the same fatigue stress of 24 kg/mm^2. It is clear
that even though the stress amplitudes were the same for all specimens, the changes in residual stress were larger for the short life
specimen than for the longer life specimens. It implies the
existence of the scattering of fatigue strength of such a large
scale component. It is noteworthy that this trend is similarly
found in the change of half-value breadth ratio during fatigue as
is shown in Figure 10.

   These are examples of nondestructive fatigue damage detection.
A number of similar works have been presented at the Committee
meetings mainly by the industry laboratories, where this sort of
rough and simple method of nondestructive fatigue damage detection
is very widely availed according to the spontaneous interest of
individual industries [4].

   The simplest method of fatigue damage detection is usually
carried out by comparing the line broadening or the value of
residual stress of the spot where fatigue damage is susceptible to
be accumulated because of cycling of high stress amplitude with
that of the locations of low stress in service.

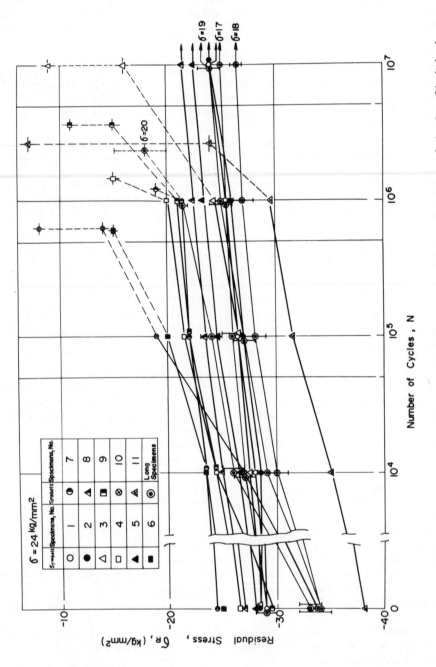

Figure 9. Change in surface residual stress during fatigue of lathe finished
large wheel shaft.

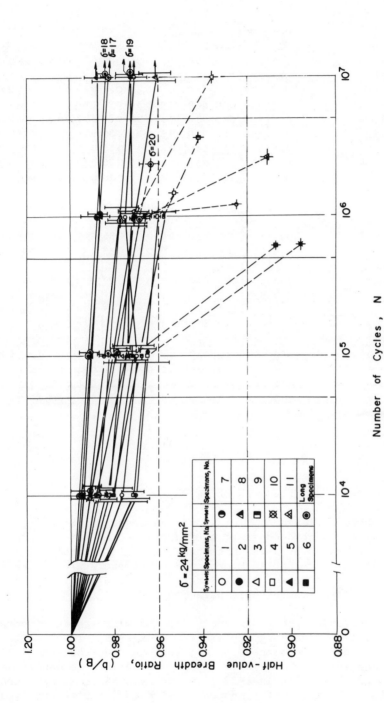

Figure 10. Change in surface half-value breadth during fatigue of lathe finished large wheel shaft.

TABLE 3

Specimen	Number of Cycles $N_f = \times 10^6$ *	Phenomena	Mechanical Properties (kg/mm^2, %)				Impact Value ($kg\text{-}m/cm^2$)	Chemical Compositions (%)					
			Yield Stress	Tensile Strength	Elongation	Reduction of Area		C	Si	Mn	P	S	Cu
79567-12	0.7046	Failure	37.7	59.4	31.0	53.0	9.2	0.34	0.28	0.70	0.010	0.008	0.13
69165-13	0.7963	"	42.9	64.6	31.6	54.0	10.2	0.32	0.30	0.69	0.010	0.009	0.15
79907-5	0.8541	"	40.3	59.4	32.8	55.9	10.3	0.32	0.24	0.64	0.012	0.008	0.14
70611-16	0.8679	"	34.4	58.1	32.8	54.9	10.4	0.34	0.29	0.64	0.014	0.006	0.13
70392-3-2	0.9036	"	38.3	57.8	32.2	53.0	12.5	0.34	0.27	0.77	0.018	0.009	0.14
79837-9	0.9056	"	39.6	58.5	33.2	55.0	10.3	0.36	0.30	0.71	0.013	0.008	0.14
79057-14	1.0639	"	44.5	67.9	33.6	54.9	8.6	0.34	0.35	0.73	0.013	0.008	0.15
79411-14	1.0639	"	41.2	59.1	33.6	56.8	10.9	0.33	0.35	0.72	0.010	0.010	0.14
71997	1.1274	"	40.3	59.1	30.4	46.9	10.4	0.32	0.30	0.65	0.009	0.011	0.10
79252-9	1.1672	"	39.3	60.4	34.9	56.8	10.9	0.32	0.30	0.73	0.011	0.007	0.14
70262-10	1.5503	"	33.8	57.8	31.2	56.8	11.1	0.34	0.33	0.67	0.013	0.008	0.14
63705-4	1.6000	"	—	—	—	—	—	—	—	—	—	—	—
79626-1	2.0431	"	38.6	59.4	30.0	49.0	10.9	0.32	0.35	0.66	0.012	0.008	0.12
78731-8	2.0600	"	41.8	62.4	30.8	53.0	—	0.34	0.35	0.70	0.012	0.007	—
76021-10	2.6071	"	—	—	—	—	—	—	—	—	—	—	—
78908-6	3.2400	"	39.3	60.4	29.4	49.0	8.2	0.33	0.27	0.73	0.010	0.008	0.18
79706-15	3.4289	"	40.9	59.1	31.2	50.0	11.8	0.34	0.31	0.62	0.009	0.008	0.13
72232-14	4.3760	"	40.3	62.0	27.2	42.7	8.6	0.36	0.33	0.74	0.013	0.008	0.17
63438-10	7.7934	"	39.0	62.4	26.0	43.8	6.3	0.36	0.31	0.69	0.008	0.009	0.10
75259-9	10.0047	No Failure	—	—	—	—	—	—	—	—	—	—	—
72079-7	10.0610	"	37.3	60.4	28.0	48.0	8.1	0.35	0.37	0.66	0.009	0.008	0.16
71738-7	10.1206	"	40.9	61.7	27.4	40.5	8.0	0.35	0.31	0.65	0.009	0.008	0.17
71599-10	10.2643	"	38.3	59.1	31.6	51.0	8.0	0.33	0.30	0.68	0.010	0.008	0.12
72523-1	10.8179	"	40.3	59.8	31.4	54.9	8.4	0.34	0.31	0.67	0.015	0.011	0.13

* Rotating Bending Stress : 24 kg/mm^2

MICROSCOPIC PARAMETERS IN RELATION TO FATIGUE PROCESS

As is shown in Table 1, microscopic parameters that are associated with submacroscopic or microscopic parameters can be numerically evaluated by x-ray microbeam analysis or profile analysis. Studies evaluating microscopic parameters are basically close in fundamental or scientific interest to the interpretation of submacroscopic or macroscopic parameters.

Higher resilving power for analyzing the microscopic structure is made possible by inducing a microbeam of x-rays through a micro-slit to irradiate it onto the surface; one can then observe sensitive changes in the microstructure. The principle was first presented by P. B. Hirsch in 1952 [13] and even since it has been applied within the field of fundamental studies. For several years,

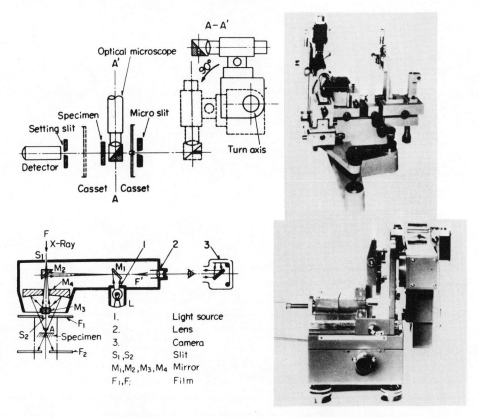

Figure 11.   X-ray micro cameras.

the characteristic features of x-ray microbeam analysis, such as
high resolving power and local observation, have attracted the
attention of researchers working on engineering problems in order
to achieve a deeper understanding of the phenomena for wider
application [14].   The development of the x-ray microcamera combined
with the x-ray microfocus generator stimulated research activities
concerned with evaluating the change in microscopic parameters which
occurs during the macroscopic mechanical behavior of engineering
materials.

Figure 11 exhibits a part of the x-ray microcamera.   In combination with an optical system, a fine x-ray beam with a diameter of
50 to 150μ can be positioned most accurately on the part to be
measured on the surface of the specimen or the machine element.   The
microbeam is irradiated on the observed spot by an optical microscope
to obtain a diffraction pattern on the film.   A certain number of
crystals in polycrystalline materials leave their image on the film
by diffraction.

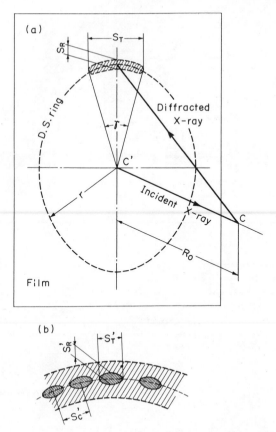

Figure 12.   Schematic pattern of x-ray microbeam Debye analysis.

Figure 12 shows a schematic pattern of the microbeam Debye analysis for a deformed metal. The haded part is the diffraction of a deformed crystal. The image of an annealed crystal is a spot and it spreads in radial and tangential direction by a deformation of the crystal [Figure 12(a)]. Measurement of the radial spread S_R and the tangential spread S_T gives the microlattice strain $\Delta d/d$ and the misorientation β by the formula:

$$\frac{\Delta d}{d} = \frac{\cos^2 2\theta}{\tan \theta} \frac{S_R}{2R_O} \quad , \tag{1}$$

$$\beta = \frac{8}{\pi} \text{ arc sin } (\frac{|\tan 2\theta|}{\sin \theta} \frac{S_T}{4R_O}) \tag{2}$$

where θ is the Bragg angle and R_0 is the distance between specimen surface and film. When the formation of substructure is the controlling mechanism of the deformation of the crystal, the diffraction image contains clear small spots on a rather dull background. In such a case [Figure 12(b)], we can count the number m of the spots, which enables us to evaluate the subgrain size t. The spread of spots (S_R^s and S_T^s) is also associated with the microscopic distortion of substructure.

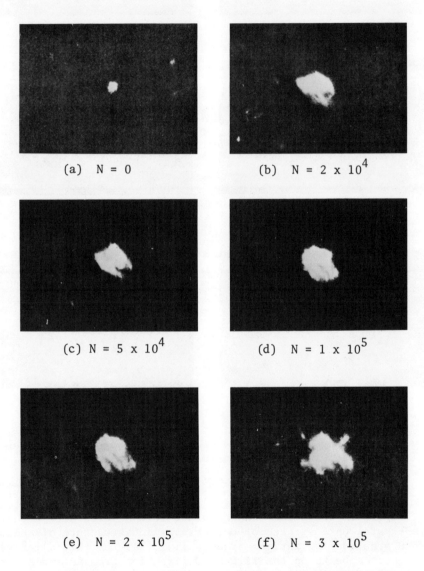

(a) N = 0   (b) N = 2 x 10^4

(c) N = 5 x 10^4 (d) N = 1 x 10^5

(e) N = 2 x 10^5 (f) N = 3 x 10^5

Figure 13.   Change in the shape of x-ray microbeam diffraction
             pattern of annealed 0.06%C steel during fatigue.

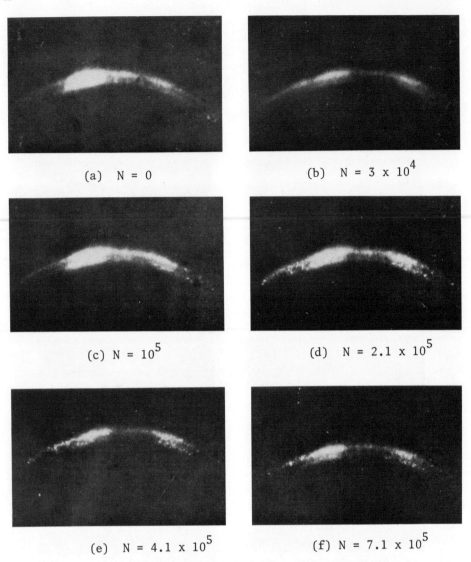

(a)   N = 0   (b)   N = 3 x 10^4

(c) N = 10^5 (d)   N = 2.1 x 10^5

(e)   N = 4.1 x 10^5 (f) N = 7.1 x 10^5

Figure 14.   Change in the shape of x-ray microbeam diffraction
             pattern of 15% cold-rolled 0.06%C steel.

        Figure 13 shows the change in the form of diffraction pattern
of a crystal on a surface of polycrystalline structure of annealed
0.06%C steel during fatigue process [15].   At the initial stage, a
small spot appears on the film by diffraction from an undistorted
crystal.   As the result of stress cycling, minute distortion occurs
in the crystal and the diffraction image spreads both in tangential

and radial directions with the increase in number of stress cycling.

    Figure 14 shows the change of microbeam diffraction pattern from
15% cold-rolled 0.06%C steel during fatigue. The initial pattern
shows an arc that is the diffraction from a distorted crystal grain
in polycrystalline aggregates. The arc is composed of a number of
diffused small spots that correspond to substructures formed by
plastic deformation. With an increase in the number of stress cycles,
the number of spots increases and individual spots become sharper.
Thus, specialty in fatigue damage is characterized by clear formation
of substructures.

    These patterns were analysed and the changes in microscopic
material parameter with number of cycles are presented as shown in
Figure 15 for annealed material, and Figure 16 for cold-rolled
material. In both figures, the fatigue damage in terms of microscopic parameter is shown for three crystal grains involved in the
area of x-ray irradiation, respectively. These figures give a good
interpretation to the mode of change in diffraction profile line
broadening for annealed and cold worked material during fatigue as
were shown before.

    An accurate profile of the object to be measured is obtained

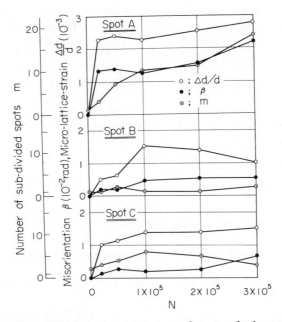

Figure 15.   Change in microparameters of annealed steel during
              fatigue.

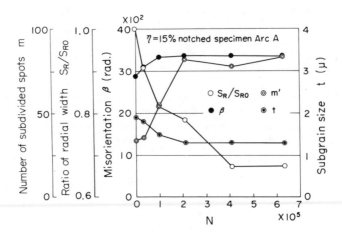

Figure 16.   Change in microparameters of cold-rolled steel during
              fatigue.

by an x-ray diffractometer which is either Fourier analyzed according to the method devised by B. E. Warren and B. L. Averbach [16] or
otherwise analyzed by integral breadth measurement according to
W. H. Hall [17] to obtain the microscopic parameter below; the
microscopic strain $<\varepsilon^2>^{1/2}$ and the particle size are separately
obtained. A detailed description of the analysis is not given here
because a number of publications are available on this subject.

     When we apply this analysis to observe the change in micro-
structure of materials subjected to mechanical processes, we have
to be careful to prepare a geometry of the specimen surface suitable
for the diffractometer to get a correct x-ray profile. An example
of engineering application is shown in Figure 17 in the case of the
observation of change in the microparameters during low cycle fatigue
of carbon steel [18].

                X-RAY STUDY ON FATIGUE CRACK PROPAGATION

     The changes in x-ray material parameters described above are
directly related to the process of fatigue until the initiation of
fatigue crack. After the fatigue crack appears, the propagation of
crack is another interesting field for application of x-ray microbeam technique, displaying the advantage of observing the change in
microstructure at a very local area around the tip of the fatigue
crack once it appears.

     Figure 18 is an example of the application of x-ray microbeam

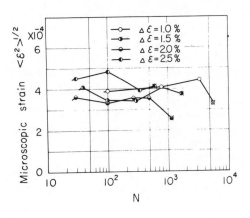 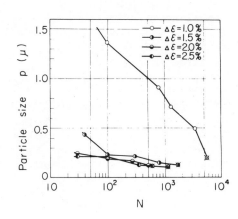

(a) Microscopic strain versus   (b) Particle size versus
     number of cycle ratio   number of cycle ratio

Figure 17.   Change in microparameter of annealed 0.15%C steel during
             low cycle fatigue.

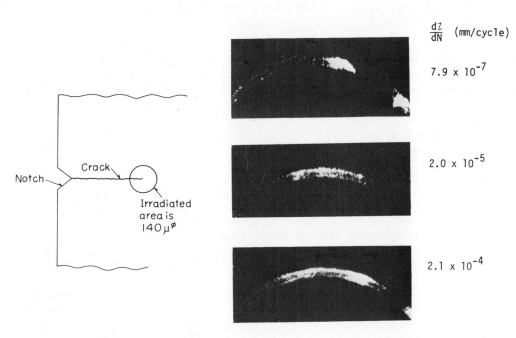

Figure 18.   X-ray microbeam observation on microstructure at a
             fatigue crack tip.

technique for the study of fatigue crack propagation [19]. Figure 18(a) is a schema for observation and (b) is examples of the x-ray microbeam patterns obtained at the tip of fatigue cracks which appeared in a 0.03%C steel and were propagating at low and high rates of propagation. Looking at these patterns, it may be clearly seen that the modes of the diffraction patterns for different rates of propagation are distinctly different, complying with the difference in the change in microstructure. Thus the change in microstructure at the tip of propagating crack is very much dependent on the rate $d\ell/dN$ of propagation. This feature of the change in microstructure at the tip of fatigue crack has been studied quantitatively and the mechanism of fatigue crack propagation is discussed in terms of x-ray microparameters measured by the x-ray microbeam technique. The results are briefly introduced below.

Figure 19 shows the measurement of the changing microparameters at the tip of fatigue crack for specimens of 0.03%C steel, plotted against the crack propagation rate $d\ell/dN$ [21]. While (a) gives the excess dislocation density at the tip of crack derived from microparameters, (b) shows the $d\ell/dN$ dependence of the subgrain size. It is seen that these values are closely connected with the propagation rate $d\ell/dN$ of the crack. Diagrams similar to Figure 19 may be drawn for the cases of other carbon steels. This approach could be used for detecting the possibility of propagation of fatigue crack appearing in structure components in service.

Crystal deformation near fracture surface was also studied by directly observing the fracture surface by means of the x-ray microbeam technique, in order to study the mechanism of fatigue crack propagation [21]. In this line of work, specimens cracked under cyclic stressing were cooled down to -196°C by liquid nitrogen and broken by a Charpy impact testing machine. The region of fatigue fractured surface thus revealed adjacent to the former tip of the fatigue crack was the target for irradiation by x-ray microbeam.

The changes in microparameters at the surface fractured under fatigue with different rates of propagation was measured by this means and example of the results is presented in Figure 20 for the case of in-plane cyclic bending of three kinds of annealed low carbon steels. The diagram shows the excess dislocation density at the fractured surface in relation to the rate of crack propagation.

The value D of excess dislocation density derived from x-ray microparameters has been studied in the case of monotonic tensile deformation by utilizing the x-ray microbeam technique, and it has been found that the value D has a close relation to true strain ε_F for individual materials. Some examples are presented in Figure 21 for the case of three kinds of low carbon steels. Combining Figures 20 and 21, the true strain is substituted for the

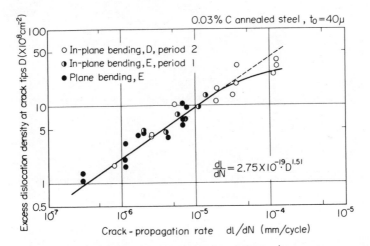

(a) Excess dislocation density at crack tip versus crack-propagation rate relation

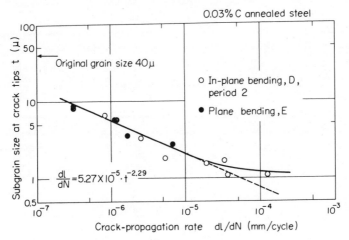

(b) Subgrain size at crack tip versus crack-propagation rate

Figure 19. Changes in microparameters, excess dislocation density D and subgrain size t, at the tip of propagating crack with the rate $d\ell/dN$ of crack propagation.

microparameter, and the diagram of equivalent true strain ε_F versus crack propagation rate $d\ell/dN$ (Figure 22) is obtained. Figure 23 is crack propagation rate versus stress intensity factor K and it is clearly found that the relations are slightly different between

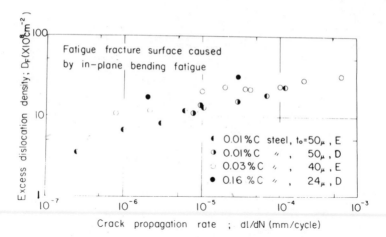

Figure 20.   Change in microparameter at fractured surface in relation with crack propagating rate $d\ell/dN$.

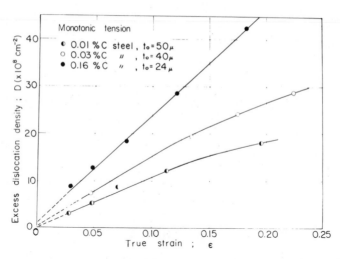

Figure 21.   True strain in relation with excess dislocation density in monotonic tension.

the kind of materials.   It is a good contrast to the case of Figure 22, where ε_F versus $d\ell/dN$ relation is represented by a single curve independent of the difference in materials.

A recent development in the x-ray microbeam technique that is worthy to note is the introduction of the oscillating beam principle.

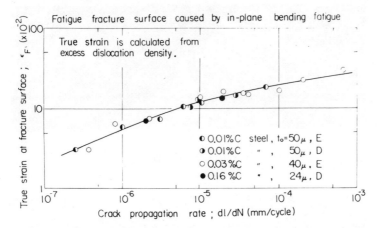

Figure 22.   Crack propagation rate $d\ell/dN$ versus true strain at
             fracture surface ε_F relation for three kinds of low
             carbon steel.

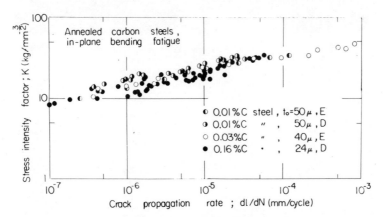

Figure 23.   Crack propagation rate $d\ell/dN$ versus stress intensity
             factor K relation for three kinds of low carbon steels.

This method has led to direct measurement of macroscopic residual
stress around the tip of propagating fatigue cracks.   Figure 24
shows an example of the continuous x-ray pattern obtained by the
oscillating microbeam technique used on a material that gives an
ordinary pattern by the stationary microbeam technique, shown in
the same figure.   Figure 25 shows the distribution of axial
residual stress near the tip of propagating crack of a quench-
tempered (at 500°C) 0.35%C steel along the crack growth direction
[22].   The influence of overstressing of 28 kg/mm^2 in gross stress

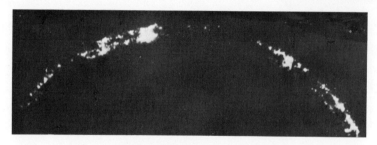

STATIONERY BEAM

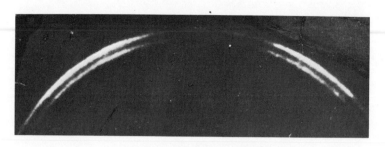

OSCILLATING BEAM

Figure 24.   X-ray microbeam patterns of stationary and oscillating
             beam method.

on the rate of crack propagation for the case of normally applied
gross stress of 14 kg/mm^2 is studied for a center cracked specimen.
As is seen in the figure, even single cycling of overstressing
causes a notable residual stress near the tip of crack and also it
should be noted that the number of cycles of overstressing has a
significant influence on the magnitude and distribution of residual
stress.   The effect of residual stress on the retardation of crack
growth is the objective of this line of work and it is clearly
substantiated by Figure 26, where it is seen that the higher value
of residual stress gives the more remarkable retardation effect
immediately after the end of application of overstressing.

CONCLUDING REMARKS

        X-ray diffraction technique are only a part of experimental
means that are available for revealing the properties of materials.
When these are applied for the studies on mechanical behavior of
materials, detection of minute deformation is the main objective

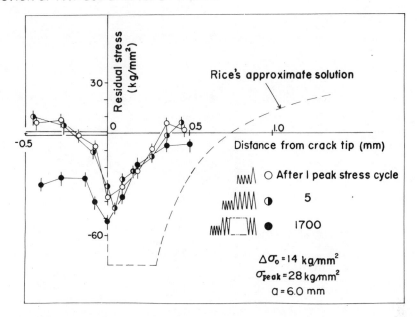

Figure 25. Distribution of axial residual stress near the tip of
            propagating crack as influenced by overstressing.

of their applications. When we consider the x-ray diffraction
techniques as tools for nondestructive detection of fatigue, the
sort of materials is limited within the range of materials with
ductility. For the case of fatigue of materials with very low
ductility, the fatigue would proceed at very localized areas and
early detection of fatigue by x-rays for such localized area con-
fronts serious difficulties in finding the location where the
x-ray should be irradiated. In the case of materials of low
ductility, the residual stresses are also rather stable and so it
is hard to take the change in residual stresses as the measure of
fatigue damage.

    However, there are a variety of engineering problems where
simple and rough methods of fatigue damage detection are occasionally
wanted. In such cases, the x-ray diffraction technqiues for nondestructive fatigue damage detection as described above would be
available efficiently.

    These measurements are now applied in a wide range of
engineering fields, from small size of machine element to the parts
of large scale structural constructions. The object of these
measurements is not confined to fatigue damage detection but also
is used as quality control technique that are such as the check of

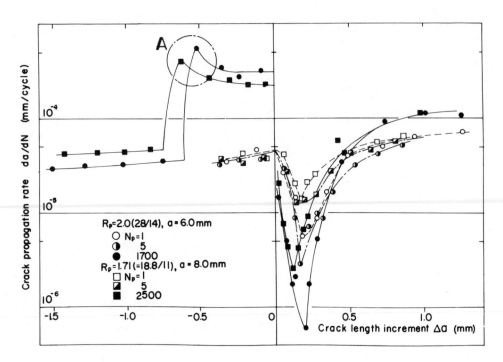

Figure 26.   Retardation of fatigue crack propagation due to the application of overstressing.

uniformity of quench hardened components, inspection of residual stress for the stress relieving treatment in preventing stress corrosion cracking of pipe and other construction, and surface damage control of rolling roll in steel makers, and so on.

REFERENCES

1. Taira, S., "X-Ray Studies on Mechanical Behavior of Materials",
    Society of Materials Science, Japan, 1974.

2. Taira, S., "X-Ray Approach for the Study on Mechanical Behavior
    of Metals", Proc. Intl. Conf. Mech. Behav. Mat., Kyoto, Special
    Vol. (1972) 111-28.

3. Taira, S., "X-Ray Diffraction Approach for Studies on Fatigue
    and Creep", Exp. Mech., Vol. 13, No. 11 (1973) 449-64.

4. J. Mat. Sci. Japan, Vol. 11 (1962), 629-718; Vol. 12 (1963)
    829-911; Vol. 13 (1964) 905-1012; Vol. 14 (1965) 923-1029;
    Vol. 15 (1966) 813-901; Vol. 16 (1967) 921-1041; Vol. 17
    (1968) 1047-80; Vol. 18 (1969) 1033-56; Vol. 20 (1971) 1239-60;
    Vol. 21 (1972) 1045-1171; Vol. 23 (1974) 1-96; Vol. 24 (1975)
    1-102; Vol. 25 (1976) 1-129.

5. Taira, S. and K. Honda, "X-Ray Investigation on the Fatigue of
    Metals", Bull. of JSME, Vol. 4 (1961) 230-37.

6. Taira, S., K. Honda and K. Matsuki, "X-Ray Investigation of
    Fatigue Damage of Metals (On the Change in Half-Value Breadth
    of X-Ray Diffraction Line, Residual Stress, Micro-Structure
    and Micro-Hardness Due to Stress Repetitions)", Trans. JSME,
    Vol. 28 (1962) 1325-43.

7. Ohuchida, H., A. Nishioka and M. Nagao, "X-Ray Detection of
    Fatigue Damage in Machine Parts", Proc. 13th Jap. Congr. on
    Mat. Res. (1970) 167-72.

8. Ohuchida, H., M. Nagao and M. Kuboki, "X-Ray Detection of
    Fatigue Damage in Wheel Shafts", Proc. 1974 Symp. on Mech.
    Behav. Mat., Vol. 1 (1974) 381-404.

9. Rowland, E., "Effect of Residual Stress on Fatigue", Proc. 10th
    Sagamore Army Mat. Res. Conf. (1964) 229.

10. Nelson, D.V., R.E. Ricklefs and W.P. Evans, "The Role of Residual
    Stresses in Increasing Long-Life Fatigue Strength of Notched
    Machine Members", ASTM STP 467 (1970) 228-53.

11. Honda, K. and T. Goto, "X-Ray Committee Report on X-Ray Study
    on Fatigue of Metal - Report 2", J. Soc. Mat. Sci., Japan,
    Vol. 19 (1970) 714-21.

12. Ohuchida, H., M. Nagao and M. Kuboki, "X-Ray Detection of
    Fatigue Damage in Wheel Shafts (Size Effect of Notched Specimens,
    Surface Area Effect and Probability Study)", J. Soc. Mat. Sci.,
    Japan, Vol. 24 (1975) 86-96.

13. Hirsch, P.B. and J.N. Keller, "A Study of Cold-Worked Aluminum by an X-Ray Micro-Beam Technique, I: Measurement of Particle Volume and Misorientations", Acta Cryst., Vol. 5 (1952) 162-67; Hirsch, P.B., "A Study of Cold-Worked Aluminum by an X-Ray Microbeam Technique, II: Measurement of Shape of Spots, and III: The Structure of Cold-Worked Aluminum", Acta Cryst., Vol. 5 (1952) 168-72.

14. Taira, S. and K. Hayashi, "X-Ray Study on Fatigue of Mild Steel (On the Feature of Fatigue Mechanism in Comparison with Other Mode of Fracturing)", Trans. Iron and Steel Inst., Japan, Vol. 8 (1968) 220-29; Hayashi, K., "X-Ray Study of Micro-Structure Change Due to Fatigue of Low Carbon Steel", Proc. Intl. Conf. Mech. Behav. Mat., Kyoto, Vol. 2 (1972) 26-38.

15. Hayashi, K., Dr. Eng. Thesis, Kyoto Univ. (1967).

16. Warren B.E. and B.L. Averbach, "The Effect of Cold-Work Distortion on X-Ray Patterns", J. Appl. Phys., Vol. 21 (1950) 595-99.

17. Hall, W.H., "X-Ray Line Broadening", Proc. Phys. Soc., A-62 (1949) 741-43.

18. Taira, S., T. Goto and Y. Nakano, "X-Ray Investigation of Low-Cycle Fatigue in an Annealed Low-Carbon Steel", J. Soc. Mat. Sci., Japan, Vol. 17 (1968) 1135-39.

19. Taira, S., K. Hayashi and K. Tanaka, "X-Ray Investigation on Fatigue Fracture of Notched Specimens (Investigation on Fatigue Process of Cold-Rolled or Quench Aged Low-Carbon Steel Specimens)", Proc. 11th Japan Congr. Mat. Res. (1968) 18-24; Taira, S. and K. Tanaka, "Study of Fatigue Crack Propagation by X-Ray Diffraction Approach", Engng. Frac. Mech., Vol. 4 (1972) 925-38.

20. Taira, S. and K. Tanaka, "Microscopic Study of Fatigue Crack Propagation in Carbon Steel", Proc. Intl. Conf. Mech. Behav. Mat., Kyoto, Vol. 2 (1972) 48-58.

21. Taira, S., K. Tanaka, T. Tanabe and Z. Ryu, "Investigation on Fatigue Crack Propagation of Low Carbon Steel", Bull. JSME, 328 (1973) 3531-42.

22. Taira, S. and K. Tanaka, to be published in Bull. Soc. Mat. Sci., Japan.

Chapter 3

A HISTORICAL EXAMPLE OF FATIGUE DAMAGE

In Memoriam: F. Regler, Vienna, Austria*

H. K. Herglotz

Engineering Physical Laboratory, E. I. du Pont de
Nemours & Co. (Inc.), Wilmington, Delaware

ABSTRACT

     X-ray diffraction methods are a well-established tool today
for the assessment of fatigue damage. It is both instructive and
fascintating to look back at the work of an early pioneer in this
field. F. Regler, Vienna, Austria, reported in the 1930's his
"live" investigations on railroad bridges which had been in heavy
use for 80-100 years. X-ray diffraction on positions subjected to
unidirectional pulsating loads from passing trains showed very fine
grain (continuous, broadened Debye-rings) and orientation, while
diffractograms from unloaded members of the same bridges showed the
fairly coarse grain of the original construction steel, without
orientation. Regler's interpretation of grain rotation and fragmentation when a load was renewed before complete relaxation from the
previous one drew strong criticism from proponents of classical
elasticity theory.

     Regler's x-ray results were later supported by impact data on
samples taken from dismantled bridges; the fine-grain samples with
orientation were brittle, while samples showing the unchanged pattern
of the construction steel displayed high impact properties. His
results were confirmed 20 years later in the laboratory by a student
who was unaware of Regler's earlier work but who had modern equipment
at his disposition.

---

* Regler, in his 76th year, died on the very day (August 25, 1976)
  that this presentation was delivered at the 23rd Sagamore
  Conference.

Figure 1.   Nondestructive XRD equipment on railroad bridge.

    (a)   Wooden platform, suspended by chains
    (b)   Mercury diffusion pump
    (c)   Demountable x-ray tube

     At first glance it may seem out of place to describe work
done around 1930 [1,2] that has been superseded by modern developments. The topic was selected because: it is very instructive,
it exemplifies the uselessness of polemics, it tells a timeless
tale about creative, young scientists and the academic establish-
ment, and it commemorates F. Regler of Vienna, a pioneer in
nondestructive testing. I chose to report it because I was Regler's
assistant in the 1950's and was involved in the confirmation of his
disputed results.

     Figure 1 shows x-ray diffraction equipment, modern in 1930,
quite obsolete today, on an Austrian railroad bridge. Modern in
those days meant an x-ray tube of the demountable type continuously
evacuated by a diffusion pump. Cooling water and high voltage were
supplied by hose and wire. Accomplishment of XRD work on a bridge,
interrupted repeatedly by rolling trains, attests to the initiative,
resourcefulness, and courage of the experimentor. The work was
stimulated by an event on the Austrian railroad. An over-sized
load grazed a member of the Stadlau bridge, near Vienna, which was
built in the second half of the 19th century and therefore was
about 80 years old. Mere touching of one of the trusses fractured
it like glass and triggered a panic. If the bridge was that fragile,

how in good conscience could passenger trains be permitted to use
it? Bridges had always been tested by Instron samples taken from
"safe" parts of the bridge, i.e., members not exposed to load.
Previous tests on the Stadlau bridge had always indicated that it
was safe. Regler, who operated an x-ray test center of his own and
had developed a reputation as an expert, was called in as a consul-
tant. He started the daring project to which Figure 1 bears witness.

Regler tested "in vivo" by nondestructive x-ray diffraction in
the way seen in Figure 1, which also identifies the components.
Regler's cone back-reflection camera, described in the literature
[3-6], was particularly suitable for the job. Figure 2 shows the
more complete and undistorted way in which the film records x-ray
patterns. Conical film has since been reinvented, e.g.[7], apparently
without knowledge of Regler's early work.

Many x-ray diffraction patterns were obtained from various
parts of the bridge. The results are typified by Figure 3. Parts
not subjected to load showed diffraction typical of the coarse
crystallites in structural steel used for bridge construction in
the 19th century. In contrast, bridge members intermittently subjected to unidirectional loads from trains corssing the bridge at
short intervals exhibited continuous Debye-Scherrer rings and often
gave indications of preferred orientation. In other words, substantial changes in the crystalline lattice of the iron had
occurred, although the loads had been in the elastic region where,
according to the definition of elasticity, nothing irreversible
should happen.

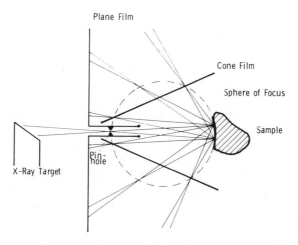

Figure 2. Focusing back reflection cone compared with plane film.

Figure 3. Cone back reflection patterns from bridge member not
           exposed to load (A); and from member exposed to uni-
           directional repetitive load (B).

           (022) reflection   FeK$\alpha_{1,2}$ θ = 72.75°
           (013) reflection   FeKβ θ = 75.67°

     Regler explained what he saw by two unconventional and, as it
turned out, controversial assumptions:

     In a polycrystalline conglomerate some crystallites can
     experience plastic deformation even though the work-piece as
     a whole is elastically deformed.

     Crystallites can be rotated by minute angles when load
     is applied; this process is reversible, but with a time delay
     (hysteresis). If the next load is applied before full
     recovery, there is an accumulation of the effects, resulting
     in the observed texture in the x-ray diagrams.

The crystallites of Figure 3(A) are nearly perfect from a crystallographic point of view, with crystallographic planes producing
well-defined points along the Debye-Scherrer circle. If subjected
to plastic deformation, these crystallographic planes are shattered,
as symbolized in Figure 4. This leads to peripheral broadening of
the diffraction spot along the Debye circle. Figure 5 from Ref. [8]

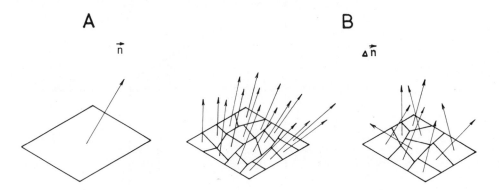

Figure 4.   Scheme of a crystallographic plane with its normal
            vector $\vec{n}$

            (A)   in an intact crystallite
            (B)   in a shattered crystallite

is Regler's original interpretation of the reason for peripheral
broadening that converts the spots on the Debye circle into a
continuous homogeneous diffraction ring. It appears as the 180°
arc of Figure 3 after the conical film is rolled off into the
planar semicircle exhibited in Figure 3.

    Figure 6 defines, in spherical projection, the geometrical
relations between the variables ρ, Θ and δ of the Polanyi formula:

$$\cos \rho = \cos \Theta \cos \delta$$

as they pertain to the back reflection cone [9]. Figure 7(A) describes the same variables ρ, Θ and δ in spherical projection. If
the vector $\vec{n}$ of Figure 4 wobbles in the way indicated because of
fragmentation, then we have the conditions of 7(B). This can be
expressed mathematically by the derivative of the Polyani equation:

$$\left(\frac{\partial \delta}{\partial \rho}\right)_{\Theta} = \frac{\sin \rho}{\cos \Theta \sin \delta}$$

in which $\partial \delta$ stands for the peripheral broadening (at constant Bragg
angle Θ) that leads to the continuous Debye circles of Figure 3(B).
In this chapter it is dealt with first because it is so apparent
and prominent on the x-ray patterns. But there is another, maybe
more significant but less obvious effect. Deformation of the
crystalline lattice resulting influctutations ∂d of the lattice

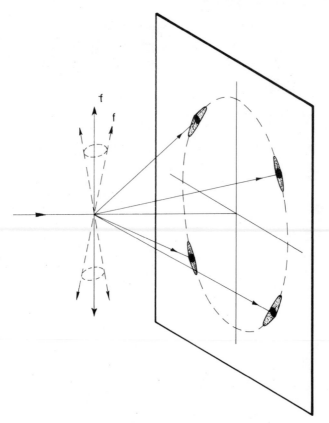

Figure 5. Peripheral broadening of Bragg reflections [8].

dimensions d(hkl) leads to radial broadening $\partial\Theta$ of the Debye-Scherrer circle according to the derivative

$$(\frac{\partial\Theta}{\partial d})_\lambda = - \frac{1}{d} \tan \Theta$$

of the Bragg equation

$$n\lambda = 2d \sin \Theta \ \ .$$

In order to obtain quantitative assessment of this radial broadening and therefore of the deformation a lattice has undergone, radial microdensitometer scans of deformed and undeformed material are usually compared. In the very lean economic conditions in Austria between the two world wars, Regler lacked access to a microdensitometer of sufficient precision. Thorough study of the

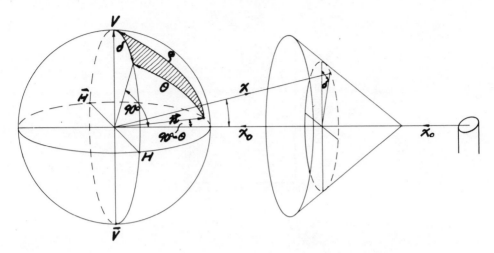

Figure 6.   Polanyi relation for the back reflection cone.

$V-\overline{V}$, $H-\overline{H}$ Vertical and horizontal direction
$X°$, X Incident and diffracted x-ray directions
$\vec{n}$ Plane vector of diffracting plane
Θ Bragg angle
ρ Angle between plane vector $\vec{n}$ and vertical
               direction $V-\overline{V}$
δ Angle between vertical on x-ray pattern and
               diffraction spot; appears also in rectangular
               spherical triangle $\Theta\rho\delta$

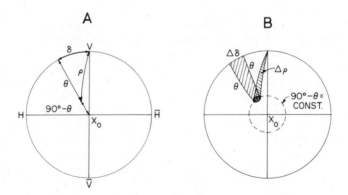

Figure 7.   Stereographic projection of back reflecting crystallite.

(A)   Intact
(B)   Shattered

All variables are the same as in Figure 6.

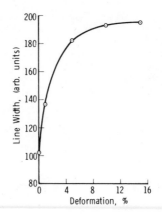

Figure 8.   Shape of line width versus deformation-curve.

diagrams, however, showed an effect which could not be explained
but gave a reliable measure of the radial diffraction line width.
Frequently, but not always, edges were observed on both sides of
the line and the distance between edges correlated very well with
the deformation a lattice had experienced. Systematic evaluation
of lattice deformation by "width between edges" (WbE) led to a
relationship described in Figure 8: WbE increases with increased
deformation and finally leads to saturation. The lattice cannot
tolerate further deformation and fracture occurs. Width between
edges (WbE) was used to measure the degree of previous deformation
and to predict when and where fracture would occur.

Such predictions were pure heresy in the eyes of some established materials testing experts and were branded as unscientific
black magic. Solids were a continuum for the representatives of
the classical elasticity theory and there was no room for discontinuities such as crystallites, crystallographic planes, and the
like. But the bridge was torn down anyway, based on Regler's
assessment of its structural strength. Samples became available
for destructive testing by the impact strength method, which fully
substantiated Regler's evaluation and predictions. Those samples
with diffractograms of the Figure 3(B)-type caused by 4(B)-type
shattering, broke very easily after a few strokes. Those exhibiting
the original steel pattern of Figure 3(A), indicating a lattice of
the 4(A)-type, did not break even after several thousand strokes.
Typical test data are shown in Figure 9.

Although these results seemed to vindicate Regler's controversial findings, there remained a good deal of ridicule, made mostly
because the measurement of WbE was highly subjective, and only
people with good eyesight and motivation were able to reproduce it.
The origin of the edges remained unknown.

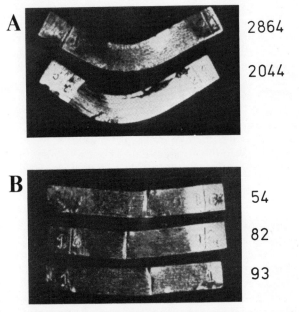

2864

2044

54

82

93

Figure 9.   Results of impact strength tests on samples from
            dismantled railroad bridge.

(A)   With structure of the type of Figures 3(A), 4(A),
      unbroken after > 2000 strokes
(B)   With structure of Figures 3(A), 4(B); broken after
      < 100 strokes

The number of strokes is indicated to the right side of
the photograph.

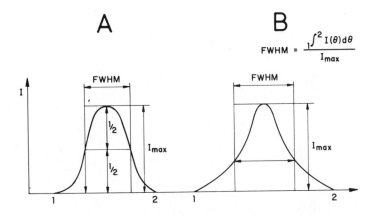

$$FWHM = \frac{\int_1^2 I(\theta)d\theta}{I_{max}}$$

Figure 10.   Two definitions of FWHM (Full Width at Half Maximum).

It was, therefore, a worthwhile and interesting project to let
a student (now Dr. E. Schiel) of the Institute of Experimental
Physics at the Technical University of Vienna measure the full
width at half maximum, a reproducible and well-defined entity
defined by Figure 10. Samples from the dismantled bridge were available at the Institute, stored there for two decades. The student
measured FWHM as defined in Figure 10(B). The samples were identified only by number and the original WbE figures were unknown to him.
After submission of his results, the old Regler values were disclosed
to him. Comparison of the two sets of numbers is shown in Figure
11. Complete correspondence between the two types of measurements
is evident, reconfirming the validity of Regler's work. The edges
on both sides of x-ray diffraction lines, whose very existence had
been disputed, were made objectively visible by microdensitometer
trace (Figure 12). F. Lihl [10] showed that the critical angle of
total reflection could quantitatively account for the edges. There
were always only some crystallites on the rough iron surface in
the proper position with respect to the primary beam to explain
the relative infrequency and limited observability of the edges.
The specific geometry of the demountable x-ray tubes, differing
from later versions of commercial tubes, was perhaps beneficial,
and responsible for the inability of some laboratories to duplicate
Regler's results.

In summary, there is no doubt that Regler's work opened up an

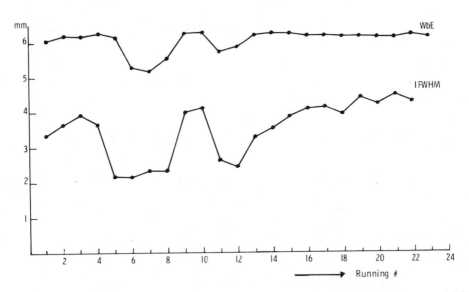

Figure 11.   Width between edges and integral FWHM [see Figure 10(B)]
of steel samples from dismantled railroad bridge.

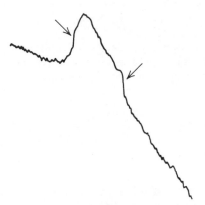

Figure 12. Edges of diffraction line (marked by arrows) recorded
by microdensitometer [10].

Figure 13. "Röntgenwagen" (X-ray Car) of Austrian Federal Railroads,
containing all equipment for "in vivo" nondestructive
testing. F. Regler stands in front of it.

era of nondestructive testing.. His successes led to the application
of the method in other cases, even outside Austria, as, e.g., on the
Margarethen bridge in Budapest. His further work was made easier by
the use of an x-ray railroad car, shown in Figure 13.

The events of this story, from early difficulties with the
materials testing experts that almost terminated Regler's scientific
career, to later successes and recognition, carry two messages apart
from the technical enlightenment:

> A talented young man has to be very cautious when
> presenting new and revolutionary views, particularly
> if his measurement methods are subjective and not fully
> understood.

> The powerful scientific establishment ought to be
> reluctant to clobber a promising and courageous young
> talent expounding new ideas, just because they are
> both preliminary and unconventional.

ACKNOWLEDGEMENTS

Facts and data for this article have been compiled from
sources readily accessible, supplemented by my best recollection
of references no longer available to me.

REFERENCES

1. Regler, F., "Röntgenographische Feingefügeuntersuchungen an
    Brückentragwerken", Z. Elektrochemie, 43 (1937) 547-57.

2. Regler, F., Verformung und Ermüdung metallischer Werkstoffe
    im Rontgenbild, C. Hanser, München, 1939.

3. Regler, F., "Über eine neue Methode zur vollständigen
    röntgenographischen Feinstrukturuntersuchung an technischen
    Werkstücken", Zeitschrift für Physik, 74 (1932) 547-64.

4. Herglotz, H., "Das Droppelkegelverfahren und seine Eignung für
    Untersuchungen der Verformung und Textur", Zeitschrift für
    Metallkunde, 46 (1955) 620-22.

5. Herglotz, H.K., "Simple Method of Determining Crystal
    Perfection", J. of the Electrochemical Society, 106 (1959)
    600-05.

6. Amorós, J.L., M.J. Buerger and M.C. de Amorós, The Laue Method,
    Academic Press, New York, 1975.

7. Arguello, C., "New Conical Camera for Single-Crystal Orienta-
    tion by Means of X-Rays", The Rev. Sci. Instruments, 38 (1967)
    598-600.

8.   Regler, F., "Neue Ergebnisse mit Röntgen-Rückstrahlverfahren,
     Teil I", Radex-Rundschau, 4 (1952) 166-80.

9.   Herglotz, H.K., "Ein Röntgenverfahren zur Bestimmung der
     Fehlerhaftigkeit des Aufbaues von Einkristallen", Acta Physica
     Austriaca, 15 (1962) 40-56.

10. Lihl, F., "Über die bei Röntgen-Interferenzlinien auftretenden
     Erscheinungen der Interferenzpunkstreuung under der kantigen
     Linienbegrenzung", Acta Physica Austriaca, 3 (1949) 156-83.

Chapter 4

THE APPLICATION OF X-RAY TOPOGRAPHY TO MATERIALS SCIENCE

S. Weissman

Rutgers University, New Brunswick, New Jersey

ABSTRACT

Basic x-ray topographic methods are described and principles of image contrast of the defect structure are briefly reviewed. Special emphasis is placed on showing the usefulness of these methods to problems in materials science and to the characterization of materials when applied in combination and in conjunction with x-ray methods of precise structural analysis.

INTRODUCTION

The term "topography" usually evokes associations with a discipline in geology that concerns itself with the survey and mapping of surface features of a locality. In analyogy, x-ray topography may be regarded as a branch of x-ray diffraction which aims to survey, to map and to characterize lattice defects in crystalline materials. When topographic methods are being used in reflection, the characterization of the defects will usually pertain mainly to surface features. On the other hand, transmission methods are capable of assessing the defect structure through relatively large crystal volumes. Early attempts in x-ray topography dealt principally with the visualization of relatively gross lattice defects [1,2]. In recent developments, however, there has emerged the aim for quantitative determination which is not solely restricted to gross defects but encompasses, also, fine features of the microstructure such as dislocations, point defects and their interactions [3-5,38,39].

The three major areas of materials research in which x-ray

topography found its greatest and most fruitful applications are:
(1) crystal growth and the study of crystal perfection, (2) deformation processes and (3) defect structure and properties of electronic
materials. On the basis of precise characterization and quantitative
analysis of microstructural features, aided by remarkable advances
in image contrast theory [6,7], x-ray topography is now not only
capable of explaining the mechanism of important physical and
chemical processes but can function also as a powerful tool for
quality control of technological processes [8].

After reviewing briefly the basic topographic methods and the
underlying principles of image contrast, the author will attempt to
show how the usefulness of x-ray topography to materials science can
be heightened by combining various topographic methods. Special
emphasis will be placed, therefore, on the synergistic interplay of
complementary methods and their application to various aspects of
materials science.

Topographic Methods

There are several ways in which various topographical methods
may be classified. One of the most useful criteria is the the type
of experimental setting employed [6]. Thus, one may distinguish
the Bragg case (reflection setting) or the Laue case (transmission
setting).

Bragg – Reflection Setting

Berg–Barrett Method: One of the earliest attempts to record
images of the diffracting crystal was made by Berg [1] by irradiating
the cleavage surface of a rock salt crystal at a considerable distance from the x-ray source and placing the film close to the crystal
surface. The Berg method was refined and further developed by
Barrett [2] and is often referred to as the Berg–Barrett or B–B
method. The incident beam is usually not monochromatized, only the
K_α lines being used. The divergence of the beam is usually much
wider than the half-width of the reflection (rocking curve) of the
crystal. The topographic images obtained by the B–B method result
from local variations of the integral reflection power of the
diffracted beam which, in turn, are caused by the structural variations on or beneath the surface. A B–B x-ray topograph of a tungsten
single crystal is shown in Figure 1(b). Due to the lattice misorientation which exists across the sub-boundaries, the substructure
becomes clearly delineated by this method [9]. Except for some geometrical distortion, the images are similar to those disclosed in
the light microscope by the etching method [Figure 1(a)]. Kingman
[10] has worked out a method by which topographic image distortions
can be analyzed utilizing the principle of differential geometry.
Kingman gave a procedure for extracting the true spatial configuration from the distorted image and showed the applicability of the

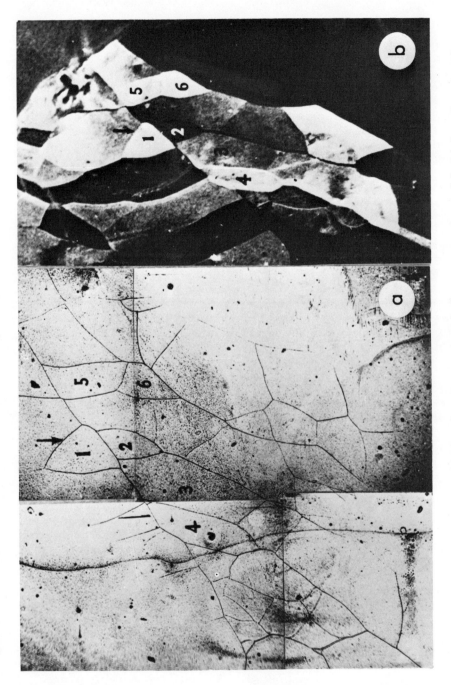

Figure 1. Correlation of substructure in tungsten crystals. (a) Light micrograph disclosed by etching with 3% boiling H_2O_2. (b) Berg-Barrett x-ray micrograph. Unfiltered radiation, copper target. Effective size of focal spot 40μ.

method by presenting a practical problem in trace analysis.

Besides disclosing the substructure of crystals, the B-B method has been employed to study twinned structures induced either by deformation or growth processes [11,12] and to study the dislocation arrangement in copper single crystals deformed to various stages of deformation [5]. Newkirk [13] shoed that single dislocations could be resolved by the B-B method and their Burgers vector experimentally determined. Newkirk concluded that the principal reason for disclosing individual dislocations by the B-B method is the diminution of primary extinction caused by the strain field surrounding dislocations. The image obtained by the B-B reflection method, however, derives from purely kinematic reflections and not from the dynamic interaction of the rays. In the kinematic region, the primary wave K_0 and the diffracted wave K_g are not coherently bound to each other as a wave field, and the coupling of K_0 and K_g is destroyed because of the large local lattice distortion. The term "extinction contrast", sometimes referred to this type of image in the literature, must be regarded as an unfortunate misnomer and should be discontinued.

Since individual dislocations can be resolved by the B-B method, it was possible to apply the method to the study of dislocation mobility. Pope, Vreeland and Wood [14] measured the $\{0001\}<11\bar{2}0>$ dislocation mobility in zinc crystals. Using a scratch technique, they introduced dislocations of predtermined line vectors and Burgers vector, and after subjecting the crystals to torsional pulses of various amplitudes, measured the displacement of the dislocations. Dislocation mobilities were also measured by the B-B method in aluminum [15], copper [16] and iron crystals [17].

Double-Crystal Diffractometer Method: The x-ray beam emerging from a line focus is reflected from a reference crystal A possessing a high degree of lattice perfection (Figure 2). The beam is then directed toward the second crystal B, which is the test crystal to be analyzed. The double-reflected beam is photographically or electronically recorded. The reflecting power of the test crystal is measured by rotating ("rocking") the crystal through its angular range of reflection [18]. If the reference crystal and the test crystal consist of the same kind of material, the same spacing of the reflecting plane can be used in both crystals and the crystals have, then, a perfectly parallel (n,-n) arrangement. For this arrangement the dispersion effect is absent, and the rocking curve thus obtained is essentially independent of the spectral distribution $\Delta\lambda$ of the radiation. The rocking curve becomes much narrower than any spectral line, and because of this property the method becomes eminently suitable to measure quantiatively small lattice tilts. When the test specimen is set at the flank of the rocking curve, corresponding to its steepest descent, tilts less than 0.1 second

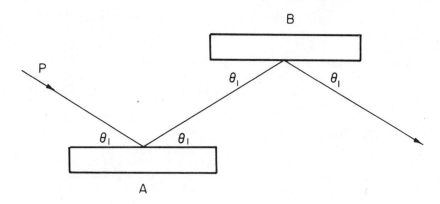

Figure 2. Double-crystal diffractometer (n,-n) arrangement.

and corresponding strains of $\Delta d/d$ less than 10^{-8} give rise to
detectable changes of the recorded intensity. Using this technique,
Bonse has shown that deformation of this magnitude occurs at distances up to 50 or 100 μm from the cores of single dislocations,
depending on the material examined [19].

   Combination of Reflection Topography with Double-Crystal
Diffractometry: Although the B-B method yields satisfactory topographical information regarding the lattice inhomogeneities of the
crystal surface, it is not directly amenable to a quantitative
evaluation of all the important substructure characteristics.
Particularly, the misorientation across the subgrain boundaries and
the nature of the boundaries - that is, whether they are pure tilt,
pure twist or a mixture of tilt and twist - cannot be ascertained
directly from the B-B topographs. To make such determinations
conveniently available, Weissmann, Evans and Intrater [20,21] combined the microscopic imaging capability of the B-B method with
the powerful analytical tool of the double-crystal diffractometer
method. In this combination method the monochromating crystal,
usually a dislocation-free silicon crystal, is retained for all the
different test crystals. The troublesome dispersion effects which
now arise, because the test specimens no longer reflect in a strictly
parallel arrangement, are removed by suppressing the K_{α_2} component
with a knife edge of the collimating system. The reflected
intensities are registered by a radiation detector and plotted as a
function of rotation angle. The width at half-maximum of the
rocking curve thus obtained is a measure of the crystal perfection.
For ideal crystals the half-width values extend only over a few
seconds of arc, while in imperfect crystals they may extend over
many minutes or even degrees and may also exhibit multiple peaks

instead of single peaks, encountered in more perfect crystals. In
order to locate the region on the crystal which contributes to each
part of the reflection curve, the entire angular range is remeasured,
using a photographic film. The specimen crystal is rotated in
discrete intervals and the film is slightly displaced following
each exposure. The film may be placed in front of the radiation
detector or close to the specimen or in both positions, and the
diffracted rays can thus be traced back to their origin on the
surface of the crystal.

This combination method was applied to elucidate the substructure characteristics and the relationship of various sub-
structural entities in deformed aluminum [20,21] and the substructure
and dislocation networks in tungsten [22]. This combination method
proved to be also very useful in determining the extent of shock-
induced plastic deformation by a small nuclear explosion and in
characterizing the induced defect structure in halite crystals [23].
The 5.3 ± 0.5 kT nuclear explosion, referred to as the Salmon event,
occurred at a depth of 827.8 m in the Tatum salt dome near
Hattiesburg, Mississippi. It was detonated in October 1964. The
salt dome consists of a predominantly halite core. The lattice
defects of halite crystal samples before and after the Salmon event
were investigated by the combination method described. At various
distances from the shot point, the half-widths of the rocking curves
of thirteen samples were measured. The cumulative distribution
curves of the half-widths at several representative distances are
shown in Figure 3(a). Figure 3(b) exhibits the dependence of the

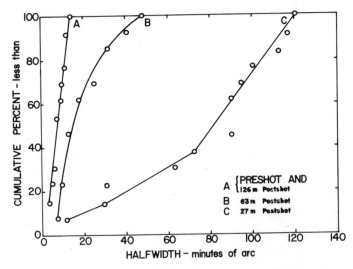

Figure 3(a).   Cumulative distribution of half-widths for preshot
and postshot specimens.

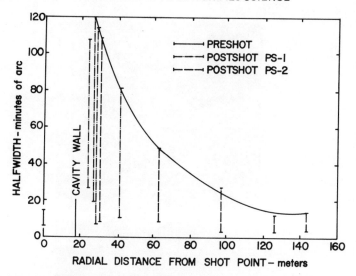

Figure 3(b).   Range of half-widths as a function of radial distance
from the shot point in Salmon event [23].

rocking curve half-widths on the radial distance from the shot point.
The results of the measurements indicated that beyond a radial distance of 127 m from the shot point the halite crystals had experienced no detectable shock-induced plastic deformation. The combination method revealed that at distances less than 126 m the strain
accommodation consisted of pronounced lattice kinking and eventually
led to a general fragmentation of the halite crystals. The breakup
of the halite lattice into small subgrain blocks and the misorientation between subgrain blocks increased when the shot point was
approached. Because of the relatively low values of the angular
range of reflection of the subgrain blocks in comparison with the
large induced misorientation between the blocks, it was inferred
that excessive accumulation of long-range stresses was absent and
that the shocked structure tended to assume a configuration of low
lattice strain.

        If the specimen is polycrystalline, which is the case encountered
most frequently when problems in materials science have to be solved,
the reflecting crystallites can still be analyzed by the combination
of x-ray topography and double-crystal diffractometry using a
method developed by the author [24,25]. Indeed, the method of
analysis is quite analogous to that of large crystals. To establish
a correlation between the surface topography of the reflecting
crystallites and their corresponding reflection curves, the images
are traced from the specimen surface to the recording film of a

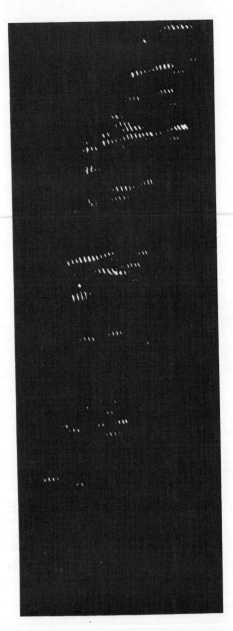

Figure 4. Detailed view of grain reflections of polycrystalline aluminum exhibiting reflection curves with multimodal intensity distribution characteristic of polygonization substructure. Specimen 96% cold-worked and annealed one hour at 300°C. Specimen rotation interval between successive spot reflections 4 minutes of arc. (422) reflection. Crystal monochromated CuK_{α_1} radiation.

cylindrical Debye-Scherrer camera. Because the beam incident on
the specimen was made parallel and monochromatic by reflection from
a silicon crystal, K_{α_2} being suppressed by a slit, relatively few
crystallites will reflect and thus the outward image tracing
can be carried out without complications. Depending on the perfection of the crystallites, the specimen is rotated in angular
intervals of seconds or minutes of arc, and the spot reflections
of the crystallites recorded for each discrete specimen rotation
are separated by film shifts. This multiple-exposure technique
gives rise to an array of spots for each reflecting crystallite,
as shown in Figure 4. Since each small crystallite may be regarded
as the test crystal of a double-crystal diffractometer, the array
of diffraction spots is analogous to the reflection curve of a
large single crystal and can be analyzed in a similar manner. Thus,
if the crystallites contain a substructure, the intensity distribution of the arrays of diffraction spots will be multipeaked in
complete analogy to the reflection curves obtained from large
single crystals containing substructure. The method was applied
to such diverse problems as the deformation substructure of nickel
and its correlation to mechanical properties [26], growth processes
in the recrystallization of aluminum [27,28], substructure formation
in iron during creep [29] and residual stress and grain deformation
in BeO ceramics [30]. This combination method proved to be also
very helpful in a recent study of a precipitation-hardening alloy
of the Ti-Al-Mo system. High-temperature strength and ductility
increases were obtained for this alloy by applying a step-aging
program based on principles of particle-dislocation interactions
[31]. Thus, precipitation of ordered Ti_3Al particles in the high-
temperature beta titanium phase field induced the moving dislocations to bypass the particles. Consequently, the slip became
dispersed in the hard β-phase and in this way gave rise to increases
in ductility. Small alpha titanium and Ti_3Al particles, subsequently
precipitated in the low-temperature phase field during the step-
aging program, were cut by the moving dislocations, thereby increasing the yield strength of the alloy. The materials science problem
involving x-ray topographic analysis was to elucidate the effect
of Ti_3Al particles as efficient agents of slip dispersal in the
high-temperature β-field. To achieve this objective, the x-ray
combination method was employed [32]. Figure 5 shows the rocking
curves for the β and Ti_3Al grain reflections prior to deformation
and after 1 and 2% tensile deformation at room temperature. The
identity of the respective phase reflections was established by
image tracing. It will be seen that after deformation of 2% the
angular range of reflection for the Ti_3Al particles did not exceed
12 minutes of arc, while that for the β-grains exceeded more than
1 degree. Hence the lattice misalignment in the β-matrix became
at least five times as large as that in the Ti_3Al particles. It
could be concluded, therefore, that room deformation of the β +
Ti_3Al phase caused the β-matrix to undergo preferentially plastic
deformation by homogeneous slip, while the undeformed, ordered

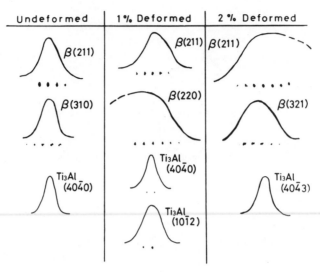

Figure 5.   Details of x-ray double-crystal diffractometer rocking
curves showing dependence of lattice misalignment of β
and Ti_3Al on deformation of Ti-7 Mo-16 Al alloy.   Step-
aging 980°C → 910°C/4 h → water quench.   Specimen
rotation 6 minutes of arc.

Ti_3Al particles functioned as hard particles controlling the slip
distribution.

Laue - Transmission Setting

Lang Method: Utilizing transmitted reflections, Lang [33-35]
developed a method by which the dislocation structure of thin
crystals can be disclosed provided the dislocation density is not
too high.   According to this method, schematically outlined in
Figure 6, a collimated x-ray beam a impinges on a crystal slab b,
which is placed approximately at right angles to the incoming
beam so that a set of transverse planes can satisfy the conditions
of Bragg reflection.   The transmitted reflected beam is recorded
on the film c, while the transmitted primary beam is prevented by
a stationary screen d from striking the film.   When the specimen
and film are kept stationary, one obtains so-called section topographs.   The distribution of lattice defects can be disclosed over
a range that extends beyond that of the volume irradiated in a
stationary crystal by coupling the crystal holder and film holder
mechanically and moving them to and fro during the exposure.   The
topographs thus obtained are called transverse topographs.

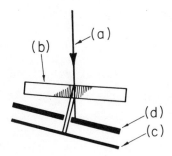

Figure 6.   Principle of Lang's x-ray transverse topographs.
            (a) Collimated x-ray beam; (b) crystal; (c) film;
            (d) stationary screen.

The maximum density for resolution of individual dislocations
depends upon the specimen thickness, and for traverse topographs it
is roughly 10^4 lines cm^{-2} for one specimen 0.1 mm thick. For section
topographs, the limiting density if 10^6 - 10^7 cm^{-2} for dislocations
and somewhat more than 10^7 for precipitates. To insure good resolution of the images, the incident beam should be carefully collimated,
and for optimum results should be monochromatized by prior reflection
from a monochromator crystal. X-rays of short wavelength such as
MoK_α or AgK_α are usually employed because these rays have a considerable penetrating power, permitting, therefore, the investigation
of thicker crystal slabs.

The Lang method has found wide applications, particularly to
the study of semiconductor crystals and devices, and a very comprehensive review up to 1969 is given by Meieran [8].

Transmission Topography in Thin Crystals by the Pendellösung
Effect [7]: In perfect crystals, dynamical interaction occurs
between the primary and secondary beams, analogous to that which
takes place between a pair of coupled pendulums whose frequencies
are nearly equal. The primary beam has all the energy at the
surface, but at a certain depth Δ this state of affairs is reversed
and the secondary beam has all the energy, and so the alternation
goes on. The depth of the layer, called extinction distance, at
which complete alternation takes place is the greater the more
nearly the velocities of the two waves approach. This closeness
is, however, limited by the diameter of the hyperbola of the dispersion surface, since the respective wave points must always lie
on two different branches of the hyperbola. Consequently, inter-
change must always take place. This periodic oscillation in
perfect crystals with the thickness of the direction of the energy

flow was first developed by Ewald and is called Pendellösung.

Experimental results to prove the existence of Pendellösung effects were first obtained by Kato and Lang [36], who studied the transmission of wedge-shaped slices of silicon, lithium fluoride and quartz crystals. Because the crystals studied had a wedge shape, both direct and reflected beams on emergence consisted of two super-imposed wave trains, of the same wavelength but differeing slightly in direction. The crystal had indeed different refractive indices for field 1, corresponding to the wave point of the α-branch of the dispersion surface, and field 2, corresponding to the respective wave point of the β-branch. Consequently, since in each beam these two wave trains were coherent, they gave rise to interference effects (beat effects) which manifested themselves as interference fringes when recorded on a photographic film. Each position in the Lang traverse topograph corresponded to a certain thickness of the crystal wedge and thus the fringe system pertaining to the reflected beam could be recorded. In agreement with the dynamical spherical wave theory developed by Kato [7], Pendellosung fringes (PF) can be dis-closed by Lang section topographs even in flat crystals. However, traverse topographs of flat crystals cannot reveal PF, since on traversing the crystal the PF images become superimposed, resulting in a diffuse background of equal intensity. The existence of small strains in the crystal become manifest in section PF topographs by the appearance of extra fringes. These appear because the strain field of the defect narrows the PF spacing, while the specimen thickness remains unaltered.

X-Ray Topography by Anomalous Transmission - the Borrmann Effect: The Pendellösung effects so far considered referred to thin perfect crystals for which the absorption was negligible. It applies to crystals for which the product of the linear absorption coefficient μ and thickness t is 1 or less. When thick perfect crystals are taken into consideration, e.g., crystals for which $\mu \cdot t \geq 10$, and when they are set to reflect a nearly parallel beam of monochromatic x-rays in transmission, the absorption coefficient of one of the two excited wave fields becomes abnormally small, while that for the other becomes abnormally large. Two standing waves, viz. A and B waves, are generated in the transverse direction to the reflection (hkl) planes, while the energy flow propagates parallel to the reflecting plances. The standing-wave field has nodes and antinodes parallel to the net planes. If the nodes, such as those of wave A, coincide with the atomic planes, the atoms are in regions of field zero and therefore remain at rest and do not eject photoelectrons. If the antinodal planes, such as those per-taining to wave B, coincide with the atoms, the absorption is enhanced beyond the normal value. Consequently, the standing-wave B will be highly absorbed by the atoms, whereas wave A will be transmitted with low attenuation. Upon leaving the crystal, the A wave splits into two components, namely, the transmitted beam R_o

and the diffracted beam R.   This effect of anomalous transmission
(AT) was first observed by Borrmann [37] on quartz and became known
as the Borrmann effect.

The observation of the Borrmann effect in crystals is indicative of a high degree of lattice perfection.   If any atoms are displaced from their normal nodes, for example, as a result of the
strain field of a dislocation, discontinuities of the anomalously
intense lines will be observed on the film.   The critical conditions
for AT to occur are being destroyed and normal absorption has set in.

Image Formation of the Defect Structure in X-Ray Transmission
Topography: Following Authier's classification [6], it is convenient
to distinguish three types of topographic images of the defect
structure, namely, direct, dynamic and intermediary.   Figure 7 may
serve to aid the visualization of the formation of these images.
The well-defined, primary x-ray beam passing through the slit S_1
enters the diffracting crystal at A and leaves it at B.   The
diffracting rays are parallel to the direction AC, and the formation
of the image of the defect structure, here exemplified by a dislocation line, takes place within the Borrmann fan ABC.   The dislocation
cuts the direct beam, and the region around the dislocation line

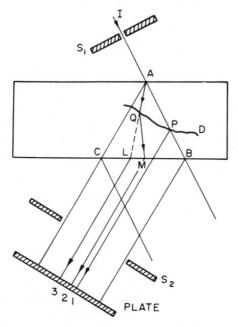

Figure 7.   Types of image formation in transmission topography [7].
        1.   Direct image.   2.   Intermediary image.   3.   Dynamical
        image.

reflects intensity from the direct beam, giving rise to the direct image at 1. This type of image results from the reflection of the deformed region, which behaves as if it were a small mosaic crystal, and the kinematical or geometric theory of diffraction is applicable.

When the crystal is sufficiently thin ($\mu t < 1$), there exists no coherence between the perfect lattice and the reflection from the distorted region. The direct image is therefore a purely kinematical image. A photographic plate placed in the path of the reflected beam will show images of the distorted regions as dark spots on the lighter background of the perfect lattice reflection. In terms of the Burgers vector $\vec{b}$ of a dislocation and the diffraction vector $\vec{g}$, minimum contrast is obtained when $\vec{g} \cdot \vec{b} = 0$ and maximum contrast is achieved when $\vec{g}$ is parallel to $\vec{b}$.

The dislocation line cutting the paths of the wave fields generated by the dynamical coupling of reflected and incident waves within the Borrmann fan casts a shadow if the crystal is sufficiently thick ($\mu t > 2$) and gives rise to the dynamical image at 3 (Figure 7). The dynamical image may be sharp or diffuse, depending on the location of the strain field relative to the surface of the crystal. If the strain field is near the beam exit surface, it will appear sharp and will become superimposed over the direct image. If the strain field is near the beam entrace surface, the entire Borrmann fan ABC will be excited, and the dynamical image will be diffuse and will lie to one side of the direct image. Depending on the sense of the lattice strain, the dynamical image can be enhanced (black) or depleted (white) with respect to the background.

The intermediary type of image results from the creation of new wave fields generated by the wave fields incident on the defect structure. The incident wave fields intercepted by the dislocation line decouple at Q (Figure 7) into their incident and reflected wave components, which on reentering the good portions of the crystal excite a new wave field. The wave point of this new wave field belongs to the other branch of the dispersion surface, which implies that a transfer of energy from one branch to the other had taken place. This process may be referred to as interbranch scattering. The new wave fields created at Q propagate along QM, giving rise to the intermediary image. As the path of the wave fields incident on the dislocation becomes closer to that of the direct beam, that is, when G moves into P, the intermediary image gradually merges into the direct image [6].

Detection of Micro Defects in Nearly Perfect Crystals: Micro defects such as point clusters and small dislocation dipole loops are usually too small to be disclosed by the Borrmann or Lang topographic techniques. For such small defects, the kinematically scattered intensity is relatively weak compared with the dynamically

scattered background and the defect structure cannot be delineated.
This difficulty was overcome by the "kinematical image technique"
(KIT) of Chikawa et al. [38]. A transmission, double-crystal
diffractometer method was applied such that the first (analyzing)
crystal and the second (test) crystal were arranged in the (n,-n)
parallel setting. An appropriate slit was interposed between the
two crystals which suppressed the dynamical background in the test
crystal by screening out all deflected rays of the Borrmann fan.
Thus, if micro defects were present in the test crystal, they could
be imaged without background intensities. Renninger accomplished
the suppression of the dynamically scattering by setting the first
monochromating crystal in asymmetric reflection (Bragg case) while
the test crystal was in symmetric transmission (Laue case) [39].
The angular difference, $\Delta\theta$, between the diffracting lattice planes
of the two crystals may be chosen arbitrarily and held constant
during the exposure. The essential feature of the method consists
of taking a topograph at an "operation point" on one of the slopes
of the rocking curve which represents the angular distribution of
the dynamically diffracted radiation, and the angular width of
this rocking curve is made very small compared with the one kinematically scattered by the micro defect core.

Application of a Combination Method Based on X-Ray Pendellösung
Fringes (PF), Double-Crystal Diffractometry and Transmission
Electron Microscopy (TEM) to Fracture [40-42]: A useful combination
method was recently developed which, applied to the study of fracture
of crystals, is capable of detecting and assessing the microplastic
regions and elastic stress fields generated by the crack tip. The
novel approach takes advantage of the disturbance of PF obtained by
x-ray transmission topography of wedge-shaped crystals. This method
is highly sensitive, since PF topography, like AT topography, is
based on the dynamical interaction of the x-ray wave fields inside
the crystal. The aspect of PF patterns most relevant to the fracture
study is the face that very small lattice disturbances have a pro-
found effect on the dynamical interaction of the wave fields and,
therefore, change drastically the PF pattern.

From the viewpoint of the materials scientist, the fracture
study of silicon crystals offers several attractive features.
First of all, these crystals can be obtained nearly dislocation-
free. Secondly, at elevated temperatures (650° to 1000°C) silicon
deforms plastically like a metal. It undergoes a ductile-brittle
transition at lower temperature and is totally brittle at room
temperature. Moreover, any lattice distortions introduced at
elevated temperatures become frozen in at room temperature, so that
the details of the defect structure which has been outlined by the
preceding x-ray investigation can then be disclosed by TEM without
fear that the defect structure was altered by the thinning process
during specimen preparation. Lastly, it sould be pointed out that
the extinction distance, ξ_g, is inversely proportional to the

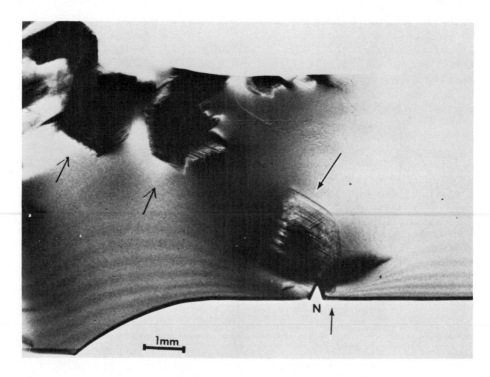

Figure 8.   Traverse PF topograph showing distribution of residual
            elastic strains constrained by microplastic zones in
            notched silicon bent at 800°C. Bending moment 6 kg mm.
            N = notch.

structure factor F of the reflecting (hkl) plane and, via F, it is
also inversely related to the atomic number Z of the crystal.   It
is expected, therefore, that perfect wedge-shaped crystals of low
Z values, namely, silicon, will yield a well-resolved PF pattern.

    Figure 8 may serve to illustrate the type of information which
can be obtained from PF patterns. The regions of plastic zones,
such as those associated with the notch N or those generated from
the specimen surface opposite the notch, are characterized by the
total destruction of the PF pattern. The direct image of the
plastic zone appears black. Adjacent to the direct image of the
plastic zone appears the depleted (white), dynamical image, indicating lattice curvature associated with the plastic strain field.
The regions of elastic strains are characterized by the bending of
PF and by the systematic narrowing of the fringe spacings when the
plastic zone is approached. Although the distortions of the PF
patterns are quite reminiscent of the optical fringe contours
encountered in photoelastic stress analysis of materials transparent to light, it must be remembered that the distortions of

the PF pattern result from displacements on an atomic scale and depend only on the transparency of perfect crystals to x-rays.

Once the plastic zones have been identified, their dislocation density can be quantitatively assessed by the method of double-crystal diffractometry. Thus, the primary beam is first reflected from a perfect silicon crystal and the monochromatized radiation is directed toward that area on the test crystal which is outlined by the destruction of the PF pattern. The reflecting power of the test crystal is measured by rotating the crystal through its angular range of reflection. The width β at half-minimum of the obtained rocking curve is a measure of the crystal perfection and can be related to the excess dislocation density of one sign D by [43] $D = \beta/3bt$, where b is the magnitude of the Burgers vector and t the linear size parameter of the crystal area investigated. Thus, the PF technique functions as a "guiding eye" to locate the microplastic zone which, subsequently, is evaluated quantitatively by the double-crystal diffractometer method.

The sensitivity of detecting microplastic zones can be gaged from the fact that a lattice misalighnment β of 45 seconds of arc was sufficient to destroy the PF pattern which corresponds to a dislocation density of $\sim 10^5$ cm^{-2}. The dislocation structure of the microplastic zone itself can be disclosed by TEM. It was shown that because the microplastic zones were strain-hardened, they were able to constrain the long-range residual elastic strains [41]. To assess the strain gradient in the vicinity of the notch tip, one may take advantage of the strain sensitivity which PF section topographs exhibit. Figure 9 shows a series of PF section topographs of a flat, notched silicon crystal taken at 0.5-mm intervals from the notch tip. It will be seen that additional fringes appeared as the notch was approached, and that at a distance of 0.5 mm away from the notch tip a total of five extra fringes were displayed.

The formation of these strain-hardened microplastic zones and the associated regions of locked-in, elastic strains explained the occurrence of the notch-brittle transition observed around 675°C. At low deformation temperatures, viz. 600°C, no microplastic zones other than the very small one at the notch tip were formed, and all the elastic strain energy at the notch tip found catastrophic release at a critical stress level, resulting in cleavage fracture. At an elevated deformation temperature, viz. 700°C, strain-hardened microplastic zones were generated, principally at the specimen surface, and these became effective agents to constrain long-range, elastic residual strains (Figure 8). Consequently, both the yield and fracture stress increased considerably. At still higher deformation temperatures, both the yield and fracture stress declined, because the dislocations in the plastic zones could rearrange

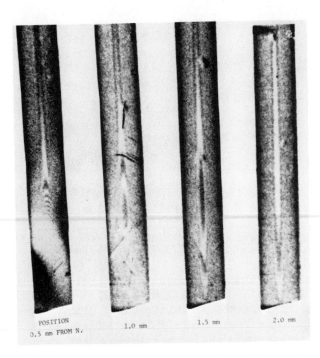

Figure 9.   PF section topographs of notched flat silicon crystal
bent at 700°C.

themselves into a configuration of lower energy by cross-slip and
climb, and thereby decrease the efficiency of the barrier effect
in constraining residual elastic strains [41].

It is empirically well known that residual elastic strains can
be removed by high-temperature annealing. By applying the combination method, one can obtain valuable information regarding the
mechanism by which this process occurs. The traverse topotgraph
of Figure 10 shows the effect of annealing at 1000°C for 1 hour on
the microstructure of the bent silicon specimen, shown in Figure 8.
Focusing attention on the regions marked by arrows (Figure 8), it
will be seen that upon annealing, dislocation loops emanated from
the microplastic zones and penetrated into the areas previously
occupied by the elastic residual strains. As a consequence of
this plastic relaxation process, shown by the dislocation sweeping,
the elastic residual strains became relaxed and largely disappeared.
This was manifested by the regained regularity of the striped PF
pattern. It is interesting to note that the regions of the
dynamical (depleted) image in Figure 8 (indicated by arrows),
indicative of lattice curvature, exhibited after annealed a three-
dimensional polygonization substructure (Figure 10). The latter

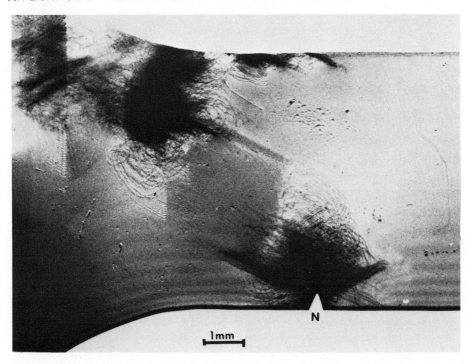

Figure 10.   Traverse topograph with polygonization detail showing
              the effect of annealing upon strain accommodation. Si
              bent at 800°C and annealed at 1000°C for 1 hour.

became apparent from the regularity of the alternating black-white
contrast effects. Thus, removal of the residual elastic strains
is brought about by a relaxation process of the microplastic zones
which constrained the elastic strains.

     To study the lattice defects in thick crystals by means of
the Borrmann (AT) effect, the x-ray divergent beam method [44] can
be effectively employed using the experimental arrangement shown
in Figure 11. This method offers the advantage that it never
"loses sight of" the elastically strained lattice planes, for if
the Bragg angles are slightly changed, some rays of the divergent
beam will still be in reflecting position and an AT pattern will
still be recorded. Furthermore, compared to the parallel beam
method, the exposure times for obtaining AT patterns are five to
ten times shorter. Figure 12 shows an AT pattern of a germanium
crystal with the specimen kept stationary. If now the synchronized
specimen and film translation is employed as shown in Figure 11,
one obtains a topographic mapping of the lattice perfection of the

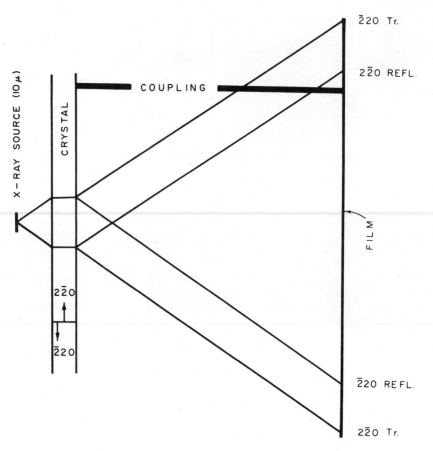

Figure 11.   Experimental arrangement for obtaining AT patterns by
             the divergent beam method.

crystal.   Thus, if adjacent areas exhibit such a high degree of
lattice perfection that AT patterns can be obtained, the pattern
of Figure 11 will continuously broaden as the crystal is traversed.
If, on the other hand, lattice defects are encountered which destroy the AT conditions, nonreflecting areas will appear on the
divergent beam pattern.   To elucidate the defect structure which
caused the elimination of AT, the divergent incident beam is transformed into a parallel beam at the corresponding location of the
specimen traverse.   This transformation is quickly performed by
increasing the distance between specimen and x-ray source and by
placing a slit aperture in front of the specimen.   It is interesting
to note that the exposure of an anomalous transmission topograph
of a 1-mm thick germanium crystal using CuK_α radiation took 20 hours

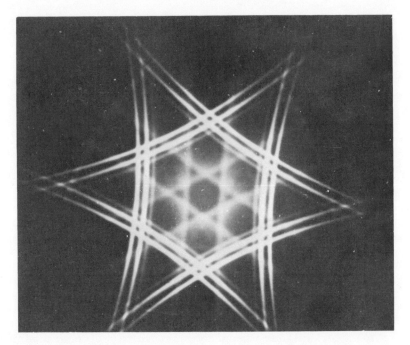

Figure 12.   Anomalous transmission of divergent beam pattern of
             unstrained germanium.

for the parallel beam technique while the entire scan of the crystal
by the divergent beam method required only two hours. Thus the
divergent beam scan functioned as an efficient, quick surveyor of
the defect structure which, once its location was established, could
be subsequently imaged by converting the incident divergent beam
into a parallel beam. It was shown that when crystals fractured
in a totally brittle manner, such as the semiconductor crystals at
ambient temperature, AT patterns could be obtained in closest
proximity to the fracture surface or at sites 0.5 mm adjacent to
the notch. By contrast, deformation in ductile materials such as
copper eliminated AT entirely when the dislocation density exceeded
10^5 cm^{-2} [44]. Minari, Pichaud and Capella [45] studied high-purity
copper crystals under increasing tensile stress at room temperature
using the Borrmann transmission x-ray topography method. Fresh dislocations were produced at a resolved shear stress of 1.5 g mm^{-2}.
Dislocations were generated at surface sites and the nucleation of
half-loops by the same sources gave rise to pileups. The local
stress resulting from the pileups was sufficient to occasion cross-
slip, and this observation formed the basis of a mode of a dislocation multiplication mechanism.

    Greenhut, Kingman and Weissmann [46] studied compressively
shock-loaded copper single crystals employing complementary methods

of x-ray topography, TEM and SEM. It was found that the defect structure introduced depended on the crystal orientation relative to the direction of shock propagation and also upon the existence or absence of a reflected tensile wave. The defect structure induced by compressive shock loading exhibited clusters of micro twins alternating with a tangled dislocation substructure. Failure zones did not occur. The defect structure exhibited a low net microscopic misorientation which amounted to less than two degrees of arc. By contrast, specimens subjected to a reflected tensile pulse exhibited a failure zone, and the defect structure in its vicinity showed a substantial macroscopic misorientation.

Combination Method of Structure Analysis Based on X-Ray Divergent Beam Methods: The x-ray diffraction method using a divergent beam as the x-ray source is well suited to provide the basis for a quantitiative strain analysis of single crystals [47] and to offer a visualization of the same defect structure as disclosed by x-ray topography and by TEM. Using this method, an electron beam originating from an electron gun is focused by means of an electromagnetic lens onto the tip of the vacuum-tight tube closed by a thin metal foil. Since the metal foil is bombarded by the electron beam, it functions as an x-ray target. By operating the tube at a suitable voltage, an x-ray beam composed mainly of characteristic radiation emerges from the tip of the tube, exhibiting a divergence of nearly 180 degrees of arc. At the point of emergence the beam size is about 10 μm in diameter. When this beam impinges on the specimen, which is placed close to the tip of the tube, diffraction patterns of the characteristic spectrum in transmission, as well as in back-reflection, may be recorded. Since the x-ray source is located outside of the specimen, the conic patterns thus obtained are referred to frequently as pseudo-Kossel patterns to distinguish them from the true Kossel patterns, which are obtained by generating the x-ray source inside the crystal.

The method offers the unique advantage that various (hkl) planes, even sets of planes of a form, which satisfy the Bragg condition for reflection of the impinging divergent beam, will be recorded as separate reflections without the necessity of rotating the crystal. The diffraction cones intersect the film in ellipse-like figures in the back-reflection region, while in transmission two types of patterns are obtained, namely, the diffraction conic and the deficiency conic patterns (Figure 13). Both types of the transmission pattern emanate from the same irradiated small crystal volume traversed by the beam.

It is the combined application of the back-reflection and transmission arrangement in the divergent beam method which jointly with TEM form the basis of the combination method of structural analysis.

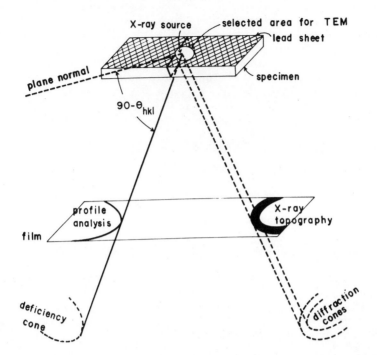

Figure 13.   Schematic diagram illustrating the origin of deficiency
and diffraction conics in transmission divergent beam
patterns from the selected area of the specimen.

The back-reflection pseudo-Kossel pattern forms the principal
basis for precision measurements of the interplanar spacings of the
crystal.   Since the exposure times of the back-reflection pattern
are very short, varying from seconds to a few minutes depending
on the diffracting power of the material, the elliptical pattern
can be repeated several times by varying, with the aid of precision
spacers, the film positions in a controlled manner.   In this manner,
the slopes of the ray paths, diffracted by the specimen, can be
determined using a method of least squares.   With the aid of these
parameters, the value of the interplanar spacing of the corresponding Bragg reflection, d, is obtained [47].

The output of d-spacing computation of the various (hkl) reflections may now be used for the computation of a complete stress-
strain analysis of the crystal [48].   If the crystal contains homogeneous, residual strains, then the changes in the interatomic
spacings Δd_{hkl} induced by the strains are manifested sensitively
by changes in the parameters of the ellipses and hence in the
observed d-spacings.   If the strained crystal is sampled from many
different directions, a collection of $\Delta d/d$ values is obtained which

characterizes the strain distribution. The strain analysis which
was developed recognizes the $\Delta d/d$ values to represent elements of
an average strain tensor, and by measuring more than six independent
(hkl) reflections the normal matrix equation is constructed and
the average tensor <T> determined. The principal strains ε_{111},
ε_{222}, ε_{333} are obtained as the eigenvalues of the matrix solution.
A computer program was written which uses as its input data the
collection of $\Delta d/d$ values, and supplies as output data the normal
principal strains ε_{111}, ε_{222}, ε_{333} and their crystallographic
directions. If the elastic constants of the crystal are known, the
complete stress-strain configuration is obtained [49]. Thus, the
maximum magnitude and the direction of the shearing strain on a
given set of crystallographic planes are obtained, and the set of
crystallographic planes on which the maximum value of the shearing
maxima occurs can also be determined.

The stress-strain analysis based on the back-reflection divergent beam method has been successfully performed for cases where
the strain inhomogeneities in the crystal were very small compared
to the residual, homogeneous elastic strains. Examples can be
found in:

(a) strain configureation of precipitation-hardened aluminum-
copper crystals where the Guinier-Preston zones of the semicoherent
θ' precipitates exert large, overall homogeneous strains in the
matrix [48],

(b) strains induced in the cubic matrix of copper-gold crystals
by order transformation to tetragonal copper-gold [49],

(c) strains induced in oxidized α-titanium crystals by the
ordering of oxygen atoms [50], and

(d) strain distribution generated during the early stages of
neutron-irradiated quartz crystals [51].

It has been shown that the stress-strain analysis is not applicable
to mechanically deformed crystals because the plastic strain inhomogeneities, caused by the induced dislocation structure, are large
and vary from area to area. Consequently, a representative sampling
of the strain distribution cannot be obtained if the $\Delta d/d$ values are
being gathered from the back-reflection divergent beam pattern.
Local strain inhomogeneities caused by mechanical deformation can
be determined, however, with the aid of divergent beam patterns taken
in transmission, and the basis of such analysis will be presently
described.

X-Ray Line Profile Analysis - Selected Area X-Ray Topography
Based on Transmission Patterns [52-55]: It may be seen with the
aid of Figure 13 that in transmission the reflection cones (also
referred to as diffraction cones), as well as the deficiency cones,
emanate from the same irradiated small crystal volume. If the

crystal is thick and of low absorption, namely, beryllium, the diffraction conics will have a large width, as shown in Figure 13, since each point along the path of the incident beam passing through the crystal functions as a vertex of a diffraction cone. Strain inhomogenieties caused by dislocations will be spread out over the diffraction conics and, therefore, it is the diffraction conics which are being used to obtain x-ray topographs. In contrast to the diffraction conics, the deficiency conics are sharp, since the vertex of the absorption cone isthe x-ray point source. Consequently, the sharp lines of the deficiency conics are being used for the analysis of the x-ray line profiles.

Before imaging the defect structure by x-ray topography, the divergent incident beam has to be transformed into a nearly parallel beam. This is accomplished by translating the specimen away from the x-ray source. The translation must be performed along the region of interest in the deficiency pattern, for only then are the identical diffraction conditions maintained. The technique of specimen translation, using a wire grid to preserve the area of interest and employing an aperture to convert the divergent beam into a parallel beam, has been described in detail [52]. Regions of special interest for analysis of the deficiency pattern are the intersections of several (hkl) lines. Such intersections pertain to a conic section which is common to the corresponding intersecting (hkl) planes. If the crystal is subjected to deformation, the line profiles will be sensitively changed. Microdensitometric measurements of the profiles carried out in the immediate vicinity of the intersection of the deficiency pattern may yield valuable information concerning the anisotropic deformation behavior of the crystal. Such information was obtained from compressive and tensile deformation of beryllium [53,54]. The corresponding topographs are obtained from the imaging of the intersection in the diffraction pattern and constitute the basis of selected area x-ray topography. Such topography affords a visualization of the defect structure of the crystal area pertaining to the corresponding line profile analysis.

To establish a correlation between the defect structure analyzed by the x-ray method and that disclosed by TEM, a thin lead sheet with a hole 1 mm in diameter is attached to the entrance surface of the specimen, as shown in Figure 13. TEM foils are prepared after the completion of the x-ray selected area studies by cutting a 2.5-mm disk centered around the hole [55].

High-Speed X-Ray Topography: Many dynamical processes in materials science such as crystal growth, phase transformation and plastic deformation occur at high speed. To study such processes by x-ray topography, a great amount of research was devoted toward improving the efficiency of the detection system of the signals. Improved speed, but without high resolution, has been obtained in the laboratory with x-ray television systems. Some investigators developed systems based on x-ray sensitive vidicons [56], on conventional television with phosphor screens, or with image intensifier interposed [57]. Meieran [58] reported a method of observing x-ray topographs by means of a secondary electron conduction vidicon system. Hashizume, Kohra, Yamaguchi and Kinoshita [59] employed a double-crystal diffractometer with an image orthicon camera, and Rozgonyi, Haszko and Statile [60] used charge-sensitive diode arrays to view x-ray topographs. With conventional x-ray sources, the exposure times of these methods varied between 0.5 second and one minute, with resolutions ranging from 10 to 30 µm over areas of a few millimeters square.

A very promising new development of high-speed x-ray topography retaining a high-resolution capability has emerged by the use of synchrotron radiation sources. Hart [61] studied perfect silicon and cleaved lithium fluoride crystals employing NINA (in Daresbury, 5 GeV and 25 mA) synchrotron radiation for x-ray diffraction topography. He found that with regard to both spatial and strain resolution, the synchrotron-source diffraction topographs were competitive with topographs obtained with conventional x-ray sources. The exposure times, however, were reduced from several hours per centimeter square to one second per centimeter square, using standard recording techniques with Ilford nuclear emulsions. With electronic detection systems, the speed is expected to be roughly proportional to the inverse square of the desired resolution and signal-to-noise ratio, so that exposure times of milliseconds are feasible. Thus, many dynamic processes which hitherto could not be studied by conventional x-ray topography will become amenable to investigations with synchrotron radiation sources.

CONCLUSIONS

In correlating the visualization of lattice defects, as disclosed by x-ray topography, to quantitative x-ray diffraction analyses of materials, such as line-profile and rocking-curve analyses, one may well have achieved a research goal that can give added confidence in interpreting the behavior of structure-sensitive properties. Nevertheless, it is important to realize that the x-rays, when passing through the material, average out details of the defect structure. It is evident that the extreme resolution achieved by TEM cannot be obtained by x-ray topography. There are, however, features in x-ray topography which offer distinct advantages over TEM. The principal disadvantage of TEM lies in the fact that only small specimen areas can be studied, and because of the limited restriction of the field of view it is frequently difficult for TEM to discern the essential features and parameters that determine and govern the structure-sensitive properties of the material. The x-ray methods, on the other hand, are capable of getting a better overall view of the defect structure. Examples were cited such as the distribution of microplastic zones and residual strains in silicon crystals, and dislocation densities and crack formation in ductile and brittle materials. X-ray and TEM methods in combination and possibly even complemented by SEM studies appear to be very useful for the characterization of materials. The combination of methods, however, becomes particularly powerful when x-ray topography is employed, not only for quick and effective location of a defect structure, but also for image contrast analyses when combined with a precise, quantitative method of structural analysis.

ACKNOWLEDGEMENT

Various parts of the work discussed in this chapter were carried out in the Materials Research Laboratory at Rutgers University with the support of the Materials Research Division of NSF; U. S. Army Research Office, Durham, North Carolina; and the Rutgers Research Council.

REFERENCES

1. Berg, W.F., "Über eine Röntgenographische Methode zur Untersuchung von Gitterstörungen an Kristallen", Naturwiss., 19 (1931) 391-96; "About the History of Load of Deformed Crystals", Z. Kristallogr., 89 (1934) 286-94.

2. Barrett, C.S., "A New Microscopy and Its Potentialities", Trans. IMD-AIME, 161 (1945) 15-64.

3. Nagata, N. and T. Vreeland, "Basal Dislocation Interaction with a Forest of Non-basal Dislocation in Zinc", Phil. Mag., 25 (1972) 1137-50.

4. Obst, B., H. Auer and M. Wilkens, "Untersuchungen zur Versetzungs anordnung in Verformten Kupfereinkristallen. II. Die Versetzungs anordnung im Bereich II", Mat. Sci. Eng., 3 (1968) 41-55.

5. Becker, C. and B. Pepel, "Dislocation Helices in Molybdenum", Phys. Stat. Sol., 32 (1969) 443-46.

6. Authier, A., "Contrast of Images in X-Ray Topography", in Modern Diffraction and Imaging Techniques in Material Science, ed. by S. Amelinckx, R. Gevers, G. Remaut and J. Van Landuyt. Amsterdam, North Holland Publishing Co., 1970, pp. 480-520.

7. Kato, N., "Dynamical Theory of Imperfect Crystals", in X-Ray Diffraction, New York, McGraw-Hill, 1974, pp. 350-438.

8. Meieran, E.S., "The Application of X-Ray Topographical Techniques to the Study of Semiconductor Crystals and Devices", Siemens Rev., 4th Special Issue 37 (1970), 39-80.

9. Nakayama, Y., S. Weissmann and T. Imura, "Substructure and Dislocation Networks in Tungsten", in Direct Observation of Imperfections in Crystals, ed. by J.B. Newkirk and J.H. Wernick, New York, Interscience, 1962, pp. 573-92.

10. Kingman, P.W., "A Method of Analyzing Image Distortions in Diffraction Topographs", J. Appl. Crystl., 6 (1973) 12-19.

11. Hirose, M., "The Strain Distribution Within and Around Deformation Twin Bands of Iron Single Crystals", Japan. J. Appl. Phys., 11 (1972) 309-18.

12. Chikawa, S. and S.B. Austerman, "X-Ray Diffraction Contrast of Inversion Twin Boundaries in BeO Crystals", J. Appl. Cryst., 1 (1968) 165-71.

13. Newkirk, J.B., "The Observations of Dislocations and Other Imperfections by X-Ray Extinction Contrast", Trans. MS-AIME, 215 (1959) 483-97.

14. Pope, D.P., T. Vreeland, Jr. and D.S. Wood, "Mobility of Edge Dislocations in the Basal Slip System of Zinc", J. Appl. Phys., 38 (1967) 4011-18.

15. Gorman, J.A., D.S. Wood and T. Vreeland, Jr., "Mobility of Dislocations in Aluminum", J. Appl. Phys., 40 (1969) 833-41.

16. Jassby, K.M. and T. Vreeland, Jr., "An Experimental Study of the Mobility of Edge Dislocations in Pure Copper Single Crystals", Phil. Mag., 21 (1970) 1147-68.

17. Turner, A.P.L. and T. Vreeland, Jr., "The Effect of Stress and Temperature on the Velocity of Dislocations in Pure Iron Mono-Crystals", Acta Met., 18 (1970) 1225-35.

18. James, R.W., The Optical Principles of the Diffraction of X-Rays, London, G. Bell and Sons, Ltd., 1948, pp. 304-17.

19. Bonse, U., "X-Ray Picture of the Field of Lattice Distortion Around Single Dislocation", in Direct Observation of Imperfections in Crystals, ed. by J.B. Newkirk and J.H. Wernick, New York, Interscience, 1962, pp. 431-60.

20. Weissmann, S. and D. Evans, "An X-Ray Diffraction Study of the Substructure of Fine-Grained Aluminum", Acta Cryst., 7 (1954) 733-37.

21. Intrater, J. and S. Weissmann, "An X-Ray Diffraction Method for the Study of Substructure of Crystals", Acta Cryst., 7 (1954) 729-32.

22. Weissmann, S., B.S. Lement and M. Cohen, "Substructure in Refractory Metals", in Refractory Metals and Alloys II, ed. by M. Semchyshen and I. Perlmutter, New York, Interscience, 1963, pp. 117-58.

23. Braun, R.L. J.S. Kahn and S. Weissmann, "X-Ray Diffraction Analysis of Plastic Deformation in the Salmon Event", J. Geophys. Res., 74 (1969) 2103-17.

24. Weissmann, S., "Method for the Study of Lattic Inhomogeneities Combining X-Ray Microscopy and Diffraction Analysis", J. Appl. Phys., 27 (1956) 389-95.

25. Weissmann, S., "Substructure Characteristics of Fine-Grained Metals and Alloys Disclosed by X-Ray Reflection Microscopy and Diffraction Analysis", Trans. ASM, 52 (1960) 599-614.

26. Weissman, S., "Quantitative Study of Substructure Characteristics and Correlation to Tensile Property of Nickel and Nickel Alloys", J. Appl. Phys., 27 (1956) 1335-44.

27. Weissmann, S., "Growth Processes in Recrystallization of Aluminum", Trans. ASM, 53 (1961) 265-81.

28. Weissmann, S., T. Imura and N. Hosokawa, "Recrystallization and Grain Growth of Aluminum", in Recovery and Recrystallization of Metals, ed. by L. Himmel, New York, Interscience, 1963, pp. 241-67.

29. Garofalo, F. L. Zwell, A.S. Keh and S. Weissmann, "Substructure Formation in Iron During Creep at 600°C", Acta Met., 9 (1961) 721-29.

30. Smith, D.K. and S. Weissmann, "Residual Stress and Grain Deformation in Extruded Polycrystalline BeO Ceramics", J. Am. Ceramic Soc., 51 (1968) 330-36.

31. Hida, M. and S. Weissmann, "High-Temperature Strength and Ductility Increases of Ti-Mo-Al Alloys by Step-Aging", Met. Trans., 6A (1975) 1541-46.

32. Hamajima, T. and S. Weissmann, "Thermal Equilibria and Mechanical Stability of Ti$_3$Al Phase in Ti-Mo-Al Alloys", Met. Trans., 6A (1975) 1535-39.

33. Lang, A.R., "Direct Observation of Dislocations by X-Ray Diffraction", J. Appl. Phys., 29 (1958) 597-98.

34. Lang., A.R., "Some Recent Applications of X-Ray Topography", in Advances in X-Ray Analysis, Vol. 10, ed. by J.B. Newkirk and G.R. MAllett, New York, Plenum Press, 1967, pp. 91-107.

35. Lang, A.R., "Recent Applications of X-Ray Topography", in Modern Diffraction and Imaging Techniques in Material Science, ed. by S. Amelinckx, R. Gevers, G. Remaut and J. Van Landuyt, Amsterdam, North Holland Publishing Co., 1970, pp. 407-79.

36. Kato, N. and A.R. Lang, "A Study of Pendellösung Fringes in X-Ray Diffraction", Acta Cryst., 12 (1959) 787-94.

37. Corrmann, G., "Über Extinktions diagramme von Quarz", Phys. Z., 42 (1941) 157-62.

38. Chikawa, J., Y. Asaeda and I. Fujimoto, "New X-Ray Topographic Technique for Detection of Small Defects in Highly Perfect Crystals", J. Appl. Phys., 41 (1970) 1922-25.

39. Renninger, M., "Topographic Observation of Micro Defects (e.g., 'Swirls') in Nearly Perfect Crystals", J. Appl. Cryst., 9 (1976) 178-80.

40. Weissmann, S., Y. Tsunekawa and V.C. Kannan, "Fracture Studies in Silicon Crystals by X-Ray Pendellösung Fringes and Double-Crystal Diffractometry", Met. Trans., 4 (1973) 376-77.

41. Tsunekawa, Y. and S. Weissmann, "Importance of Microplasticity in the Fracture of Silicon", Met. Trans., 5 (1974) 1585-93.

42. Tsunekawa, Y. and S. Weissmann, "Dislocation Generation Associated with Crack Growth of Silicon Containing Precipitate Defect Structure", Mat. Sci. Eng., 17 (1975) 51-56.

43. Hirsch, P.B., "Mosaic Structure", in Progress in Metal Physics, Vol. 6, ed. by B. Chalmers and R. King., London, Pergamon Press, 1956, p. 282.

44. Weissmann, S. and Z.H. Kalman, "Anomalous Transmission in Strained Ductile and Brittle Crystals by the Divergent X-Ray Beam Method", Phil. Mag., 15 (1967) 539-47.

45. Miniari, F., B. Pichaud and L. Capella, "X-Ray Topographic Observation of Dislocation Multiplication by Cross-Slip in Cu Crystals", Phil. Mag., 31 (1975) 275-84.

46. Greenhut, V., P.W. Kingman and S. Weissmann, "The Response of Copper to High Strain Rate Deformation as Disclosed by Electron-Optical and X-Ray Methods", in Microstructural Science, Vol. 3, ed. by P.M. French, R.J. Gray and J.L. McCall, New York, American Elsevier, 1975, pp. 475-90.

47. Ellis, T., F.L. Nanni, A. Shrier, Weissmann, S., G.E. Padawer and N. Hosokawa, "Strain and Precision Lattice Parameter Measurements by the X-Ray Divergent Beam Method", J. Appl. Phys., 35 (1964) 3364-73.

48. Imura, T., S. Weissman and J.J. Slade, Jr., "A Study of Age-Hardening of Al-3.85% Cu by the Divergent X-Ray Beam Method", Acta Cryst., 15 (1962) 786-93.

49. Slade, J.J., S. Weissmann, K. Nakajima and M. Hirabayashi, "Stress-Strain Analysis of Single Cubic Crystals and Its Application to the Ordering of CuAuI. Paper II", J. Appl. Phys., 35 (1964) 3373-85.

50. Weissmann, S. and A. Shrier, "Strain Distribution in Oxidized Alpha Titanium Crystals", in The Science Technology and Applications of Titanium, New York, Pergamon Press, 1970, pp. 441-51.

51. Weissmann, S. T. Imura, K. Nakajima and S.E. Wisnewski, "Lattice Defects of Quartz Induced by Fast Neutron Irradiation", Proceedings of Crystal Lattice Defects, J. Phys. Soc. of Japan, 18, Supplement III (1963) 179-88.

52. Glass, H.L. and S. Weissmann, "Synergy of Line Profile Analysis and Selected Area Topography by the X-Ray Divergent Beam Method", J. Appl. Cryst., 2 (1969) 200-09.

53. Glass, H.L. and Weissmann, S., "Application of the X-Ray Synergy Method to Analysis of Room-Temperature Compression of Beryllium Crystals", Met. Trans., 2 (1971) 2865-73.

54. Kannan, V.C. and S. Weissmann, "Deformation Substructure in Beryllium After Prism Slip", J. Appl. Phys., 42 (1971) 2632-38.

55. Weissmann, S. and V.C. Kannan, "Deformation Studies by a New Method Combining Analyses of X-Ray Line Profile and X-Ray Topography with Transmission Electron Microscopy", J. Mat., 7 (1972) 279-85.

56. Chikawa, J. and I. Fujimoto, "X-Ray Diffraction Topography with
     a Vidicon Television Image System", Appl. Phys. Lett., 13
     (1968) 387-89.

57. Lang. A.R. and K. Reifsnider, "Rapid X-Ray Topography Using a
     High-Gain Image Intensifier", Appl. Phys. Lett., 15 (1969)
     258-60.

58. Meieran, E.S., "Video Display of X-Ray Images, I. X-Ray
     Topographs", J. Electrochem. Soc., 118 (1971) 619-28.

59. Hashizume, H.A., K. Kohra, T. Yamaguchi and K. Konoshita,
     "Application of an Image Orthicon Camera Tube to X-Ray
     Diffraction Topography Utilizing a Double-Crystal Arrangement",
     Appl. Phys. Lett., 18 (1971) 213-14.

60. Rozgonyi, G.A., S.E. Haszko and J.L. Statile, "Instantaneous
     Video Display of X-Ray Images with Resolving Capabilities
     Better Than Fifteen Microns", J. Appl. Phys. Lett., 16 (1970)
     443-46.

61. Hart, M., "Synchrotron Radiation: Its Application to High-
     Speed, High-Resolution X-Ray Diffraction Topography", J. Appl.
     Cryst., 8 (1975) 436-44.

Chapter 5

RESIDUAL STRESS MEASUREMENT BY X-RAYS: ERRORS, LIMITATIONS
AND APPLICATIONS

C. O. Ruud, University of Denver, Denver, Colorado

and

G. D. Farmer, Jr., U.S. Army MERDC, Fort Belvoir,
        Virginia

INTRODUCTION

In a recent Air Force Materials Conference on residual stress
analysis the fact that x-ray diffraction was the only nondestructive method presently available which could measure
residual stresses in most crystalline materials, i.e., metals and
ceramics, was acknowledged. However, it was also recognized that
for the many successful applications there were also a number of
mis-applications due to inexperience, over-optimism of just plain
poor procedures on the part of the engineer or technician performing the measurements. Finally it was concluded that the
greatest impediment to greater applications of the x-ray stress
method was the limited field adaptability and portability of
existing instrumentation.

This chapter briefly explains the principles of x-ray diffraction stress analysis, discusses the most frequently encountered
sources of error, and finally summarizes the results of a program
(Contract DAAK02-75-C-0081, U.S. Army MERDC, Ft. Belvoir, Virginia)
to develop a truly portable x-ray stress analyzer.

PRINCIPLES

A rather elementary illustration of the effect of a uniaxial
stress in the plane of the surface of a material, σ_x, upon the
interplanar spacing, d, of the set of planes for which the normal
is at an angle ψ to the surface normal is shown in Figure 1. The
relation of d to this stress as derived in the first reference [1]

101

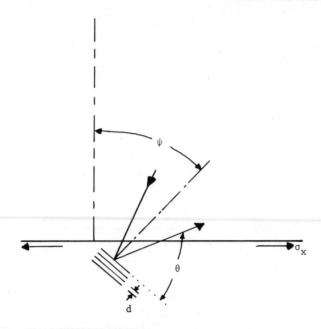

Figure 1.   Stress and interplanar spacing, d, relation.

from elemental stress-strain relationships and x-ray diffraction
principles for the double exposure technique is:

$$\sigma_x = \frac{E}{(1 + \nu)}\ \frac{1}{\sin^2 \psi}\ \frac{d_\psi - d_1}{d_1} \tag{1}$$

where   E = modulus of elasticity

ν = Poisson's Ratio

ψ = angle between surface and plane normals

d_ψ = interplanar spacing of those planes for which the
       normal is at an angle ψ to the surface normal

d_1 = interplanar spacing of those planes for which the
       normal is at $\psi = 0°$

Under uniaxial stress in the x direction d_ψ increases under tension
and decreases under compression.

                    Techniques for Measuring Residual Stresses

      The most familiar techniques are the two that apply the
diffractometer with an x-ray detector and the two that apply x-ray

film for determination of the d space, or actually two theta
degrees [1]. The d space is related to two theta by the Bragg
equation:

$$\frac{1}{d} = \frac{2}{\lambda} \sin \theta$$

where λ = wavelength of the x-radiation

            θ = Bragg angle

     By and large the diffractometer techniques use commercial
instruments with gas proportional or scintillation x-ray detectors.
Teh detector is scanned over a few 2θ degrees and these data are
used to establish the position of the diffracted peak by curve
fitting or finding the centroid of the peak. The two are described
as the double exposure technique and the sine square psi technique.

     The two exposure technique, or double exposure technique (DET),
measures the d of two sets of planes. One with a $\psi = 0$ and the
other with $\psi \neq 0$ and often $\psi = 45$ or $60°$. These exposure angles are
often chosen because these $\sin^2\psi$ values are 0.50 and 0.75, and this
simplifies calculations. However, the larger the ψ angle the more
sensitive are the stress measurements that can be made, e.g., the
sensitivity in going from 45 to 60 degrees means half again improvement in the sensitivity of d_ψ to stress.

     A number of researchers have described this technique applied
to the analysis of stresses in metals and alloys [2-5].

     The sine square psi technique requires that d, or really $\Delta d/d_o$
be plotted versus $\sin^2\psi$ for a number of ψ angles, usually ranging
from 0 to 70 degrees. Then the stress in the surface in the
direction phi, ϕ, is equal to:

$$\sigma_\phi = \text{slope} \cdot \left(\frac{E}{1 + \nu}\right)$$

where slope = $(\partial\ \Delta d/d_o)/(\partial\ \sin^2\psi)$

This is not a method limited only to using a diffractometer, as was
demonstrated by Macherauch [6] when he used film methods to obtain
a $\sin^2\psi$ curve. However, with the long exposure times required with
film this procedure could be painfully slow.

     Film techniques do not require that the two theta position of
the diffracted x-ray peak be scanned. The principle here is to
place photographic film at an appropriate position to intercept
the diffracted peak of interest. The position of the peak is then
measured with respect to a reference mark [1]. Densitometer methods

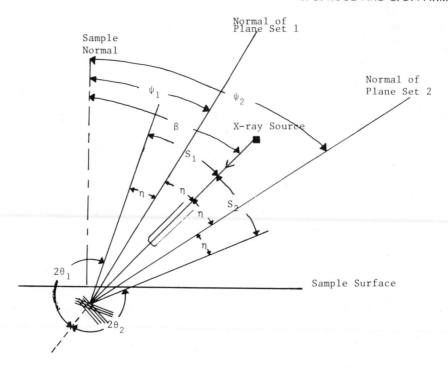

Figure 2.   Geometry of single exposure technique.

have been used to obtain film darkening versus distance plots
from exposed stress films; however, a skilled reader using an
x-ray film reading device can perform determinations which are
nearly as accurate. As mentioned previously, there are two film
techniques commonly used, one being the double exposure technique
(DET) and the other the single exposure technique (SET).

The double exposure technique essentially is the same as
applied in diffractometry, only the position of the diffracted
beam is measured on film rather than by a moving detector tube [1].
Equation (1) shows the applicable relation for this film technique.

The single exposure technique (SET) is usually preferred over
DET because it is simpler, faster and almost as accurate. Norton
[7] has compared SET and DET and found that the latter gave only
slightly more accurate results. The SET is possible since the
incident x-ray beam is diffracted from two sets of planes, of
nearly the same d space, the normals of which it bisects. Figure
2 illustrates the geometry of the SET. As shown in the figure,
diffraction occurs at $\beta - 2\eta$ and $\beta + 2\eta$ where β is the angle
between the surface normal and the incident x-ray beam and η =
$\theta - 90°$. Then to calculate the stress in the surface of the plane:

$$\sigma = (\frac{E}{1 + \nu}) \frac{1}{4R_o \sin^2\theta \sin^2\beta} (S_2 - S_1)$$

where R_o = the camera radius, i.e., distance from the specimen to the film, and S_2 and S_1 are distances from a reference point as shown in Figure 2. The derivation of this equation is shown in SAE J784a [1].

SOURCES OF ERROR

There are a number of sources of error which may frequently be encountered in the applications of x-ray diffraction to stress measurement. Realizing this is important because too many cases of x-ray stress applications have been made where errors have been overlooked only to be spotted as anomalous data by an engineer perusing the data. This has caused x-ray diffraction techniques to be labeled as unreliable in many circles.

The most common causes of error in stress measurement by x-rays are listed below. These do not include considerations such as counting statistics which must be considered in all x-ray diffractometry applications.

    Stress Constant Selection

    Diffractometer Focusing Geometry

    Diffracted Peak Location

    Cold Working

    Texture

    Grain Size

    Microstructure

    Surface Condition

Papers appearing in the Proceedings of the Denver Conferences on Applications of X-Ray Analysis will be cited which illustrate these error sources, their measurement and implications. This reference source was selected because it was judged to be the most comprehensive source available in most major science and engineering libraries.

Stress Constant Selection

Most of the experimental work accomplished to date has demonstrated that the stress constant, i.e., proportionality constant of stress, σ, and the measured change in d, $\Delta d/d_o$, should be measured experimentally for each alloy and thermomechanical condition. This constant K is a function of $E/(1 + \nu)$ and using bulk, or theoretical, values to calculate K has been shown by a number of researchers [7-9] to produce errors in measured stress as high as thirty percent. The discrepancy between bulk elastic constant and the x-ray stress constant is basically due to selecting a single crystallographic set of planes for bulk strain measurement in crystals that are elastically anisotropic.

Diffractometer Focusing Geometry

Errors from this cause are often the most subtle sources of error to quantify or even identify. Short and Kelly [10] observed errors ranging from 4 to 100 percent in measured tensile and com-pressive stresses in steel using diffractometer techniques. They cited the usual correction factors as vital considerations in obtaining reliable data, those factors being: Lorentz, polariza-tion, absorption with respect to ψ angle, atomic scattering and temperature. Larson [3] noted errors in stress determination as due to lack of parafocusing, and Jatczak and Boehm [11] cited a number of error sources related to focusing. Finally, Zantopulos and Jatczak [12] characterized the error trend caused by surface curvature on the specimen.

Diffracted Peak Location

The measurement of the diffracted peak position is usually the most significant source of error common to both film and diffractometer techniques. Peak location is determined from the intensity versus two theta data generated by the diffractometer. Essentially it is a matter of selection of the method to obtain the data, fitting a curve to the data and calculating a reference position on that curve. Data may be obtained from 2, 3, 5 or more two theta positions on the diffracted curve. A curve, usually Gaussian, is fit to these data, usually using a least square method. Then the two theta position representing the diffracted peak position is established. Figure 3 illustrates half width at half height, maximum intensity and the centroid positions.

Norton [7] demonstrated the significance of peak location error with both DET and SET film techniques. Jatczak and Boehm [11] studied peak locations with diffractometry techniques and concluded that peak resolution was the most important factor affecting peak location and that both para and stationary focusing methods would give similar results. They demonstrated with a number of illustrations effects of slits, focusing geometry and

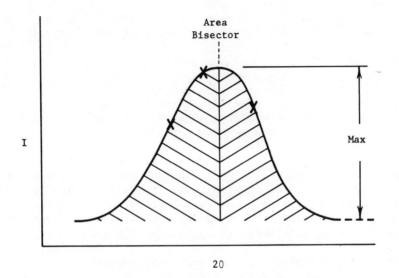

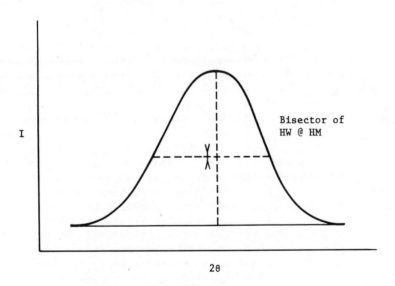

Figure 3. Peak position location method.

psi angle on the peak location. Baucum and Ammons [13] studied
methods of peak location using conditions on a diffractometer to
give the highest resolution. These conditions included an elaborate
method of specimen centering. They concluded that calculation of
the centroid was a more accurate method of peak location than parabolic fit. This was in agreement with Jatczak and Boehm [11].

Cold Working

The proportionality of stress versus the two theta angle, or
change in d space, does not seem to be affected by cold work alone.
This has been demonstrated by Macherauch's [6] results which showed
that the slope from the $\sin^2\psi$ technique, which varies with stress
and which is the quantity used to calculate the stress from the
elastic constants, increased as a straight line relation of stress.
However, the residual d space shift he observed in unidirectional
tension and attributed to a surface compressive layer has been
labeled as pseudo-macrostresses by Cullity [14] in a recent paper.
Cullity indicated that these are actually microstresses.

Although cold work alone is relatively innocuous to the proportionality necessary for x-ray stress readings, it does produce line
broadening that can make establishing the peak positions more
difficult. Cold work does cause preferred orientation, texture,
and may affect grain size and these can have a marked effect upon
x-ray stress readings, as will be discussed subsequently. In spite
of this, Ricklefs and Evans [15] observed that in most practical
applications of metals such small amounts of cold work are present
that there is no significant effect on the stress versus d space
shift proportionality.

Texture

Anomalous stress results can be displayed as a non-linearity
in the d spacing changer versus $\sin^2\psi$ curve. This condition has
been shown by Marion and Cohen [16] to vary with texture. They
showed results from several metals including Armco ireon, low alloy
steels, and a copper-gold alloy all cold worked from 14 to 35
percent. Each displayed texture and anomalous stress results.
Moreover, they proposed a procedure to enable x-ray stress measurements to be performed in spite of the anomalous stress results.

The presence of texture can also affect the intensity of the
diffracted x-ray peak, depending upon the relative direction of
the preferred orientation, to such an extent that the peak cannot
be detected.

Grain Size

The effect of an extremely fine grain size, less than a
micrometer, is to broaden the diffracted x-ray line; however, such
small grain sizes are seldom encountered. The more common problem
is too large a grain size, i.e., grains so large that the volume
irradiated does not contain sufficient grains to produce a continuous back reflected cone of x-rays. This problem can be circumvented,
at the expense of line resolutions, by increasing the irradiated
area through enlarging the collimator or oscillating the sample
with respect to the x-ray source.

Microstructure

If the microstructure of a metal or alloy in which stresses
are to be measured consists of more than one major phase, e.g.,
ferrite and martensite in steels, the diffracted x-ray line from
the phase chosen for stress analysis may be severely affected by
microstresses resulting from differing or even opposing stresses
in the two phases. This may also be a concern where a minor
phase causes hardening stresses, e.g., precipitation hardened
aluminum alloys [14].

Surface Condition

This source of possible error is usually obvious in the form
of corrosion products (oxide layer), abraded or dirty surface. The
most practical technique to remedy such problems are variations
upon the electrochemical polisher described by Bolstad and Quist [8].

FIELD LIMITATIONS

In spite of these many possible sources of error the four
classic techniques mentioned have been used successfully to measure
stresses in metals and alloys for at least three decades. These
applications span from stress measurements in nuclear fuel rods [2]
to aircraft landing gear [8].

However, there are still severe limitations to the practical
field applications of stress measurement by x-ray diffraction. The
classic instrumentation, film cameras and laboratory diffractometers,
suffer from a number of shortcomings which may include lack of speed,
portability, accuracy, ruggedness, etc. Except for a few field trip
applications [8] the film camera techniques have been considered too
slow for wide application. The laboratory diffractometers usually
require the same to be mounted upon a sample holder and several
minutes up to an hour are required for accurate readings. Moreover,

these instruments are far from portable, due to their bulk and delicateness. A number of innovative designs to bring the diffracto- meter to the field, which do not require that the sample be small, have been marketed over the past few years. These have been by Siemens Corporation, Rigaku Denki and American Analytical. However, none of these possess the combination of speed, accuracy and versa- tile portability.

DEVELOPMENT OF ADVANCED PORTABLE INSTRUMENTATION

There is an obvious need for a highly portable, rapid and accur- ate instrument to exploit the well-established techniques of x-ray stress analysis in the laboratory, shop and field. There are two possible sources of improvement in x-ray instrumentation which could pave the way for such a development. These are smaller, lighter and higher intensity x-ray sources and smaller, lighter, more sensitive and higher spatial resolution position-sensitive x-ray detectors. There are a number of researchers working on development of a port- able x-ray stress analyzer from the standpoint of incorporation of improved detector systems and this is a logical first step since this direction is most likely to lead to major improvements over present technology.

Recently James and Cohen [17] reported the application of a position-sensitive gas proportional detector to x-ray stress analysis by mounting this device on a laboratory diffractometer. They showed greatly improved speed over that possible by scanning with a standard detector. The spatial resolution of the counter which they reported was 183 micrometers. A French company (Compagnie Generale De Radiologie Dept., Recherche et Industrie, 48, Boulevard Gallieni, 92133 Issy Les Moulineaux) is presently marketing what they describe as a portable x-ray stress analyzer using a similar position-sensitive gas proportional detector. Moreover, James [18] at a recent confer- ence presented a paper suggesting the possible applications of this type of detector to a portable x-ray device. However, Ruud et al. [19] reported that upon evaluation of a position-sensitive gas proportional detector, from the same manufacturer as James and Cohen's, they found it quite suitable for replacement of the standard detectors on laboratory diffractometers but not suitable for a port- able field device. They cited the size, shape, gas supply and gas flow regulation limitations as well as poor spatial resolution as reasons for these conclusions.

Spatial resolution is an important consideration because it affects speed, accuracy and versatility of application. For example, a detector with a spatial resolution of 180 micrometers would have to be placed three times farther away from the sample to produce the same two theta, or d space, reading accuracy as a detector with a

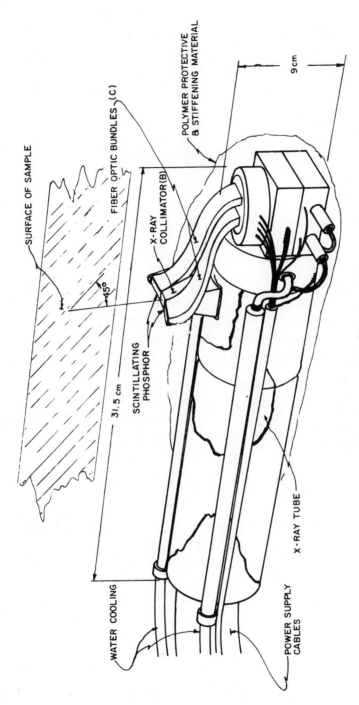

Figure 4. Sensing head and x-ray source.

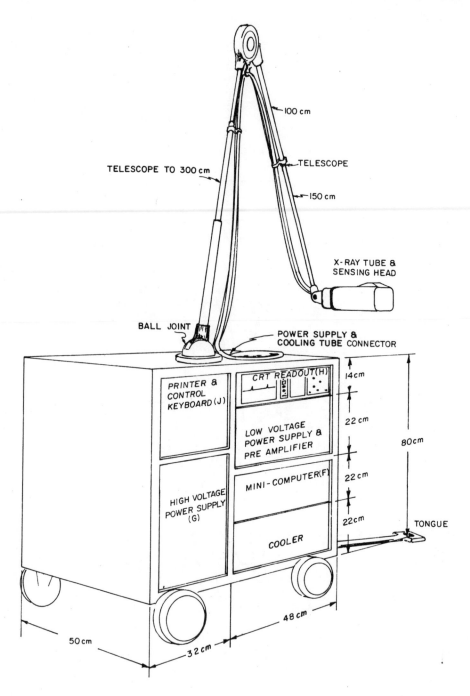

Figure 5. Power supply and data console.

spatial resolution of 60. If the 180 resolution detector were placed closer the accuracy of reading would diminish. Furthermore, the intensity of the diffracted x-rays incident on a given area of detector surface would be 1/9 that of the higher resolution detector, neglecting factors such as absorption. These dimensional considerations also imply that the x-ray sensitive surface of the poor resolution detector would have to be three times longer than that of the detector. Furthermore, many more restrictions would be found in applications under conditions of confined space.

An entirely new position sensitive detector has recently been developed and built on U.S. Army MERDC contract [19], the principle of which is based upon electro-optics and solid state self-scanners. This development has been found to be small, light, versatile and demonstrates a spatial resolution of approximately 60 micrometers. It is considered versatile because the x-ray sensing surface can be placed remotely from the electronics and the electronic amplification and scanner package is only about two inches cubed. The possibility of a portable field x-ray stress device with an x-ray source and detector head, stress head, the size and configuration shown in Figure 4 seems feasible [19]. The stress head depicted in Figure 4 is about the size of a quart thermos bottle and weighs approximately 10 pounds. The console from which the device obtains power is controlled and the data are reduced and displayed as shown in Figure 5. This console is an estimated weight of 300 pounds and approximately 20 x 28 x 32 inches.

CONCLUSION

In spite of the many sources of error inherent in x-ray stress analysis it has been proven to be a rapid, accurate, non-destructive technique for the measurement of residual stress.

The major impediment to wider applications, expecially field use, has been speed in conjunction with accuracy and portability.

Recent developments in advanced position-sensitive detectors will soon make possible the building of a highly portable, rapid, accurate and versatile x-ray stress analyzer.

REFERENCES

1. SAE J784a, "Residual Stress Measurement by X-Ray Diffraction", Handbook Supplement, Society of Automotive Engineers, Inc. (1971).

2. Wooden, B.J., E.C. House and R.E. Ogilvie, "Precision X-Ray Stress Analysis of U and Zr", Vol. 3, Advances in X-Ray Analysis, Plenum Press, New York, 1960, p. 331.

3. Larson, R.C., "The X-Ray Diffraction Measurement of Residual Stresses in Aluminum Alloys", Vol. 7, Advances in X-Ray Analysis, Plenum Press, New York, 1964, p. 31.

4. Woehrle, H.R., et al., "Experimental X-Ray Stress Analysis Procedures for Ultrahigh-Strength Materials", Vol. 8, Advances in X-Ray Analysis, Plenum Press, New York, 1965, p. 38.

5. Esquivel, A.L., "X-Ray Diffraction Study of the Effects of Uniaxial Plastic Deformation on Residual Stress Measurements", Vol. 12, Advances in X-Ray Analysis, Plenum Press, New York, 1969, p. 269.

6. Macherauch, E., "Lattice Strain Measurements on Deformed FCC Metals", Vol. 9, Advances in X-Ray Analysis, Plenum Press, New York, 1966, p. 103.

7. Norton, J.T., "X-Ray Stress Measurement by the Single-Exposure Technique", Vol. 11, Advances in X-Ray Analysis, Plenum Press, New York, 1968, p. 401.

8. Bolstad, D.A. and W.E. Quist, "The Use of a Portable X-Ray Unit for Measuring Stresses in Al, Ti and Fe Alloys", Vol. 8, Advances in X-Ray Analysis, Plenum Press, New York, 1965, p. 26.

9. MacDonald, B.A., "Application of the X-Ray Two-Exposure Stress Measuring Techniques to Carburized Steel", Vol. 13, Advances in X-Ray Analysis, Plenum Press, New York, 1970, p. 487.

10. Short, M.A. and C.J. Kelly, "Intensity Correction Factors for X-Ray Diffraction Measurements", Vol. 16, Advances in X-Ray Analysis, Plenum Press, New York, 1973, p. 379.

11. Jatczak, C.F. and H.H. Boehm, "The Effects of X-Ray Optics on Residual Stress Measurements in Steel", Vol. 17, Advances in X-Ray Analysis, Plenum Press, New York, 1974, p. 354.

12. Zantopulous, H. and C.F. Jatczak, "Systematic Errors in X-Ray Diffractometer Stress Measurements Due to Specimen Geometry and Beam Divergence", Vol. 14, Advances in X-Ray Analysis, Plenum Press, New York, 1971, p. 360.

13. Baucum, W.E. and A.M. Ammons, "X-Ray Diffraction Residual Stress Analysis Using High Precision Centroid Shift Measurement Techniques Application to U-0.75 Ti Alloy", Vol. 17, Advances in X-Ray Analysis, Plenum Press, New York, 1974, p. 371.

14. Cullity, B.D., "Some Problems in X-Ray Stress Measurements",
    Vol. 20, <u>Advances in X-Ray Analysis</u>, Plenum Press, New York,
    1977.

15. Ricklefs, R.E. and W.P. Evans, "Anomalous Residual Stresses",
    Vol. 10, <u>Advances in X-Ray Analysis</u>, Plenum Press, New York,
    1968, p. 273.

16. Marion, R.H. and J.B. Cohen, "Anomalies in Measurement of
    Residual Stress by X-Ray Diffraction", Vol. 18, <u>Advances in
    X-Ray Analysis</u>, Plenum Press, New York, 1975, p. 466.

17. James, M.R. and J.B. Cohen, "The Application of a Position-
    Sensitive X-Ray Detector to the Measurement of Residual
    Stresses", Vol. 19, <u>Advances in X-Ray Analysis</u>, Kendall/Hunt
    Publ. Co., Dubuque, 1976, p. 695.

18. James, M.R. and J.B. Cohen, "Study of the Precision of X-Ray
    Stress Analysis", Vol. 20, <u>Advances in X-Ray Analysis</u>, Plenum
    Press, New York, 1977.

19. Ruud, C.O., R.E. Sturm and C.S. Barrett, "A Nondestructive
    X-Ray Instrument to Instantly Measure Residual Stresses",
    Type II Interim Technical Report on Phase I, Prepared for U.S.
    Army Mobility Equipment Command, Fort Belvoir, Virginia.

Chapter 6

FLASH X-RAY

Q. Johnson, Lawrence Livermore Laboratory, University of
    California, Livermore, California

and

D. Pellinen, Physics International, San Leandro,
    California

ABSTRACT

The complementary techniques of flash x-ray radiography (FXR)
and flash x-ray diffraction (FXD) provide access to a unique domain
in nondestructive materials testing. FXR is useful in studies of
macroscopic properties during extremely short time intervals, and
FXD, the newer technique, is used in studies of microscopic properties. Although these techniques are similar in many respects,
there are some substantial differences. FXD generally requires
low-voltage, line-radiation sources and extremely accurate timing;
FXR is usually less demanding. Phenomena which can be profitably
studied by FXR often can also be studied by FXD to permit a complete
materials characterization.

INTRODUCTION

Flash x-ray (FX) is a borderline nondestructive testing (NDT)
technique. Although x-rays are, for the most part, nondestructive,
the fact that FX methods require less than one μs per exposure
makes them ideally suited to the task of probing material that is
moving or changing very rapidly, and the acceleration or decelera-
tion necessarily involved in such tests often destroys the material.
FX can thus be viewed as a NDT technique for materials on their way
to being destroyed!

Flash x-ray radiography (FXR) is nothing more than the x-ray
equivalent of optical strobing. It is used to probe the inside of
fast-moving objects and to examine detail in cases where object

117

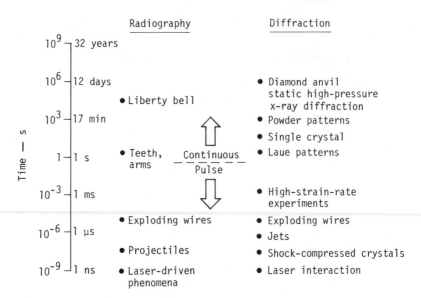

Figure 1.   Time requirements for various x-ray experiments.   The
dashed line indicates the approximate boundary between
exposures requiring continuous sources and exposures
requiring pulsed sources.   A radiograph of the Liberty
Bell required 7.5 hours [1].

self-illumination presents a problem.   Often, such events cannot
be viewed by other techniques.   FXR is especially valuable in
revealing macroscopic changes in volume or density.   A typical
system will use a pulse 10^{-8} to 10^{-7} s in length.   The blur will
be equal to the product of the object velocity and the exposure
time.   For example, if a bullet from a 220 Swift, which has a
muzzle velocity of 1.2 km/s, were radiographed with a relatively
slow 10^{-7}-s pulse system, the motion blur would be only 1.2 x 10^{-2}
cm.   FXR can be used to define accurately the bullet's location
as it moves through opaque media (the barrel, the muzzle brake, or
the target).

Flash x-ray diffraction (FXD) is a more recent development
which makes use of coherent elastic scattering.   It is sensitive
to microscopic changes (simple atomic displacements or complete
rearrangements as in a crystalline structure transformation).

Combining these two techniques makes it possible to access a
unique domain in NDT.   The type of phenomena observable by the two
FX methods and the exposure times required are shown in Figure 1.
Exposure times for some continuous x-ray studies are also shown.
Most of the short-lived phenomena have already been explored in

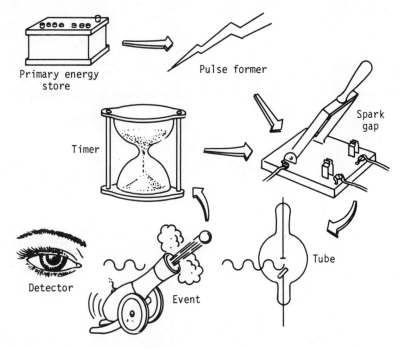

Figure 2.   Seven elements of a flash x-ray experiment.

some detail.   One exception, however, is the interaction of lasers with various materials.   It is anticipated that as shorter pulse machines and techniques become available, studies of laser phenomena will begin.

ELEMENTS OF A FLASH X-RAY EXPERIMENT

Energy Storage, Pulse Shaping

As shown in Figure 2, FX experiments typically involve seven basic elements in addition to the power supply.   The first of these is the energy storage unit.   (The others will be discussed in turn.) The simplest and most common way to store energy is in a low-inductance capacitor.   If this energy-storage unit is connected directly to an evacuated x-ray tube, as shown in Figure 3, a simple FX device results.   Charging the capacitor C through the isolation resistor R leads ultimately to a flash discharge in the x-ray tube resulting in a pulse of x-rays.   (It is assumed that the circuit is completed by connection of the left- and right-hand parts of the circuit and that the anode/cathode gap in the x-ray tube is small enough to break down before the capacitor does.)   This is not a very

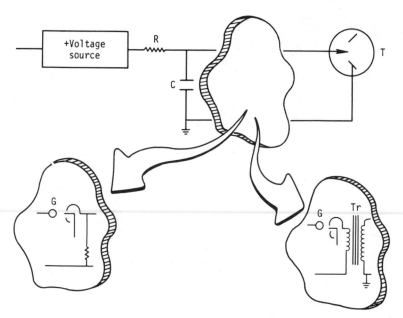

Figure 3. Simple flash x-ray unit consisting of a voltage source,
           charging resistor R, storage capacitor C, and flash tube
           T. Addition of lower-left circuit element containing
           a triggerable spark gap G, permits timing. Insertion
           of lower right element containing a pulse transformer,
           Tr, permits pulse shaping.

sophisticated system; it lacks provision for both timing and pulse
shaping. The incorporation of a simple triggered spark gap (shown
in the lower left-hand side) corrects the first of these deficiencies.
The right-hand circuit component, a spark gap and pulse transformer,
addresses both.

A more sophisticated approach is shown in Figure 4. This is a
schematic drawing of a Marx gnerator [2], which combines energy-
storage and pulse-shaping functions in a deceptively simple circuit.
Such generators allow one to charge the capacitors in parallel and
to discharge them in series. This permits erection of a fast, high-
voltage pulse from relatively simple circuit components. After the
capacitors are charged to a suitable voltage, discharge is initiated
through the triggered spark gap shown on the left. In simple Marx
systems, the following gap then breaks down as a result of an over-
voltage transient. The avalanche thus started leads to a fast-rising
voltage pulse on the x-ray tube equal to and in some cases greater
than the charging potential times the number of stages. In more

complicated hybrid triggering systems, a number of gaps are triggered and the remainder are capacitively coupled to ensure fast erection with low jitter [3].

Rather large Marx generator arrays are used to produce the very high voltages required for state-of-the-art FX devices. An example is the array used in a recently fabricated 9-MeV device [4] and shown in Figure 5. In the photograph, it is possible to see

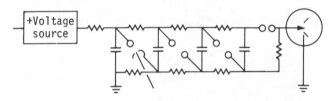

Figure 4.   Schematic of Marx generator.

Figure 5.   Marx generator room of the 9-MeV flash radiographis unit. Vertically stacked capacitors are interconnected by spark gaps and isolation resistors, consisting of solutions of $CuSO_4$ in plastic tubes.

Figure 6.   The 9-MeV, 200-ka, 75-ns flash x-ray unit for radio-
             graphic studies.

individual capacitors, connecting spark gaps, and clear plastic
tubes filled with a $CuSO_4$ solution, which function as the isolation resistors. In operation, the whole room is filled with oil
to prevent high voltage breakdown. The complete x-ray device is
shown in Figure 6. The box-shaped unit at the upper right of the
figure is the tank containing the Marx generator. This is connected to a cylindrical unit containing a Blumlein pulse transformer.
The actual electron-accelerator x-ray head is at the lower left.

At the other end of the scale of complexity from the 9-MeV
device described above is the Blumlein FX device used in FXD studies
at the Lawrence Livermore Laboratory (LLL). Shown schematically in
Figure 7, it combines energy-storage and pulse-forming components.
Basically, it consists of a low-impedance, triaxial transmission
line approximately 20 cm in diameter and 7 m long. The indicated
conductor is charged by a power supply to 50 to 60 keV. To indicate
a discharge, this conductor is connected to the outer conductor via
a triggered spark gap. The Blumlein circuit employed in this FX
unit enables a voltage gain on the x-ray head over the charging
value according to the following:

$$V/V_0 = 2Z_1/(Z_0 + Z_1) \ ,\qquad\qquad (1)$$

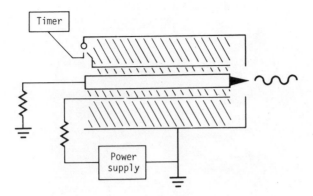

Figure 7.   Schematic of low-voltage, high-current FXD device.

where Z_1 is the load (anode/cathode-gap) impedance and Z_0 is the
characteristic impedance of the Blumlein [5].   Since it is desirable to keep the voltage relatively low while increasing the
electron current to enahnce the component of characteristic radiation [6], this unit is usually charged to only 50 to 60 keV.   During
pulsing, the voltage across the anode/cathode gap reaches a maximum
value of approximately 85 keV, while the current reaches 10 to 15
kA [7].

Spark Gap

A schematic of a low-impedance, triggered spark gap, the next
element in an FX system, is shown in Figure 8.   The gap is charged
to a voltage near self-breakdown; then, to initiate switching, a
trigger pulse is fed in via the trigger electrode.   This creates an
over-voltage condition that initiates spontaneous breakdown.   In

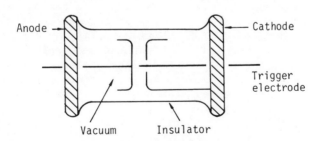

Figure 8.   Schematic of triggerable spark gap.

Figure 9.   Triggerable spark gap.   Trigger electrode is the cylin-
            drical device at midplane in the gap.

some gaps, the trigger electrode is a concentric ring positioned
midway between the anode and cathode elements as shown in Figure 9.
This particular gap is capable of carrying a current of 140 kA at a
voltage of 120 keV. During operation, the gap is normally filled
with SF_6 and argon. Nitrodgen is used for lower voltage applications

X-Ray Tube

    The output voltage pulse from the pulse-forming unit is carried
to a vacuum diode. Two typical diodes are shown in Figure 10.   On
the left-hand side is a transmission cathode diode used in low-
voltage machines; on the right-hand side is a transmission anode
diode used at high voltages. If the voltage pulse is negative, it
is applied to the cathode; if positive, it is applied to the anode.
At higher voltages, the anode and cathode are isolated from each
other by an isnulator or a stack of insulators. Spacings between
electrodes are set so that the negative surface begins to field-
emit electrons. After a few runs a plasma is formed, which will
emit freely with low electric fields at the cathode.

    The voltage/current relationship for diodes in planar geometry
with circular cathodes and low magnetic fields can be described
approximately by an equation originally developed by Langmuir [8]

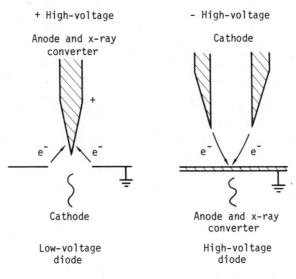

Figure 10.   Flash x-ray diodes.

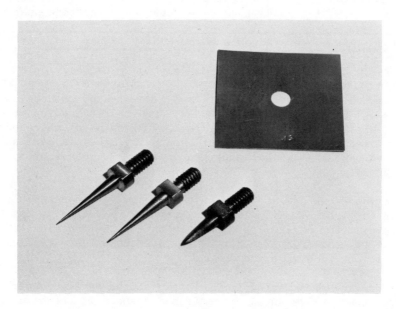

Figure 11.   Anodes (lower left) and cathode (upper right) from a
demountable flash x-ray diode used for diffraction.

for vacuum tubes:

$$I = \frac{V^{3/2}}{k} (\frac{r^2}{d^2}) \; , \tag{2}$$

where k is a constant, r is the cathode radius, and di is the electrode spacing. This expression applies only to diodes where magnetic fields are not important. At higher voltages and currents, the magnetic field of the beam tends to focus the electrons. Current for a planar diode under these conditions is given by

$$I_p = \frac{r}{d} I_A \gamma_0 \; \ell n[\gamma_0 + (\gamma_0^2 - 1)^{1/2}] \; , \tag{3}$$

where

$$\gamma_0 = \frac{eV + m_0 c^2}{m_0 c^2} \; . \tag{4}$$

In this expression, I_A is the Alfen current = 8.5 kA, V is the voltage across the diode, m_0 is the rest mass of the electron, c is the velocity of light, and I_p is the para-potential current [9].

In Equation (3), consideration has been given to the effect of plasma motion, which leads to a time-dependent electrode spacing d.

Diode assemblies available commercially are as small as 10 cm in diameter for voltages under 100 keV and as much as a meter in diameter for units such as the 9-MeV machine previously described.

A cathode and three anodes from a transmission cathode diode are shown in Figure 11. These anodes, used in FXD studies at LLL, have a conical 1.5-cm long active region. Erosion of the anode under intense electron bombardment eventually makes it useless. The shorter anode shown has experienced about 20 pulses of 85-keV, 12-kA current.

The anode/cathode area of the 9-MeV machine discussed earlier is shown in Figure 12. The actual x-ray active region is inside the 5-cm-diameter conical cathode structure. The function of these diodes is to convert the energy of the electrons into x-radiation through interaction with the anode.

The x-radiation useful in FXD, is produced by fluorescence. Beam electrons remove an electron from its shell in the atom, and an x-ray photon with a discrete energy is emitted as fluorescent radiation as a higher shell electron drops into that vacancy. Fluorescent radiation is isotropic and discrete. Most of the

Figure 12.   Anode/cathode area of 9-MeV flash x-ray unit.

x-ray energy, however, especially in high voltage systems, is a
result of the process of bremsstrahlung. A beam electron passes
by a nucleus and is accelerated toward it; an x-ray photon is
then emitted to conserve momentum and energy. This x-radiation will
have a continuous spectrum from close to zero to near the total incident electron energy. The mean value of this spectrum occurs at
1/5 to 1/3 of the incident electron energy. At low voltages (<500
keV), the radiation will be essentially isotropic; at higher voltages, relativistic effects lead to a maximum in the forward
direction. Calculated spectra for two diodes are shown in Figure
13 [10]. Note the line radiation in the lower voltage diode shown
in Figure 13(a).

For a thick target, efficiency of conversion of electron
energy into x-ray energy is given by:

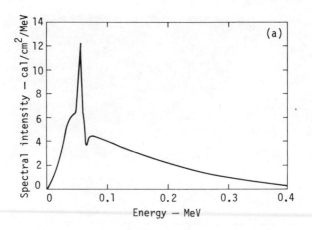

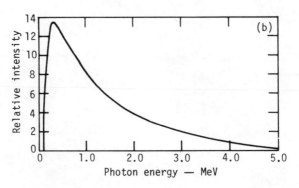

Figure 13.   Calculated x-ray spectra for two flash x-ray diodes.

$$\varepsilon = \frac{(6 \times 10^{-4} \ ZV)}{(1 + 6 \times 10^{-4} \ ZV)} \quad , \tag{5}$$

where Z is the atomic number of the target and V is the electron
energy in MeV [11]. With practical FXR sources in the 0.3- to 1.0-
MeV range, we have measured efficiencies between 0.5 and 2.0%
(somewhat lower than claculated by the above expression).

Event, Timing

FX techniques can be used to study an exploding wire or foil,
development of a jet by the action of a shaped charge, the behavior
of a projectile as it moves through air or as it penetrates some
material, shock-induced, microscopic transformations in materials,
or any of a number of other high-speed phenomena. Specific examples

of several of these and their study by FXR and/or FXD methods will
be detailed shortly. Each event poses a timing problem which may
be trivial or quite formidable. Sometimes it is sufficient to
initiate the event and delay firing of the FX unit by a calculated
amount using an electronic delay unit. Often it is necessary to
arrange the start of this delay by sensing a prior event, such as
the interruption of a light or x-ray beam [12] by the passage of
a projectile. The timing problem for events requiring synchroniza-
tion to µs is trivial. Slightly greater levels of sophistication
are needed to position events within an x-ray "window" which is
between 1000 and 100 ns. Very sophisticated techniques are
required below the 100-ns level. In such cases, it becomes necessary
to sense the actual shock wave, determine (or if necessary, estimate)
its velocity, and erect the high voltage pulse rapidly on the FX
unit in time to coincide with the passage of the phenomena through
the x-ray window. Few FXR applications require this level of sophistication, whereas many, if not most, of the more interesting FXD
applications do.

Detectors

Detectors, like timing devices, range from the simple to the
exotic. FX images often require enhancement, and this is the
principal complication. Figure 14 shows three kinds of detectors

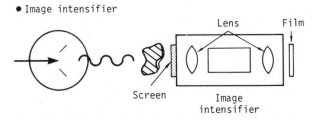

Figure 14. Three basic methods of detection used in FX.

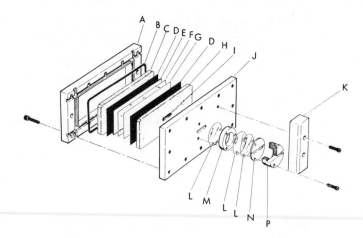

Figure 15.   Blast cassette used in FXD studies.

using film: film without a screen, film with an intensifying
screen, and film with image enhancement [13].   Scintillation
detectors, arrays of detectors [14], or films with gated image
enhancement [15] are less frequently used.

      Since the event under study is often of a very destructive
nature, it may also be necessary to provide elaborate safeguards
so that the film or detection device is not destroyed.   Figure 15
shows a blast cassette used in FXD studies [16].   Here it was
necessary to allow access to the film by the 8-keV x-rays and yet
to deny access to blast products generated by the detonation of
150 g of a high explosive only 10 cm distant.   The film and
intensifying screen are indicated by G and F, respectively.   These
are contained in a light tight plastic bag, D, protected from
major abuse by foam cushions C and H and steel J, and shielded from
small fragments by the slit assembly L-P.

FLASH X-RAY RADIOGRAPHY

      In FXR, the object selectively absorbs some of the x-rays, and
the remaining x-rays form a corresponding image.   The intensity of
the transmitted beam is given by

$$I = I_0 e^{-\mu t} , \qquad\qquad (6)$$

where I_0 is the incident intensity, I is the transmitted intensity,
μ is the absorption coefficient, and t is the path length.   The

absorption coefficient is dependent on energy and material. X-rays
are absorbed in three ways: through the photoelectric effect,
through scattering, and through pair production. As x-ray photons
collide with atoms of the absorbing material, they remove electrons
and transfer energy to them. This is the photoelectric effect. It
is an important process at low photon energies and is strongly
dependent on atomic number. Scattering can be either coherent or
incoherent. The former, resulting from the back-and-forth acceleration of an electron by the incident x-ray, is the process utilized
in FXD. The latter, termed Compton scattering, occurs when an
x-ray photon gives up part of its energy to an electron in the
absorbing material and thereby has its trajectory modified.
Scattering is dependent upon electron density, and it slowly decreases with electron energy. Pair production occurs when an x-ray
photon disintegrates in the vicinity of a nucleus to form an
electron-positron pair. Approximately 1 MeV is required for this
process, and the corss section increases with energy.

Absorption curves that consider these three effects are shown
in Figure 16 for three different elements. Data for these curves
were obtained from Grostein [17]. More extensive data are available
in Plechaty, Cullen, and Howerton [18]. Three observations can be
made from Figure 16:   (1) low-energy x-rays can be used very
effectively to show differences in atomic number for materials;
(2) absorption is relatively independent of energy above about
1 MeV and it shows a minimum of about 5 MeV depending on the atomic
number of the absorber; high-intensity beams in this energy range
are optimal for radiography of thick or dense objects; (3) photon
energy above about 10 MeV may be counterproductive for higher-Z
material because of increasing absorption due to pair production.

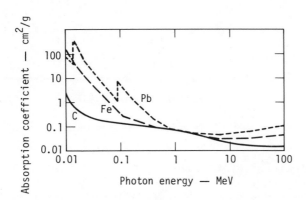

Figure 16.   Absorpition coefficient for C, Fe, and Pb versus x-ray
             photon energy.

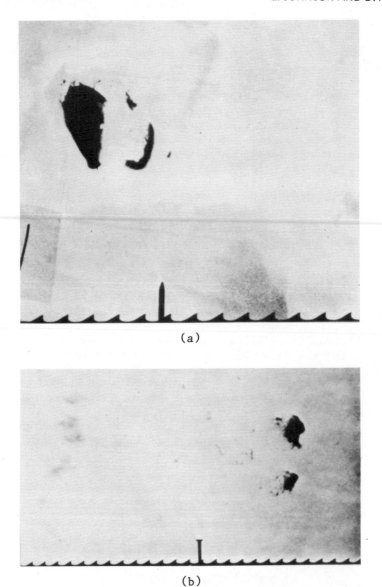

(a)

(b)

Figure 17. Flash radiographs of projectile breakup.

A practical FXR system has to take into account all of the
previous considerations to provide sufficient x-ray flux at the
target to expose the detector and provide sufficient contrast to
view the object. In fixed-pulselength Blumleins, some measure of
flexibility with respect to exposure can be obtained by changing
the dc charging voltage and the diode impedance.

Several FXR events are shown in Figures 17–22 to illustrate some of the types of materials study possible and to point out some of the limitations. Most of these examples have been chosen to coincide with FXD analogues that will be discussed later. The macroscopic characterization of the event by FXR can thus be contrasted with the microscopic characterization by FXD.

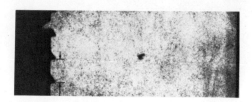

Figure 18.   Flash radiograph of lithium–magnesium projectile flying in air at 12.2 km/s. Fiducial mark is 1.27 cm.

Figure 19.   Heavy rod penetrator undergoing tensile fracture upon exit from a steel target.

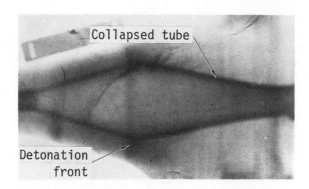

Figure 20. Collapse of a conical gun barrel imploded by a high
            explosive.

    An explosively produced shock wave is an obviously destructive
phenomenon. Material experiencing the full effects of such an
event comes apart as complex wave interactions lead to stresses
exceeding tensile strengths. Figure 17 shows the breakup of a
projectile fired from an explosively driven launcher [19] and
travelling at a velocity slightly below 10 km/s. FXR photographs
permit a determination of the effect of changes in gun design and
materials parameters. The changes are made to try to achieve projectile acceleration without breakup.

    Figure 18 shows the successful launch of a projectile moving
at the exceptional velocity of 12.2 km/s [20]. (To put this in
perspective, we should note that the detonation velocity of the
high-explosive PETN is only 8.3 km/s [21].) The projectile, fabricated from a lithium-magnesium material, is 0.66 cm in diameter

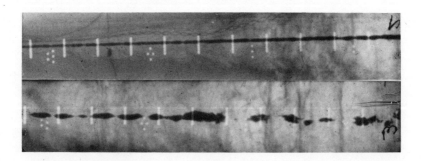

Figure 21. Jet formed by a shaped charge. The bottom half of the
            photograph is a continuation of the top half. The jet
            is travelling from right to left.

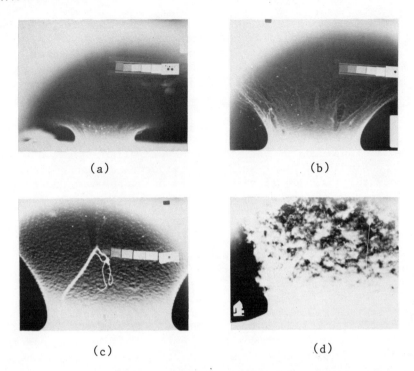

(a)   (b)

(c)   (d)

Figure 22. Flash radiographs of cratering by an explosive sphere
            half buried in plasticene clay. Pictures were taken
            at 50, 200, 500, and 3000 µs after detonation.

and 0.36 cm long. Because of its velocity and the length (100 ns)
of the FXR pulse, the radiograph is motion-blurred by over 1 mm.

The behavior of a material experiencing high dynamic stresses
is shown in Figure 19. This FXR photograph, taken with a 300-keV
FX unit having a 20-ns pulse length, shows a penetrator which was
travelling at 1 km/s, exiting from a target of mild steel [22].
Macroscopic motion of parts of the material can be inferred as the
penetrator comes apart under tensile failure. The object of such
FXR studies is to tailor penetrator design and materials properties
to avoid such failure.

Macroscopic changes in volume are readily observed by FXR,
provided that there is a materials density contrast. Figure 20
shows a gas container in the process of being imploded by a high-
explosive-driven liner of steel [20]. The collapsed region of the
container can be seen easily in this photograph because of the
density contrast between the compressed gas and the liner. An

additional feature of interest here is the location of the detonation front. This photograph demonstrates one of the important features of FXR - its ability to probe the location of materials buried inside objects. (In this case, the steel liner and the compressed gas volume are contained within a high-explosive jacket.)

The object of this particular study was to determine implosion parameters so as to ensure that the gas was properly contained. Later, it was to be used to accelerate projectiles to hypervelocities. Too weak an implosion is bad, since the propellant gases will not ever be confined and compressed. Too vigorous an implosion is also bad, since the imploded liner will rebound and permit the gases to escape.

Another interesting event to study is a high-explosive-produced jet. If a high-explosive charge is shaped as an inverted cone and lined with metal, it will be capable of forming that liner into an extremely high velocity jet, very effective in drilling holes in materials. Unfortunately, the jet velocity is not constant. There is a high-velocity leading portion capable of penetration and a low-velocity tail that can plug the hole just drilled. Figure 21 shows a macroscopic FXR characterization of a jet [23]. The slower slug-forming portion is shown on the bottom half of this photograph. FXR studies of this nature permit tailoring of the shaped charge so as to eliminate the hole-plugging tail.

Time dependent phenomena can be probed both by FXR and by FXD. Figure 22 shows the development of a crater as a function of time. The crater is being formed by explosion of a 3.5-cm-diameter sphere of C-4 explosive buried in plasticene clay [24]. The experimentally measured size and shape of this crater is used to normalize computer calculations of this phenomenon. Notice the increase, as a function of time, of the particle size of the crater debris. This series illustrates one of the special advantages of FXR over high-speed optical photography. The light from the explosion would prevent optical photography from recording details of the initial cratering. Examples of applications of FXR techniques to other interesting time-dependent phenomena, such as exploding wires, are shown in Jamet and Thomer [25].

FLASH X-RAY DIFFRACTION

The examples shown above of events studied by FXR and many more in the literature (see, for example, [25]), serve to point out that FXR techniques are suited to macroscopic characterization and that they are relatively mature in development. By contrast, FXD techniques, although dependent on much the same technology, are in the early stages of development and are particularly suited to microscopic characterization.

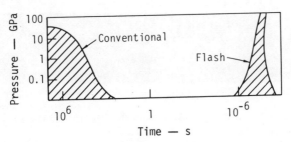

Figure 23.   Pressure/time domain for conventional and flash x-ray
             diffraction experiments.

    FXD is conventional x-ray diffraction conducted on a very
short time scale. Making use of FXR hardware and techniques, FXD
has opened a radically new region of materials study. Figure 23
shows qualitatively the regions of pressure/time space reachable
by means of conventional versus flash x-ray diffraction. The upper
limit of static-pressure x-ray measurement is around 50 GPa [26],
although recent experiments suggest that 100 GPa may soon be
attainable [27]. Such experiments require extremely long exposures,
however, lasting in some cases a week or more. FXD is well suited
to events occurring in times less than 1 μs but more than 1 ns.
Perhaps 50- to 200-ns events are optimal at present. By taking
advantage of the very high pressures attainable via shock waves,
we should some day be able to use FXD to study materials at 100 GPa
and beyond. The current upper limit is below 50 GPa.

    Several historical reviews of FXD have been written [25,28].
Currently, FXD work is carried out in only a handful of laboratories,
of which LLL has been one of the most active. Much of the work we
will describe was done at LLL.

    To conduct a diffraction experiment in the very short time
periods described, it is necessary to replace conventional dc
sources with flash units having very much higher instantaneous
brilliancy factors. Two approaches have been used. In one of
these, conventional FXR units were adapted to diffraction requirements [28]. There is a problem in this approach, however. FXD
requires low-voltage, high-current systems, whereas FXR usually
stresses voltage at the expense of current. The necessary compromises result in very weak images and high backgrounds. A more
successful approach pursued at LLL has been to adapt a research
FXR unit especially built for radiographic studies requiring a
high-current, low-voltage source. This device, already described
in Figure 7, has characteristics similar to those needed for FXD,
i.e., a small focal spot, moderate voltage, high current, a short

Figure 24.   Pinhole photographs of ten consecutive fires from a
             FXD source.

and easily modified pulse length, and a short (<1/2 μs) interval
between trigger pulse and x-ray pulse. Focal spot diameter is
~2 mm and x-ray intensity is moderately reproducible. By contrast,
the focal spot of the 9-MeV FXR unit has been described as between
3 and 10 mm, depending upon x-ray energy.

     A series of pinhole photographs of the LLL FXD anode is shown
in Figure 24. An interesting but bothersome phenomenon is observed
on the second photograph. The tip of the conical anode has a
"blossom" appearance due to the fracturing of the tip by the impact
of the intense electron bombardment on the first fire. This feature
disappears in subsequent fires as the material erodes. The fifth
fire is particularly weak as a result of premature switching in the
triggering circuits.

     As mentioned earlier, the instantaneous brilliancy of such
devices surpasses that of standard dc microfocur units by several
orders of magnitude. It even surpasses state-of-the-art anode
machines by 2 or 3 orders of magnitude. An approximate comparison
is shown in Figure 25. In spite of this advantage, it is found
that compromises must be made to carry out diffraction investigations with FX units. One quick remedy is to place the material
close to the source, sacrificing resolution for intensity. Typical,
sample-to-source distances are ~10 to 12 cm and equivalent sample-
to-film distances as required. A schematic of a conventional FXD
experiment is shown in Figure 26. With the arrangement shown, it
is possible to obtain diffraction records of the quality shown in
Figure 27. Both records were made using a single pulse of copper
radiation on a LiF sample. The upper one shows a portion of the
powder pattern obtained from a polycrystalline sample, and the
lower one shows diffraction from the 200 reflection of a single
crystal of LiF. The discontinuity of the image resulted from a
lead foil mask placed over the center of the crystal for alignment
purposes.

     An actual FXD experiment using a high explosive to study the
effect of high dynamic pressure on a crystalline material is shown
in Figure 28. At the bottom of this photograph is the armor plate

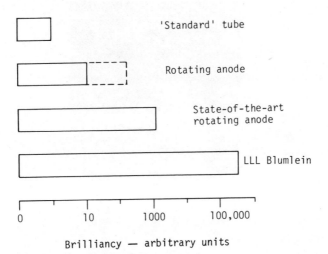

Figure 25.   Source brilliancy for continuous x-ray sources and the FXD source used at LLL.

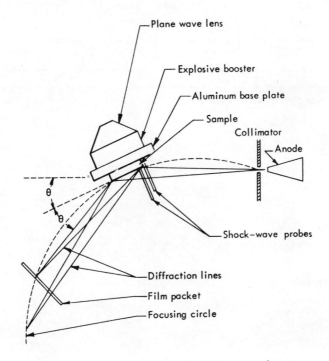

Figure 26.   Geometry of FXD experiment.

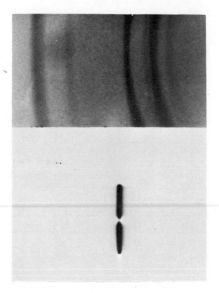

Figure 27. FXD photographs taken of LeF using a Cu anode. Upper
            record is from a polycrystalline sample, lower one is
            from a single crystal.

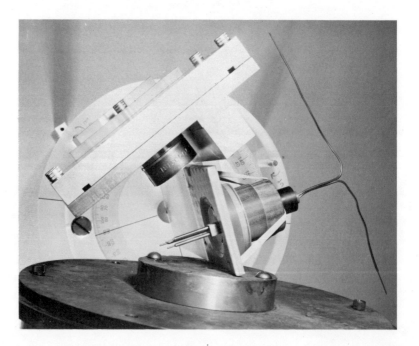

Figure 28. Actual experimental configuration of a FXD experiment.

used to protect the anode area of the x-ray machine from damage due to the high-velocity fragments produced by the shock-wave generator. The generator, consisting of a detonator, a shaped charge, and a high-explosive booster, is shown in the central part of this photograph. It is supported by an inexpensively fabricated goniometer device, which permits orientation of the sample at the correct Bragg angle with respect to the source, and of the camera, shown at the top of the photograph, with respect to the sample. The camera is an assembled version of that shown in Figure 15. Protruding from the front of the sample holder are shock-wave probes used to sense the location of the shock wave and to obtain its velocity. The x-ray source is triggered by a pulse from one of these probes.

This apparatus has been used to answer a variety of questions. Perhaps the simplest and certainly the most fundamental concerns order. Is there crystalline order behind the shock front, or is there rather a state of frenzied atomic motion and chaotic confusion of atoms? Underlying this question is the concern that the obviously violent and destructive nature of the shock-wave process might be inimical to crystalline order. It is conceptually difficult, upon first consideration, to imagine how the process can do more than disorganize the crystal lattice into a nearly amorphous entity. As the FXR study of projectile breakup illustrates, disintegration is possible on the macroscopic scale.

The answer to this important question came out of FXD studies on polycrystalline materials. It was shown that it is possible to obtain a diffraction record from a shock-compressed polycrystalline sample. An example of these results is shown in Figure 29 [29]. The top record, obtained before shock-wave compression, shows the first two diffraction lines of the powder pattern of LiF. The bottom record shows the pattern obtained during shock-wave compression to approximately 18 GPa. This record tells us that not only is the material behind the shockfront ordered, it is actually experiencing approximately hydrostatic pressure conditions. This follows from a measurement of the $\Delta 2\theta$ shift of the powder lines, which indicate directly how much the atoms have moved as a result of the shock pressure.

These results led to a study of the behavior of single crystals under shock-wave compression. The object was to shed some light on the question of atomic motion. Exactly how is it possible for apparently hydrostatic conditions to be obtained at the atomic level, when the sense of the shockwave is unidirectional? We contrast this study to the FXR study of the penetrator, which implied macroscopic motion transverse to the initial penetrator velocity vector.

Results from two different single-crystal materials are shown in Figure 29(b) [30]. The top record was obtained from a single

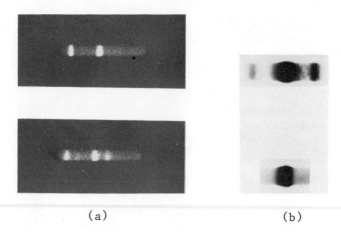

(a)   (b)

Figure 29.   FXD records from shock-wave experiments:
             (a)   Obtained from polycrystalline LiF; top image is
                   before shock compression; lower is during shock
                   compression.
             (b)   Obtained from single crystal undergoing shock
                   wave compression; the top record is from a LiF
                   crystal; the bottom record is from a graphite
                   crystal.

crystal of LiF shock-compressed to about 30 GPa. Features from
left to right are $K_\beta 200$ and $K_\alpha 200$ from the uncompressed lattice,
and $K_\alpha 200$ from the shock-compressed LiF. The bottom record shows
diffraction from the 002 reflection of pyrolytic graphite shock-
compressed to about 20 GPa. From considerations of the intensity
of the diffracted reflection obtained from the shock-affected portion of the crystal, it can be deduced that the crystal does not
disintegrate into a lot of randomly oriented cyrstallites, but
rather, to a first approximation, behaves as though it is still a
single crystal with orientation unchanged. The results of these
experiments suggest that models of microscopic motion behind the
shock-front must preserve crystal order and orientation, while
accounting for motion transverse to the shock vector to effect the
hydrostatic condition demonstrated by the earlier polycrystalline
experiments.

     FXR studies of volume changes have an analogue in the powder
and single crystal FXD studies. In addition, it will now be shown
how FXD can help to characterize materials undergoing discontinuous volume changes.

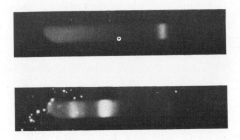

Figure 30.   FXD records of the phase transformation in BN.

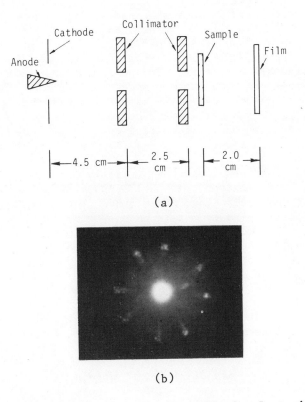

(a)

(b)

Figure 31.   (a)   Schematic for FXD transmission Laue photograph;
             (b)   Laue photograph of thin copper foil using a Mo
                   anode FXD unit.

A pivotal question in shock-wave studies concerns phase transformations behind the shock front. Bancroft, Peterson, and Minshall's study of shock-compressed iron [31] using conventional shock-wave techniques led to the conclusion that phase transformations could occur in the very short time intervals involved in a shock-wave experiment. Unfortunately, such measurements are indirect and do not establish that the phase transformation hypothesis is the unique and correct description of the observed phenomena. X-ray diffraction offers a direct test. Upon transformation to a different structure, the diffraction pattern changes so that, were it possible to observe the pattern from material presumably undergoing a transformation behind a shock wave, it should be clear that a new crystalline modification had formed.

In FXD studies conducted with BN, Hohnson and Mitchell [32] showed that it is possible to say conclusively that a phase transformation has occurred, to say what the new structure is, and to establish a mechanism for the transformation. The records obtained in that study are shown in Figure 30. The top record, taken before the shock compression, shows a partial pattern of the graphite-like normal low-pressure structure. The bottom record was taken as the sample was experiencing a shock pressure of about 25 GPa. The new features in this record establish that the material has transformed to the Wurtzite (diamond-like) structure. This type of study holds promise of answering many interesting questions on shock-induced pressure transformations.

A diffraction technique only partially exploited in FXD studies to date is that of Laue photography. Figure 31(a) shows the geometry for a transmission Laue experiment. The results obtained from a single pulse of 40-μs Mo x-rays onto a thin sample of copper are shown in Figure 31(b). This is a very weak image, and it would probably require enhancement if this particular material were used in an experiment under dynamic conditions. Considerably better results were obtained by Jamet [33] using this method on a single crystal of aluminum exploded by a capacitor discharge.

One of the weaknesses of the high-explosive method of shock generation in connection with the FXD studies it that only small quantities of explosives can be used if the film record is to be preserved. This means that simultaneous diffraction and shock-velocity measurements are difficult to obtain with great accuracy. One way around this problem is to do the separate measurements on separate but equivalent samples. A better way would be to devise a method for carrying out the diffraction experiment on a gas gun, where a precise determination of pressure is possible and where higher pressures can be obtained. Research in Japan is proceeding in this direction [34].

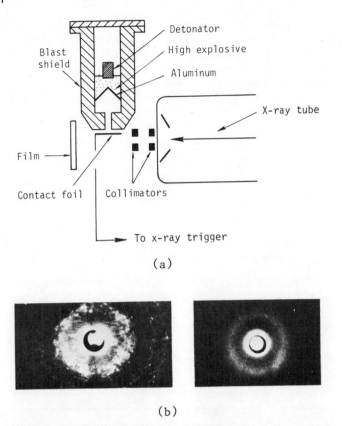

(a)

(b)

Figure 32. FXD experiment of an explosively-produced jet:
           (a) Schematic of experiment;
           (b) Powder pattern before (left) and during (right).

        In independent experiments, Green [35] and Jamet and Thomer
[36] have attempted to characterize the crystallite size of material
comprising an explosively driven jet. Figure 32 shows the experimental arrangement employed by Jamet and Thomer as well as their
results. A shaped charge, confined in a blast container, produces
a jet of aluminum. This jet contacts the foil and turns on the
x-ray unit in time to obtain a diffraction powder pattern of the
polycrystalline aluminum in the jet. The studies of Green show
that the jet was composed of particles between 0.01 and 1.0 mm in
size. The results of Jamet and Thomer indicate a smaller average
crystallite size. The left-hand photograph was taken before the
jet was formed, and the right-hand photograph was taken 5 μs after
initiation of the shaped charge. These studies clearly complement
the earlier described macroscopic characterization of jets.

The time-dependent process encountered in the explosion of a
gold foil by a capacitor discharge has been studied by Jamet [37].
The transmission powder pattern taken with copper K_α radiation is
shown in Figure 33.   The upper left picture shows the foil at rest;
the upper right picture was taken 1.3 µs after the beginning of the
capacitor discharge; the lower left and right pictures were taken at
1.5 and 1.8 µs, respectively.   The vanishing of the diffraction
pattern 1.8 µs after the discharge gives evidence that the vapor
state has been reached.

One of the foremost problems encountered in FXD is that of
synchronizing the x-ray pulse to the event of interest.   In some of
the foregoing experiments, this was not a major concern because the
event had a reasonably long lifetime.   Both the jet and exploding
foil experiments fall into this category.   The studies of the response of crystalline material to shock waves, however, requires much
more accurate timing.   Some of these events last 10 ns or less.

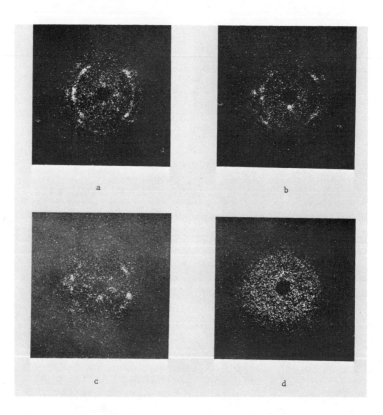

Figure 33.   Transmission powder patterns of an exploding gold foil.

Methods to increase lifetimes are thus of obvious interest. Egorov, Nitochkina, and Orekin [38] increased the lifetime of the shock-compressed state of aluminum by sandwiching a semi-transparent anvil of lithium over it. This greatly extends the time period over which it is possible to view the shock-compressed state, but it has an undesirable drawback. Due to the impedance mismatch between the sample and the anvil, the pressure drops considerably as soon as the shock wave reaches the sample/anvil interface. Nevertheless, a shock pressure of 14 GPa was attained.

The wedge method of Johnson, Mitchell, and Evans [39] is one way of increasing the event lifetime without producing the problem described above. A wedge-shaped base under the sample causes the shock wave to enter the sample at an angle. As a result, different parts of the sample are at different states, and a slit-shaped beam of x-rays can then probe the sample before, during, and after compression to give a time history of the event on a single record.

SUMMARY

The examples given here of the use of FXR and FXD in materials testing show the wide variety of problems that may be tackled. By contrasting FXR examples with approximately equivalent FXD examples, some appreciation of the types of questions each technique addresses has been obtained. These are complementary techniques; FXR is a tool for macroscopic characterization of materials, and FXD is a tool for microscopic characterization. Together, they permit access to a unique domain of NDT by virtue of the extremely short pulse lengths available from field-emitting flash x-ray devices.

ACKNOWLEDGEMENTS

We wish to thank A. Mitchell, L. Evans, J. Scudder, M. Van Thiel, K. Seifert, and F. Jamet for providing some of the material used in this manuscript. The spectrum in Figure 13(a) was calculated by S. Gair of Sandia Livermore Laboratory and used by permission.

The work was performed under the auspices of the U. S. Energy Research and Development Administration under contract N. W-7405-Eng-48.

REFERENCES

1. Turner, R.E., "Big Shot", Industrial Res., (August 1976) 43-45.

2. Marx, E., "Versuche uber die Prufung von Isolatoren mit
    Spannungsstossen", Elektrotechnische Zeitschrift, 45 (1925) 652.

3. Bernstein, B. and I. Smith, "Aurora, an Electron Accelerator",
    I.E.E.E. Trans. Nucl. Sci., NS-20, No. 3 (1973) 294-300.

4. Champney, P. D'A. and P. Spence, "PULSERAD 1480 - A 9 MV Pulsed
    Electron Accelerator with Intensely Focused Beam", I.E.E.E.
    Trans. Nucl. Sci., NS-22, No. 3 (1975) 970-74.

5. Martin, J.C., "Nanosecond Pulse Techniques", United Kingdom
    Atomic Energy Association, Berkshire, Rept. SSWA/JCM/704/49
    (1970).

6. Motz, J.W., C.E. Dick, A.C. Lucas, R.C. Pacious and J.H. Sparrow
    "Production of High Intensity K X-Ray Beams", J. Appl. Phys.,
    42, No. 5 (1971) 2131-33.

7. Pellinen, D.G., Q. Johnson and A. Mitchell, "A Picosecond
    Risetime High Voltage Divider", Rev. Sci. Inst., 45, No. 7
    (1974) 944-46.

8. Langmuir, I. and K.T. Compton, "Electrical Discharges in Gases,
    Part II. Fundamental Phenomena in Electrical Discharges", Rev.
    Mod. Phys., 3, No. 2 (1931) 191-257.

9. Creedon, J.M., "Relativistic Brillouin Flow in the High ν/γ
    Diode", J. Appl. Phys., 46, No. 7 (1975) 2936-55.

10. Personal communications, S. Gair, 1974.

11. Koch, H.W. and J.W. Motz., "Bremsstrahlung Cross-Section Formula
    and Related Data", Rev. Mod. Phys., 31, No. 4 (1959) 920-55.

12. Long, J.R. and A.C. Mitchell, "A dc X-Ray Fiducial Pulse Genera-
    tor for Light-Gas Gun Work", Rev. Sci. Inst., 43, No. 6 (1972)
    914-16.

13. Green, R.E., Jr., "Electro-Optical Systems for Dynamic Display
    of X-Ray Diffraction Images", in ADVANCES IN X-RAY ANALYSIS,
    Vol. 14, Proceedings of the 19th Annual Conf. on Applications
    of X-Ray Analysis, Denver, Aug. 5-7, 1970. Ed. by C.S. Barrett,
    J.B. Newkirk and C.O. Rund, New York, Plenum Press (1971)
    311-37.

14. Johnson, Q., A. Mitchell and L. Evans, "X-Ray Detector for
    Dynamic Diffraction Studies", Rev. Sci. Inst., 42, No. 7 (1971)
    999-1001.

15. Kusa, G.W. and J. Chang, "Nanosecond Time Resolved X-Ray Diag-
    nostics of Relativistic Electron Beam Initiated Events", in
    ADVANCES IN X-RAY ANALYSIS, Vol. 18, Proceedings of the 23rd
    Annual Conf. on Application of X-Ray Analysis, Denver, Aug. 7-
    9, 1974. Ed. by W.L. Pickles, C.S. Barrett, J.B. Newkirk and
    C.O. Ruud, New York, Plenum Press (1975) 107-116.

16. Mitchell, A.C., Q. Johnson and L. Evans., "A Film System for
    Flash X-Ray Diffraction Studies of Shock-Compressed Materials",
    Rev. Sci. Inst., 44, No. 5 (1973) 597-99.

17. Grostein, G.W., "X-Ray Attenuation Coefficients from 10 kV to
    100 MV", National Bureau of Standards, Report 583 (1957).

18. Plechaty, E.F., D.E. Cullen and R.J. Howerton, "Tables and
    Graphs of Photon Interaction Cross Sections from 1 kV to 100 MV",
    Lawrence Livermore Laboratory, Livermore, Calif., Report UCRL-
    50400, Vol. 6, Rev. 1 (1975).

19. Seifert, K. and A. Crispino, "Feasibility Study of Explosively
    Driven Hypervelocity Projectiles", Physics International Company,
    San Leandro, Calif., Report PIFR-559 (1974).

20. "Testing of Large Shaped Charges", Physics International Com-
    pany, San Leandro, Calif., Report PIP-1140B (1972).

21. Dobratz, B.M., "Properties of Chemical Explosives and Explosive
    Stimulants", Lawrence Livermore Laboratory, Livermore, Calif.,
    Report UCRL-51319 (1972) 8-1.

22. Personal communication, J. Scudder, 1976.

23. Van Thiel, M., M. Wilkins and A. Mitchell, "Shaped Charge
    Sequencing", Int. J. Rock Mech. Min. Sci. & Geomech. Absts., 12,
    No. 9 (1975) 283-88.

24. Seifert, K. and D. Maxwell, "Experimental Interim Report of
    Cratering Displacement Ejecta Processes", Physics International
    Company, San Leandro, Calif., Report PITR-659-1 (1974).

25. Jamet, F. and G. Thomer, FLASH RADIOGRAPHY, New York, American
    Elsevier Publishing Co., Inc. (1976).

26. Personal communication, G. Piermarmi, 1976.

27. Mao, H.K. and P.M. Bell, "High Pressure Physics: The 1 Mbar
    Mark on the Ruby/R_1 Static Pressure Scale", Science, 191, No.
    4229 (1976) 851-52.

28. Carbonnier, F.M., "Proposed Flash X-Ray System for X-Ray
    Diffraction with Submicrosecond Exposure Time", in ADVANCES IN
    X-RAY ANALYSIS, Vol. 15, Proceedings of the 20th Annual Conf.
    on Applications of X-Ray Analysis, Denver, Aug. 11-13, 1971.
    Ed. by K.F.J. Heinrich, C.S. Barrett, J.B. Newkirk and C.O. Ruud,
    New York, Plenum Press (1972) 446-61.

29. Johnson, Q., A. Mitchell and L. Evans., "X-Ray Diffraction
     Evidence for Crystalline Order and Isotropic Compression During
     the Shock-Wave Process", Nature, 231, No. 5301 (1971) 310-11.

30. Johnson, Q., A. Mitchell and L. Evans., "X-Ray Diffraction
     Study of Single Crystals Undergoing Sock-Wave Compression",
     Appl. Phys. Lett., 21, No. 1 (1972) 29-30.

31. Bancroft, D., E.L. Peterson and S. Minshall, "Polymorphism of
     Iron at High Pressure", J. Appl. Phys., 27, No. 3 (1956) 291-98.

32. Johnson, Q. and A. Mitchell, "First X-Ray Diffraction Evidence
     for a Phase Transition During Shock-Wave Compression", Phys.
     Rev. Lett., 29, No. 20 (1972) 1369-71.

33. Jamet, F., "Diagramme de Laue en Radiographic Instantanée",
     C. R. Acad. Sci., Ser. B, 271, No. 14 (1970) 714-17.

34. Kondo, K., A. Sawaoka and S. Saito, "A Trial Construction of
     the Flash X-Ray Diffraction System for Shock Wave Studies",
     PROC. 4TH INT. CONF. ON HIGH PRESSURE., Kyoto, 1974 (The
     Physico Chemical Society of Japan, Kyoto, 1975) 845-51.

35. Green, R.E., Jr., "First X-Ray Diffraction Photograph of a
     Shaped Charge Jet", Rev. Sci. Inst., 46, No. 9 (1975) 1257-61.

36. Jamet, F. and G. Thomer, "Diagramme de Poudre d'un Jet de
     Charge Creuse", C. R. Acad. Sci., Ser. B, 279, No. 9 (1975)
     501-03.

37. Jamet, F., "Recording of X-Ray Diffraction Patterns Using Flash
     X-Rays in Connection with an Image Intensifier", J. SMPTE, 80,
     No. 11 (1971) 900-01.

38. Egorov, L.A., E.V. Nitochkina and Yu.K. Orekin, "Registration
     of Debyegram of Aluminum Compressed by a Shock Wave", JETP.
     Lett., 16, No. 1 (1972) 4-6.

39. Johnson, Q., A. Mitchell and L. Evans, "Time-Resolved X-Ray
     Diffraction Study of Crystals Undergoing Shock-Wave Compression",
     to be published.

Chapter 7

NEUTRON RADIOGRAPHY FOR NONDESTRUCTIVE TESTING

J. John

IRT Corporation, San Diego, California

ABSTRACT

Neutron radiography is similar to x-ray inspection in that both depend upon use of radiation that penetrates some materials and is absorbed by others to provide a contrast image of conditions not readily available for visual inspection. But an important difference is the type of materials that absorb each of the two kinds of radiation. X-rays are absorbed by dense materials, such as metals, whereas neutrons readily penetrate metals but are absorbed by materials containing hydrogen.

The neutron radiography technique has been successfully applied to a number of inspection situations. These include the inspection of explosives, advanced composites, adhesively bonded structures and a number of aircraft engine components. With the availability of Californium-252, it has become feasible to construct mobile neutron radiography systems suitable for field use. Such systems have been used for in-situ inspection of flight line aircraft, particularly to locate and measure hidden corrosion.

INTRODUCTION

Neutron radiography is a nondestructive inspection technique that is similar in principle to x-ray, in that a penetrating radia-tion (neutrons) is used to obtain a visual image of the internal form of an object. The transmitted neutron beam is detected and permanently recorded on a suitable imaging system (usually film). The recorded image represents a shadow of the object with the lower

151

density on the image corresponding to portions of the object that are more effective in attenuating the direct neutron beam.

Although neutron radiographic and x-ray techniques are similar in principle, they actually complement each other. This is because the relative absorption characteristics of most elements are essentially reversed; thermal neutrons are highly attenuated by light elements (principally hydrogen), whereas x-rays are attenuated more by heavy elements. Thus with neutron radiography, nondestructive inspection can be made of light elements or certain defects encased in or behind heavy elements. It can detect imperfections in thick samples through which x-rays cannot penetrate, or in samples containing hydrogeneous matter or contaminants whose neutron absorbing and scattering properties differ significantly from the base material.

The mass absorption coefficients of naturally occurring elements for neutrons and x-rays are displayed in Figure 1. When plotted as a function of atomic number, the mass absorption coefficient for x-rays is a smoothly varying function increasing with atomic number. On the other hand, the corresponding values for neutrons fluctuate from elements to element, and on the average decrease with increasing atomic number. Hence, in general, the lightest elements (particularly hydrogen), are the most opaque and the heavier elements are the least opaque to thermal neutrons. This characteristic is the opposite of the attenuation properties of x-rays in which the heavy elements, such as lead, are the most opaque and the light elements are quite transparent. Hence by complementing x-ray inspection the neutron radiographic technique extends the capability of radiography to a much broader spectrum of applications.

The majority of neutron radiography is performed with neutrons having an energy of 0.3 eV or less. The neutrons are in thermal equilibrium with the surrounding material so that there is no net transfer of kinetic energy between the neutrons and the surrounding medium. The average energy of thermal neutrons is usually taken to be 0.025 eV. Thermal neutrons are more widely used because of their higher probability of interaction with most materials. This makes them easier to detect and gives the best contrast between the elements of interest and the surrounding medium.

In the past, the availability of suitable neutron sources has limited the growth and application of neutron radiography. Nuclear reactors and accelerators which are commonly used are expensive to acquire, troublesome to maintain, and lack portability. However, the recent availability of the man-made isotope Californium-252, a copious emitter of enutrons, has made it feasible to construct ^{252}Cf-based neutron radiographic systems, which because of their lower cost and portability, promise to significantly expand the use

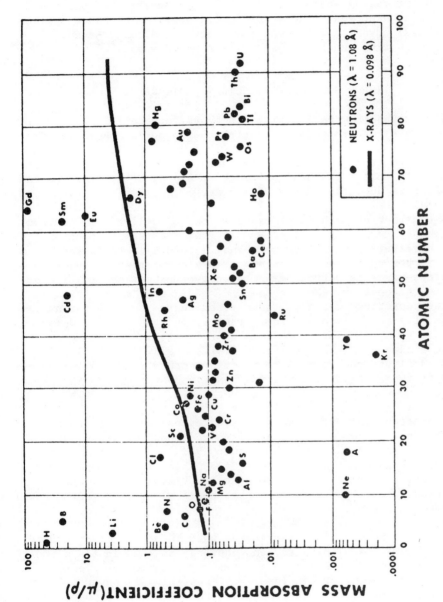

Figure 1. Comparison of x-ray and neutron radiography mass absorption coefficients.

of neutron radiography for nondestructive inspection. The neutron yeild of 2.3×10^{12} n/sec/g and its radioactive half-life of 2.65 years, make ^{252}Cf uniquely suited for use as a practical source of neutrons for radiographic applications. In addition to the cost savings over a nuclear reactor or accelerator, ^{252}Cf-based systems have several advantages in that they require no maintenance, elaborate control and safety system, or power supply.

IMAGE SYSTEMS FOR NEUTRON RADIOGRAPHY

The purpose of the system is to photographically detect neutrons. Unlike x-rays, neutrons do not interact with normal photographic films; therefore, a converter screen must be used with a conventional x-ray film to produce a permanent image. The role of the converter screen is to change the neutron image to alpha, beta, or gamma radiation which is detectable by a wide range of x-ray films.

Two different methods of exposure may be utilized in neutron radiography, depending on the type of converter used. Some of the metal converters and all of the scintillators emit their radiation promptly upon capturing a neutron. These converters are used in the direct-exposure method, with the converter screen and film exposed to the neutron beam together. The transfer method exposes only the converter screen to the neutron beam. Upon capturing the neutron, the metal screen becomes radioactive and emits a delayed radiation in a time period that is characteristic of the specific isotope produced in the metal screen. The delayed radiation is used to activate the film. The transfer method is useful when the object is radioactive, the neutron beam is contaminated with gamma rays, or the very best contrast is desirable.

Converters fall into two general classes: metal screens that emit low-energy gammas or x-rays often accompanied with electrons, and scintaillators that are loaded with ^{6}Li or ^{10}B that captures the neutron and emits an alpha particle, which in turn excites the scintillator material (ZnS) into emitting visible light. As one would expect, metal converters use standard x-ray film and scintillators use light-sensitive film.

The most common converter screen used in neutron radiography is gadolinium. Gadolinium screens can only be used in the direct-exposure method as it emits soft gamma rays and 70-keV electrons promptly upon capturing neutrons. Using a thin Gd converter about 0.001 inch thick, very high-resolution radiographs are possible. Resolutions of less than 0.0005 inch have been achieved using Gd converters in neutron beams having low angular divergence.

X-ray films having a range of speeds from 100 to 1600 (GAF Ref.) are routinely used with the converter screens mentioned above.

Resolution of cracks or opaque inclusions with one-mil dimentions are possible with GAF 100 or EK SR film when good irradiation geometry is present.

Combinations of scintillator material and photographic film have been produced to obtain increased speed by Gavaert, Polaroid, and 3M Company. Resolution suffers, but some interesting applications are possible.

As more sensitive imaging systems become available, fewer total neutrons are required to produce the desired image. However, a quantum limit exists that is determined by the statistical accuracy necessary to define an object over a given area. This limit is approximately 10^6 n/cm^2. Fewer neutrons than this per unit area will produce a fuzzy image no matter how sensitive the converter screen.

Other imaging techniques have been tried or are being developed. The tract-etch method uses cellulose nitrate screens to detect hydrogen atoms that have been scattered by the neutron beam. Neutron television systems are being studied using a scintillator material. Multiwire spark counters, similar to those used in high-energy particle experiments, are also being developed for neutron radiography.

NEUTRON RADIOGRAPHY CAMERA

The basic function of a neutron radiographic camera is to achieve as large a thermal neutron flux as possible at the object and imaging system with a minimum of gamma rays. In addition, the thermal neutron flux must arrive at the imaging plane in a beam to produce a sharp, clear picture of the object.

^{252}Cf is a compromise on the ideal neutron source for neutron radiography. It is true that ^{252}Cf has several of the attributes of an ideal source. It is a very intense source (essentially a point source) with relatively few gamma rays. Unfortunately, the neutron energy distribution is similar to a fission spectrum with an average neutron energy of slightly over 2 MeV. Consequently, to do neutron radiography, it is necessary to reduce the average energy of the neutron spectrum. The moderation of the neutron energy distribution is accomplished by placing the ^{252}Cf source in a material that has a high density of hydrogen. Hydrogen has the unique ability of moderating the neutron energy in the fewest number of scattering collisions and thus in the smallest volume.

Now that we are forced to use a neutron source with a finite volume to generate thermal neutrons, we must define an aperture

through which the neutron beam must go to get to the imaging plane.
To do this efficiently, a hole is introduced into the moderator
region that extends into the middle of the material next to the
source. This is the region with the highest thermal-neutron flux
density.

The aperature looks at the imaging plane through a divergent
collimator. The collimator is made divergent to maximize the
neutron flux over as wide an area as possible. The function of the
collimator is to capture (with a minimum of gamma activity) all
neutrons that might possibly scatter into the imaging plane.

The image plane is the location of the object. The film must
be place as close as possible to the object to produce clear, high-
resolution radiographs. The geometric unsharpness of a radiograph
is inversely proportional to the L/D ratio, where L is the distance
from the aperture of dimension D to the imaging plane. It is
usually convenient to have a varialbe L/D ratio by allowing for a
variable L and/or D.

The bulk of a neutron camera consists of biological shielding.
The cheapest shielding material for a neutron source is water.
Unfortunately, water can be inconveneint in that it has a tendency
to leak and provides poor gamma-ray shielding. We use a solid
material that contains a large amount of water and can also be
mixed with other neutron capturing material to reduce the amount of
secondary gamma rays that are produced. This material is called
water-extended polyester (WEP). It is an efficient neutron and
gamma shield when mixed with material containing boron.

In principle, a neutron camera with proper shielding can be
operated in the same room with personnel. However, for safety
reasons, it is best to have controlled access to the camera area.
A shutter and interlock system can be provided that essentially
eliminates the remotest possibility of a radiation accident. The
access door and a personnel plug board are interlocked to the
shutter. Break either interlock and the shutter automatically
closes. Loss of power will also cause the shutter to close.
Additional safety features include a flashing light when the shutter
is open, a radiation monitor with a threshold setting, a shut-off
from the cell, and a console lock key.

Portable Neutron-Radiography System

A standard neutron-radiography system using ^{252}Cf weighs far
too much to enable its use as a convenient portable source to inspect
an aircraft or an aircraft component without first removing it from
the aircraft. Fortunately, most of the weight resides in the
biological neutron and gamma-ray shielding necessary if personnel

are to be in close proximity to the radiography system. The camera
itself consists only of the source, moderator, and thermal-neutron
collimator, and weighs approximately one tenth of a standard fixed
system. Hence, a portable system, consisting essentially of only
the camera components, has been shown to be feasible [1] if the need
for personnel to be in close proximity during exposure is eliminated.

    Such a protable camera, improvised for aircraft maintenance
inspection, is shown in Figure 2. This system is shown readied to
inspect the wing tank of an E-2 aircraft [2]. Although this system
has not been specifically designed for mobile operation, its performance in this less-than-optimum configuration helps to demonstrate
the feasibility of a portable neutron-radiography system (see Ref. 3
for additional discussion). The camera and the biological shield
are constructed as separate entities, with the source residing in

Figure 2.   Portable neutron radiography camera positioned to radio-
            graph lower skin of aircraft fuel tank. The camera is
            mounted on top of the wing, with the neutron beam aimed
            downward. The Californium-252 source normally resides
            in the biological shield shown on the right-hand side of
            the photograph. The operator initiates an exposure by
            moving the source from the cask into the camera through
            a flexible hose using a remotely operated cable-driven
            mechanism.

the shield except during actual exposure. The portable neutron-radiographic camera is located on top of the aircraft wing, with the beam aimed downward. The cylindrical biological shield is located on the right-hand portion of this figure. The two subsystems are connected by a polyethylene hose through which the source is transferred between the camera and the shield by a flexible cable.

The process of making an individual exposure consists of positioning the film cassettes on the lower surface of the wing under the camera. The exposure is initiated by transferring the ^{252}Cf source from the storage container by means of a crank-driven cable. The cable arrangement allows the operator to stand some 25 feet away from the camera during source transfer, as shown in Figure 2. When the crank is driven forward, the flexible cable with a magnetic tip moves from the neutron camera through the polyethylene hose, which guides the cable into the source storage container. The magnetic tip of the cable is driven down against the source, and with contact a magnetic latch is established. The crank mechanism is then reversed, withdrawing the cable, with the source magnetically attached, from the storage container back through the polyethylene tube into the camera body. This operation requires about four seconds. The exposure is terminated by reversing this process, thereby transferring the source from the camera into the source container.

Using 5.8 mg of ^{252}Cf, it was possible to inspect 6-1/2 square feet of the lower skin of the E-2 aircraft integral fuel tank in a single exposure, lasting approximately 2-1/4 hours. The radiographic geometry allowed the resolution of most spherical clusters of corrosion with a diameter of 0.010 inch, which was considered adequate.

To comply with radiation safety regulations, an exclusion area having a radius of approximately 140 feet is necessary. The total radiation dose at the perimeter of this area is 2 mrem/hr during exposure. When necessary, partial radiation shields consisting of borated paraffin positioned between the camera and maintenance personnel can reduce the exclusion area to permit access to other portions of the aircraft during an exposure.

A series of programs [4-12] have been carried out to evaluate applications of neutron radiography to solve a number of inspection problems. A selected set of examples is described below.

METAL JACKETED EXPLOSIVES

Neutron radiography is quite effective for inspecting metal-jacketed explosive density variations, voids, and foreign materials. This application provides excellent examples of the ability of neutrons to effectively image light materials within heavy ones.

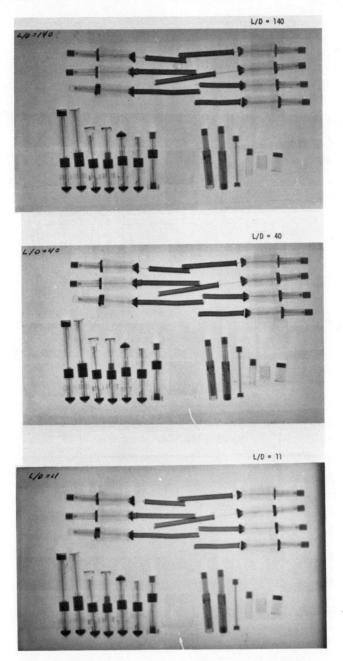

Figure 3. Prints of neutron radiographs of explosive train devices
having known defects (such as missing end tips) taken with
L/D ratio (resolution) of 140, typical of reactor-based
radiography, and L/D ratios of 40 and 11, more typical of
^{252}Cf-based neutron radiography.

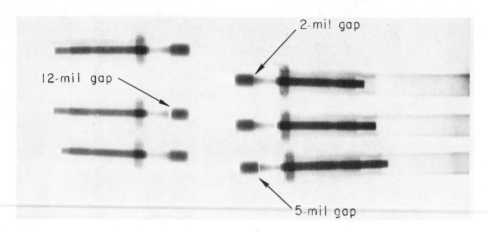

Figure 4.   Enlarged print of neutron radiograph of shielded mild
            detonating cords with known defects (gaps of 12, 6, and
            2 mils) identified by the arrows.

Examination of the same components with x-rays only reveals the
metallic claddings. Aircraft depend on propellant-activated
devices and linear-shaped charges for emergency escape systems;
here fast and reliable performance is demanded. Figure 3 shows
three neutron radiographs taken with varying L/D ratios of small
explosive train devices with known defects, such as missing end
tips. These pictures establish that although radiographs with an
L/D of 140 obtain the best resolution, an L/D of 11 provided radiographs with adequate resolution to discern many defects. This
reduction of L/D from 140 to 11 is accompanied by a reduction in
the exposure time by a factor of about 100.

Figure 4 shows three shielded mild detonating cords with
known gaps of 2, 6, and 12 mils; these cords are sealed in stainless steel cases. Neutron radiographs taken with the ^{252}Cf camera
successfully detected all of these gaps. Here again, it should be
noted that x-ray inspection images only the stainless steel casings
and provides no information on defects in the explosive.

A number of pyrotechnic devices are used in the F-111 aircraft
system. These are widely used in emergency escape mechanisms. One
such device, consisting of the end fittings and a line charge enclosed in a metallic sheath, is shown in Figure 5; this is a photograph of the device. The radiograph shown in Figure 6 clearly
images the explosive charge, since the metallic sheath is transparent
to neutrons. In this particular case, the pyrotechnic device is in
good condition and is devoid of any defects.

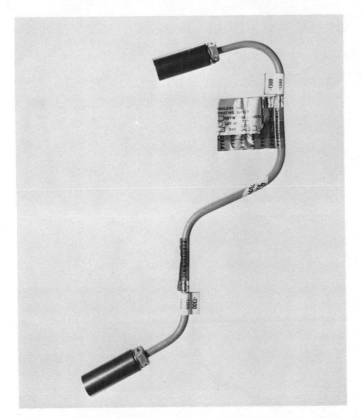

Figure 5.   Photograph of pyrotechnic device used in the F-111
            aircraft.

ADVANCED COMPOSITE MATERIALS

     The development and widespread use of advanced composites in
aircraft structures calls for a parallel effort in the development
of new NDI techniques for inspecting these components. Conventional
NDI techniques, developed primarily for metallic components, will
become of limited use as components fabricated from these advanced
composites begin to replace the all-metal ones. Neutron radiography is capable of inspecting these composites, an example of
which is shown in Figure 7. This is a single layer of tungsten-
boron fibers in an aluminum matrix. The filaments consist of 0.5-
mil tungsten core coated with boron to make 4-mil fibers. This
neutron radiograph is able to see single filaments for detailed
inspection. Cracks and other defects are easily observed.

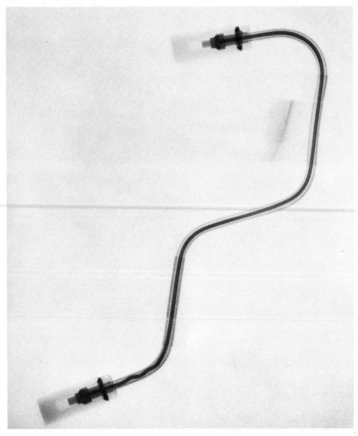

Figure 6.   Neutron radiograph of pyrotechnic device. The explosive
            material is clearly imaged through the metallic sheath.

INSPECTION OF ADHESIVE BONDED STRUCTURE

        Neutron radiography is quite effective in detecting common
defects in adhesive bonded honeycomb structure. Since neutron
radiography highlights the organic adhesive material instead of the
metallic core seen by x-rays, it is possible to detect variations
in adhesive density which can lead to inadequate bonding. This is
illustrated in Figure 8. Shown here is a neutron radiograph of a
test panel approximately 12 inches long, 7 inches wide, and 3-5/8
inches thick. The end panels are approximately 1/16 inch thick.
The honeycomb structure visible in this radiograph shows the distribution of the organic adhesive around the metallic honeycomb
core. This radiograph clearly shows four circular areas where no
adhesive was applied to one of the end panels. Careful examination
reveals that the edges of two of these circles are better defined
than those of the other two, implying that two of these voids are
on one side of the panel, while the other pair is on the opposite

Figure 7.   Enlarged print of neutron radiograph of tungsten-boron
            composite material consisting of 0.5 mil tungsten and
            4 mil boron in aluminum matrix.

side. Since neutron radiography images the adhesive, rather than
the metallic core, it is speculated that a technique involving
neutron radiography may be possible to detect defective bonds resulting from the failure of the adhesive to wet the honeycomb.

A second example for inspecting adhesive bonded aircraft
structure is seen in the case of the F-111 aircraft. The vertical
stabilizer of the F-111 aircraft consists of phenolic honeycomb
bonded to aluminum skin. The leading edge of the vertical stabilizer is an aluminum strip shaped in the form of an arrowhead. The
adhesive bond lines between the aluminum arrowhead and the aluminum
skin have been suspected of being deficient, leading to the existence of minute leakage path ("worm holes") connecting the honeycomb
cells to the outside environment. It is surmised that water
accumulation in the honeycomb cells results from these leakage paths.
During supersonic flights, considerable heat is produced in the

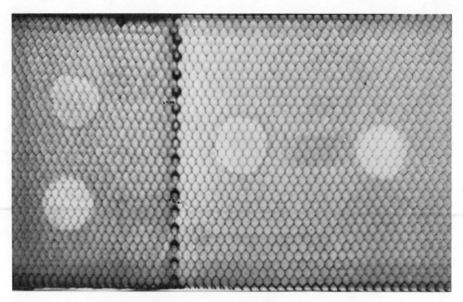

Figure 8.   Neutron radiograph of an adhesively bonded test panel
            approximately 12 inches only, 7 inches wide, and 5/8
            inches thick. The four circular areas seen in this
            figure are regions where no adhesive was applied to one
            of the end panels. The two circles with clearly defined
            edges are on the side of the panel close to the film,
            and the other two are on the opposite side of the panel.
            The honeycomb structure visible in this radiograph
            results from the adhesive around the metallic core rather
            than the core itself.

region of the leading edge, causing the water to boil and resulting
in the rupture of the skin and loss of portions of the vertical
stabilizer in flight.

    A neutron radiographic examination of an F-111 sample was per-
formed and the resulting radiograph is displayed in Figure 9. This
radiograph clearly shows the adhesive bond line between the arrowhead
and the aluminum skin. A careful examination of the radiograph
reveals adhesive deficiencies, and a number of possible paths linking
the honeycomb cells to the outside environment can be identified.
In addition, this radiograph reveals great nonuniformity in the
application of the adhesive in this region.

    Several models of helicopters are equipped with main rotor
blade systems that are bonded by adhesives. Failures of these
bondings have been attributed to insufficient amounts of adhesive
in the bond area, voids or air bubbles in the adhesive, improper

preparation of the bond surface, corrosion of the surfaces under
the adhesives, or a variety of other possible causes. Such defects
are known to result in catastrophic failures, particularly when the
disbonding occurs in the closure of a C-shaped spar of the main
rotor blade system.

The feasibility of using the neutron radiographic technique to
detect these disbonds has been established through a program [8]
conducted at IRT Corporation. Selected test blades were subjected
to the standard neutron radiography procedure. Neutron beam from the
camera was directed toward the sample at an angle so that the adhesive region on one side of the C-spar, between the spar and the spar
closure, was clearly imaged on the film. The orientation of the
blade was such that the adhesive region on the opposile side of
the C-spar did not interfere with imaging of the adhesive bond close

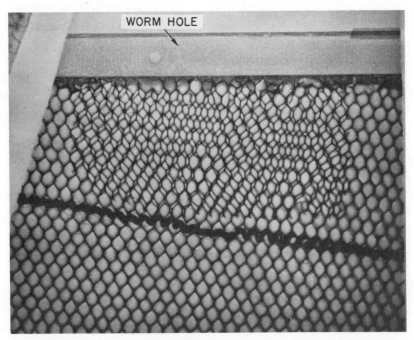

Figure 9. Neutron radiograph of a portion of the leading edge of
           the vertical stabilizer of an F-111 aircraft. The ad-
           hesive bond line between the aluminum arrowhead and the
           aluminum skin shows adhesive deficiencies providing
           possible leadage paths between the honeycomb cells
           and the outside environment. One such "worm hole" is
           labeled.

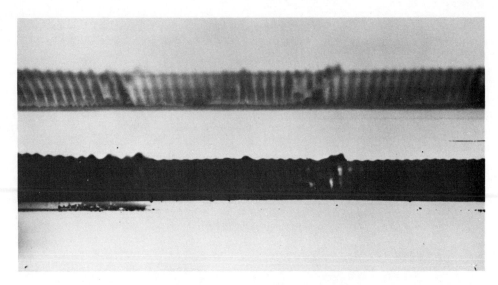

Figure 10.   A composite of two neutron radiographs of a portion of
             a 540 helicopter blade system. The one on the top is
             a standard neutron radiograph showing the bond line
             between the C-spar and the spar closure clip, as well
             as the honeycomb. The bottom half shows the result
             after treatment with an enhancing agent. The penetra-
             tion of the enhancing agent into the disbonded area
             helps clearly identify the entent of disbond.

to the film. These radiographs showed areas of non-uniform adhesive
distribution, but were naturally unable to detect disbonded areas
that were properly bonded prior to use.

    A suitable enhancing agent was then used to wet the area near
the spar closure. The orientation of the blade was such as to en-
courage the penetrant to seep into areas where any disbonding occurs.
After a short period of time the excess penetrant was removed, and
the sample was subjected to a second inspection. The penetrant, a
good absorbed of neutrons, helped highlight the disbonded or unbonded
area.

    Figure 10 shows a composite of two neutron radiographs obtained
with a blade. At the top is a standard radiograph clearly showing
the honeycomb structure in the trailing portion of the blade. The
bond line is the region between the two parallel lines below the
honeycomb. The lower radiograph was taken after treatment with the
penetrant. The disbonded portion of the spar closure area allows
the penetrant to enter the space between the spar and the closure

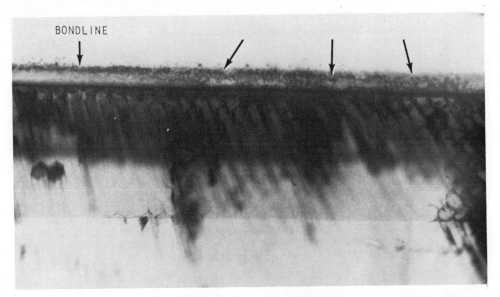

Figure 11.   Neutron radiograph of a helicopter blade with fretting
corrosion.   The uneven nature of the bond line results
from metal-to-metal contact due to inadequate quantities
of adhesive.

clip, highlighting the area of disbond.   The extent of disbond is
clearly identified.

A second type of defect in helicopter blade detected by neutron
radiography is illustrated in Figure 11.   The bond line, in this
case, appears uneven and discontinuous, indicating the location where
fretting corrosion has occurred.   When two metal surfaces are allowed
to rub against each other due to insufficient amount of adhesive,
fretting often results which leads to the initiation of cracks
followed by catastrophic failure.

After completion of the neutron radiographic inspection, these
blades were disassembled and subjected to visual inspection.   The
disbonds detected by this destructive inspection process correlated
exactly with the defects identified by neutron radiography.

INSPECTION OF AIRCRAFT ENGINE COMPONENTS

Turbine engine blades and vanes for high-temperature applications are generally cooled by internal passages formed by investment
casting processes.   Air is supplied at the base of the blade, which
flows through the passageways cooling the blade and exits through a

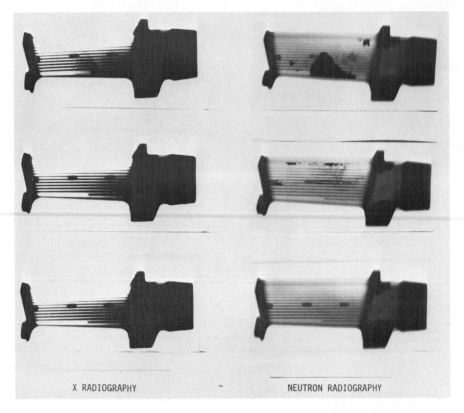

X RADIOGRAPHY   NEUTRON RADIOGRAPHY

Figure 12.   Comparison of x-radiograph and neutron radiograph for
              for inspecting turbine blades for residual ceramic core.
              The set on the left are x-rays, and the one on the right
              was obtained with neutrons.

series of holes. Cooling is aided by the use of pins and fins
within the blade, which increase the surface area to improve heat
transfer efficiency.

    A ceramic core is used to form the cooling passages when the
blade is cast, and after casting this core is chemically leached
out. The high operating temperatures necessary to achieve higher
engine efficiency demand that all the ceramic core be removed. Any
amount of residual core remaining in the blade can cause the cooling
passages to be blocked, which will reduce durability and can lead
to blade failure. The amount of residual core remaining which could
cause serious damage is bleow the resolution limits of x-radiography.

    Figure 12 shows an x-radiograph of a set of TF-41 turbine
blades on the left and the corresponding neutron radiograph on the
right. The x-radiograph is able to detect the large amounts of

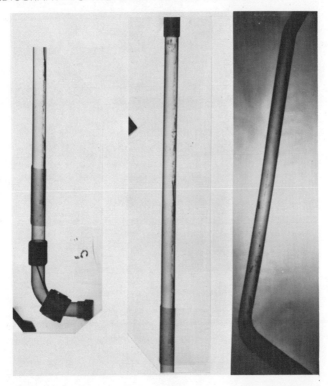

Figure 13. Neutron radiograph of portions of oil lines in aircraft
engine showing coke deposits on the inner walls. A
large piece of coke, dislodged from the tube wall, is
seen resting near the shoulder of the J-shaped tube.

residual core which is blocking almost the entire cooling passage.
This residual core is seen with much greater contrast in the
neutron radiograph. As the amount of core is reduced, x-radiography
is unable to detect it, but it is still clearly imaged in the
corresponding neutron radiograph.

Turbine blades, fuel manifolds, and oil lines in aircraft
engines are often clogged by coke deposits. The high temperatures
at which these components operate cause the hydrocarbons in the fuel
and oil to dissociate, depositing carbon on their surfaces. Figure
13 is a composite of three neutron radiographs showing sections of
oil lines of a CF-6 aircraft engine. Coke deposits are visible on
the inner surfaces of these tubes. A large piece of coke, dislodged
from the tube wall, is seen resting near the shoulder of the 'J'-
shaped tube.

The TF-30 engine used on high-performance military aircraft
has been known to fail due to fatigue cracking of compressor and

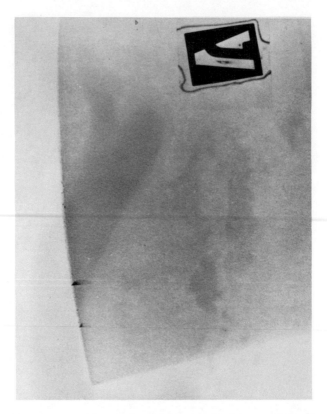

Figure 14. Neutron radiograph showing cracks at the tip of com-
            pressor blade of a TF-30 engine. The enhancing agent
            used in this case assists in defining the depth and
            orientation of the cracks.

turbine blades. An example is shown in Figure 14 which contains a
neutron radiograph of the tip of a compressor blade. The cracks
are clearly imaged in the radiograph. The enhancing agent used in
this case assists in defining the orientation of the cracks.

DETECTION OF CORROSION IN AIRCRAFT STRUCTURE

Corrosion within the cross sections of aircraft structures can
affect the structural integrity, resulting in damages that range
from fuel leakage to catastrophic failures. The timely detection
of corrosion is essential if corrective measures are to be taken
to arrest corrosion and avoid subsequent expensive rework.

The primary technique at present for identifying corrosion is
visual detection of external manifestations of corrosion, such as

appearance of blisters of the swelling of skin blanket. This technique completely fails when the corrosion is at the interface between
two structural members, or is concealed behind a heavy coating of
sealant.

Wing Tank Corrosion

The wing fuel tanks of aircraft have been known to be susceptible
to in-service corrosion and subsequent damage. This has been attributed to moisture seepage under the sealant and between the skin and
ribs, spar flanges, and fasteners. Corrosion damage identified by
de-skinning, or cutting out skin portions so corroded that fuel
leaked through, has not shown a pattern related to either flight
hours or months in service. Visual inspection through available
access panels covers only a small fraction of the area, making the
identification of wings requiring priority rework and definition of
safe damage limits unreliable.

The extreme effectiveness of the neutron radiographic technique
to identify corrosion is illustrated by two examples in the following
subsections.

Wing Tank Corrosion in a Commercial DC-9 Aircraft: Several
instances of exfoliation corrosion in the outboard flange of
station Xcw = 58.500 lower tee cap in DC-9 aircraft have been
reported. Exfoliation corrosion is attributed to fuel contamination
or water or both, allowing corrosion to originate between the stringer
and the tee cap. Exfoliation corrosion in the tee cap has been
reported on aircraft having as few as 10,000 flight hours.

The recommended inspection procedure called for visual inspection, followed by the addition of a radio-opaque oil and subsequent
x-ray inspection. The special oil mixture is forced into the area
between the stringer and the tee cap using a cacuum method prior to
x-ray inspection. The results obtained with x-rays are unsatisfactory. Since the stringer is nearly 2 inches thick, x-rays merely
display the outline of the stringer and fail to reveal corrosion
(see Figure 10 in Ref. 4).

Under a program sponsored by Douglas Aircraft Company, a series
of radiographs has been taken simulating in-situ inspection of a
DC-9 aircraft. A set of stringers, received from Douglas Aircraft
Company, were inspected using the portable neutron-radiography
system, and the resulting neutron radiograph is shown in Figure 15.
It displays the outline of the thick portion of the stringer, the
tee cap, and the bolts attaching the stringer to the tee cap. In
addition, the radiograph clearly reveals the extent of corrosion
through the thick stringer, and presents no difficulty in interpreting the results. A composite photograph showing two views of

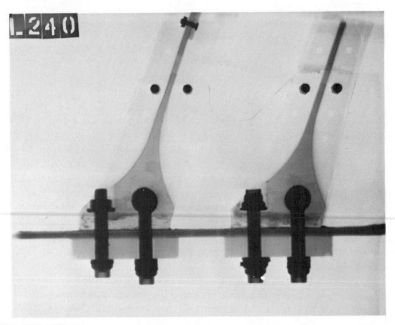

Figure 15. Neutron radiograph of a portion of the DC-9 wing tank
            showing the outline of the stringer, tee cap, and a set
            of bolts. The corroded area is visible through the
            stringer approximately 2 inches thick.

the tee cap surface is given in Figure 16. Tee caps marked 13 and
14 correspond to the radiograph displayed in Figure 15. It is clear
that the surface has been subjected to a major corrosion attack.
It should be noted that Figure 16 displays only the corrosion
visible on the surface, whereas the radiograph in Figure 15 images
both surface and subsurface corrosion, and hence the apparent dis-
parity in the extent of corrosion seen in these two figures.

     The main factor in favor of the neutron radiographic technique
is that it does not require visual inspection, the application of
radio-opaque oil, or intrusion of the oil into a corroded area for
detection. Also, it does not require removal of the barrel nut or
evacuation of the barrel nut hole. It could also locate corrosion
between the wing skin and the tee cap if it exists.

     Corrosion in C-130 Aircraft Wing: The C-130 Hercules aircraft
has all-metal wings, with two-spar stressed-skin structure, with
integrally stiffened, tapered, machined skin panels. Several instances of extensive surface and intergranular corrosion have led
to fuel leaks in this aircraft. This problem is particularly
difficult to identify when the corrosion is located within the lap
joint.

Figure 16.   Photograph showing two views of the tee cap surface from
             the DC-9 wing tank. Tee caps identified as 13 and 14
             correspond to radiographs displayed in Figure 15.

        This wing-tank problem is illustrated in Figure 17. This is a
photography of a portion of the lower wing-tank structure of a C-130
aircraft. It clearly shows the origin of the corrosion attack. An
inspection of the upper surface of the skin does not provide any indication of the severity of corrosion damage. Figure 18 shows two
neutron radiographs of this sample. The radiograph on the right
half of this figure is a standard neutron radiograph. It shows
the corroded region to originate at the locations identified in
Figure 17, and the intergranular corrosion to entend to one rivet
hole. This could clearly cause a fuel leak, although the rivet
itself may appear to be in good condition. When a small quantity
of enhancing fluid is used, the corrosion feature is considerably
enhanced, as shown in the neutron radiograph on the left half of
Figure 18. This radiograph indicates that the intergranular corrosion has enveloped several rivet holes, clearly explaining the
cause of the fuel leak.

        Stress Corrosion in A-7 Nose Landing Gear: In the course of a
program [6] sponsored by the Navy, a number of A-7 nose landing

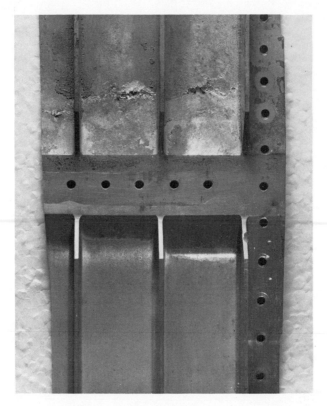

Figure 17. Photograph of a portion of the lower wing skin structure
            of a C-130 aircraft, showing the origin of the corrosion
            attack. The extent of corrosion is not clear from such
            visual inspection.

gear struts were screened using the neutron radiographic technique.
This investigation was carried out to detect corrosion on the inner
surface of the trunnion, at the interface between the aluminum
housing and the bronze bushing. Disassembly of the gear for this
inspection without damaging the component is very difficult. Over
52 landing gears were inspected, and nearly half of them were found
to have been subjected to corrosion attack to some degree.

     One of the most critical specimens examined during this study
was an A-7 strut with a stress-corrosion defect. The defect appeared
as a hairline crack in the trunnion region, barely detectable by
visual inspection, even after the location was detected by an eddy
current probe. However, the eddy current technique was unable to
provide any additional information regarding the extent or shape
of the stress-corrosion crack.

     This strut was subjected to neutron radiographic inspection,

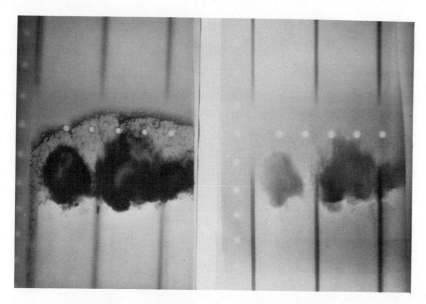

Figure 18.   Neutron radiograph of C-130 wing structure. The neutron
radiograph on the right is obtained in the usual way,
and the one on the left with the use of a penetrant.
The intergranular corrosion is seen to extend beyond
the rivet holes, clearly explaining the cause of fuel
leak from these holes.

and the resulting neutron radiograph is shown in Figure 19. The
stress-corrosion crack is seen to originate at a point on the inner
surface of the trunnion bore and extend almost to the outer surface,
clearly showing the extent, shape, and nature of the crack. This
appears to be the first experimental work in which a stress-corrosion
crack has been imaged by neutron radiography.

The sample was then sectioned; a photograph of the sectioned
trunnion area is shown in Figure 20. The extent of the stress-
corrosion crack is clearly visible, and the correlation with neutron
radiography is excellent.

Quantitative Determination of Corrosion: In order to take
full advantage of this nondestructive inspection technique, it is
desirable not only to identify the location of the corrosion, but
also determine the extent of the corrosion damage. Hence, a program
has been conducted to evaluate the potential of using the neutron
radiographic technique as a quantitative tool to measure the depth
of corrosion so that selective work of the aircraft can be carried
out. This work is reported in detail in Ref. 9.

The results of the investigation show that an experimental

Figure 19.   Neutron radiograph of an A-7 nose landing gear strut
             showing the extent, shape, and nature of stress-
             corrosion crack.   This crack was barely visible, even
             after its location was confirmed by the eddy current
             technique.

relationship can be formulated which will quantitatively determine
the corrosion depth from the film density measured from a neutron
radiograph.   This is graphically represented in Figure 21, showing
the data obtained from a corroded torque-box panel of an F-4 aircraft.   The deviation of the photographic density at a corroded
point from the corresponding value at the undamaged portions
(plotted along the ordinate) is shown to be proportional to the
depth of the corrosion (plotted along the abscissa).   Such linear
relationships have been shown to exist for a variety of aluminum
alloys sued in structure of aircraft that have operated in
different environments.

     This study has further established that the thickness of
corrosion products and the thickness of aluminum remaining can be

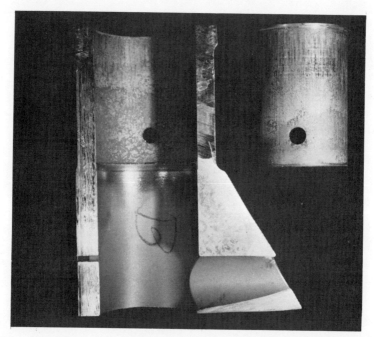

Figure 20. Photograph of the sectioned trunnion area showing the
extent and shape of stress-corrosion crack.

determined with a precision of 2 to 3%. This ability to precisely
determine the amount of structural material remaining (particularly
the aircraft skin in the case of integral fuel tanks) enables
rework decisions to be based on actual structural conditions.

COMPARISON TO TO OTHER INSPECTION TECHNIQUES

In order to compare the corrosion detection capability of the
neutron radiographic technique with those of other inspection
techniques, a careful experiment was carried out on a flight-line
aircraft. For this investigation, a particular E-2 aircraft was
selected as a test vehicle due to its past history of in-service
corrosion susceptiblity. The lower skin of the wing tank was
first surveyed with neutron radiography. The wing was later subjected to x-ray examination, ultrasonic scan, visual inspection
after the upper skin of the wing tank was removed, and a second
visual inspection after the sealant and paint were removed from
the inner surface of the tank. The major conclusions of this
study are:

1. The neutron radiographic technique, unlike x-ray and

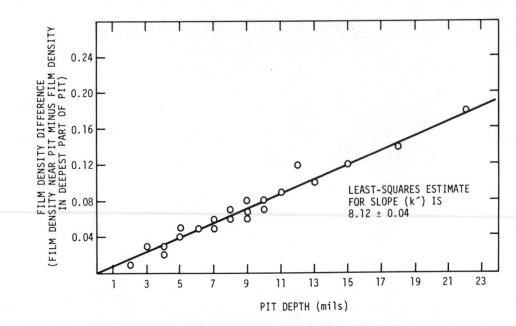

Figure 21.   Relationship between photographic density and extent of
corrosion providing a quantitative means of determining
corrosion using neutron radiography. The deviation of
the photographic density at a corroded point for the
corresponding value at the undamaged portion (ordinate)
is proportional to the depth of corrosion (abscissa).

ultrasonic techniques, detects corrosion by the presence of
corrosion products. Hence, this technique identified the
actual condition of the skin, and therefore the interpretation requires no knowledge of its prior maintenance or
rework history.

2.   Neutron radiography permits inspection of 95% of the lower
internal skin area. It can detect corrosion hidden under
sealant, behind ribs, spars, spar flanges, and fasteners.
On the other hand, x-ray, ultrasonic, and visual inspection
technqiues are severely restricted by the internal structure
of the wing, as well as the uneven nature of the skin. Only
80 to 85% of the surface can be covered by a combination of
these three conventional techniques.

3.   Of the 419 regions of corrosion identified by neutron radiography, 284 were confirmed by ultrasonic inspection and a
total of 309 were confirmed by a combination of all three
conventional techniques. Most of the uncorrelated areas
are accounted for by the limitations of the conventional

techniques resulting from inaccessibility or from interfering structure. Only five discrepancies were attributable
to poor interpretation of neutron radiographs.

COST EVALUATION

The E-2 aircraft inspection also provides a means of evaluating
the cost savings to be achieved with neutron radiography. A complete
inspection of the wing tank, as it is carried out now, calls for the
de-skinning of the upper wing surface, followed by the removal of
the sealant and paint by chemical stripping or shell blasting. With
the wing tank fully open from the top, visual access is now available
to the entire exposed portion of the upper surface of the lower wing
skin. However, even this elaborate and time-consuming operation
does not provide a means of inspecting the region between the skin
and ribs where corrosion is very likely to occur. Only complete
disassembly of the wing structure can provide any sure way to con-
duct visual inspection of the entire wing tank for corrosion detection.

It is estimated that approximately 5,000 manhours are expended
in reworking the integral fuel cell areas of a single E-2 aircraft
for corrosion. This consists of de-skinning, removal of paint and
sealant, cleaning, visual inspection for corrosion, chemical conversion coating, application of new sealant, painting, re-skinning,
and checkout. The actual inspection portion of this procedure takes
428 manhours.

The experience gained with the E-2 aircraft shows that a team
of two can carry out the neutron radiographic inspection, including
setup, film developing, and interpretation. Allowing 16 manhours
for initial preparation, and an equal amount for shutdown operations,
the entire neutron radiographic inspection of the lower skin of
the E-2 aircraft can be accomplished with an expenditure of 152
manhours. This is a substantial savings over the 428 manhours
estimated to perform the inspection using the currently recommended
practice. In addition, by eliminating the need for de-skinning,
paint removal, and re-skinning required for visual inspection, it
appears that the cost of acquiring and operating a neutron radiography system can be repaid in a short time, even if its use is
restricted to the E-2 aircraft.

CONCLUSION

These studies have helped to establish that neutron radiography
as a nondestructive inspection tool has a definite role to play in
a number of inspection situations. Its capability complements those

of other inspection techniques currently in use. The ability of neutron radiography to identify surface and intergranular corrosion, and to detect them quantitatively, is unique. The feasibility of performing neutron radiographic inspection in a maintenance environment is now clearly established.

ACKNOWLEDGEMENTS

A large number of persons have contributed to making this program possible, and any attempt to name them would only result in an incomplete list. The author would like to acknowledge their assistance. Special mention should be made of the assistance received from Naval Air Systems Command, Air Force Logistics Command, and Army Aviation Systems Command.

This work would not have been possible without the loan of ^{252}Cf sources from the San Diego Californium-252 Demonstration Center. The assistance of Mr. F. P. Baranowski, Director, Division of Nuclear Fuel Cycle and Production in the Energy Research and Development Administration, who made this Center possible, is gratefully acknowledged.

REFERENCES

1. John, Joseph, "Mobile Neutron Radiography System for Aircraft Inspection", text of paper presented at Air Transport Assoc. of America Nondestructive Testing Subcommittee Meeting, Houston, Texas, Sept. 11-13, 1973, IRT Corp. Technical Report RT-TB-151.

2. Larsen, J.E., J. Baltgalvis, F. Patricelli, M. Gallardo and Joseph John, "Evaluation of a Portable Neutron Radiographic System for the Detection of Hidden Corrosion in the Wing Fuel Tank of the E-2C Aircraft", IRT Corp. TEchnical Report INTEL-RT 6082-001, February 25, 1975.

3. John, Joseph, D.E. Rundquist and R. Sharp, "Application of Neutron Radiography Techniques for Nondestructive Detection of Corrosion in Naval Aircraft and Aircraft Components", IRT Corp. Technical Report INTEL-RT 6044-002, March 15, 1974.

4. John, Joseph, V.J. Orphan, D.E. Rundquist and R. Sharp, "Feasibility Study of Applying Neutron Radiography for Improved Maintenance Inspection of Naval Aircraft and Aircraft Components", IRT Corp. TEchnical Report INTEL-RT 6044-001, November 1, 1973.

5. Rundquist, D.E., R. Sharp and Joseph John, "Feasiblity Evaluation of Real Time Imaging for Neutron Radiographic Inspection of Naval Aircraft", IRT Corp. Technical Report INTEL-RT 6072-001, August 1, 1974.

6. Larsen, J.E., D.E. Rundquist, J. Baltgalvis and Joseph John, "Investigation of Neutron Radiography Techniques for Corrosion Inspection of A-7 Nose Landing Gear", IRT Corp. Technical Report INTEL-RT 6071-001, August 16, 1974.

7. Larsen, J.E., R. Parks, J. Baltgalvis and Joseph John, "Investigation of Neutron Radiographic Techniques for Maintenance Inspection of Air Force Aircraft", IRT Corp. Technical Report INTEL-RT 6081-001, March 3, 1975.

8. Patricelli, F., J.E. Larsen, J. Baltgalvis, R. Parks and Joseph John, "Californium-252 Based Neutron Radiography for the Detection of Disbonds in the Spar Closure Area of 540 Series Helicopter Blades", IRT Corp. Technical Report INTEL-RT 6085-001, February 28, 1975.

9. Patricelli, F., J.E. Larsen, J. Baltgalvis and Joseph John, "Experimental Evaluation of Neutron Radiography for Quantitative Determination of Corrosion in Aircraft Structure", IRT Corp. Technical Report INTEL-RT 6093-001, January 28, 1976.

10. John, Joseph, "Californium-based Neutron Radiography for Corrosion Detection in Aircraft", Practical Applications of Neutron Radiography and Gauging, H. Berger, Ed., published by American Society for Testing and Materials, January 1976, pp. 168-180.

11. John, Joseph, J.E. Larsen, F. Patricelli, H. Harper,
     J. Baltgalvis, M.J. Devine and A.J. Koury, "Neutron Radio-
     graphy for Maintenance Inspection of Military and Civilian
     Aircraft", Proceedings of International Symposium on Califor-
     nium-252 Utilization, Fontenay-aux-Roses, France, April 26-28,
     1976.

12. Harper, H., J. Baltgalvis, H. Weber, J.C. Young and Joseph John,
     "Evaluation of Californium-252-based Neutron Radiography and
     Photon Scattering Techniques for the Inspection of Hot Isostati-
     cally Pressed Components of T-700 Aircraft Engine", AVSCOM
     Report No. TR76-24, August 1976.

Chapter 8

NONDESTRUCTIVE COMPOSITIONAL ANALYSIS

L. S. Birks

Naval Research Laboratory, Washington, D. C.

ABSTRACT

X-ray spectroscopy is about the only nondestructive method
for quantitative chemical analysis. Portable x-ray machines can
be carried directly to the material to be analyzed but laboratory
x-ray machines are more common in industrial use. Historically,
x-ray spectroscopy began with Moseley's 1913 demonstration that
each chemical element emits characteristic x-ray lines when excited
by high-energy electrons. Early difficulties in quantitiative
analysis were overcome by better instrumentation and mathematical
methods for relating x-ray intensity and composition. Today many
kinds of samples including solids, liquids, powders, pollution,
etc., can be analyzed rapidly for 20-30 elements for less than
$ 10 per sample. Computer programs make the data interpretation
simple and accurate.

INTRODUCTION

Although x-rays were discovered by Röntgen in 1895 it was not
until 1913 that Moseley's historic experiments [1] demonstrated that
each chemical element emits a very simple x-ray spectrum. Figure 1
is taken from Reference 1 and shows the Kα and Kβ lines (plus weak
impurity lines) from elements between Ca and Zn. The relationship
between wavelengths, λ, and atomic number, Z, is simply

$$\lambda \propto 1/Z^2 \qquad\qquad (1)$$

Nowadays we can both calculate and measure the characteristic and
continuum spectrum very accurately as shown in Figure 2 from a

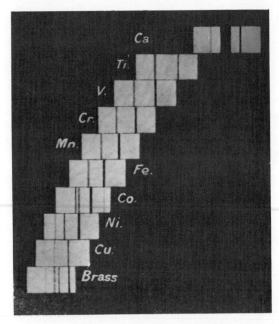

Figure 1.   Moseley's original x-ray spectra [1].

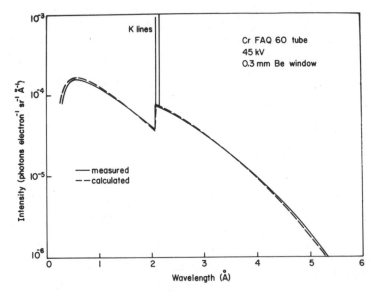

Figure 2.   Characteristic lines and continuum from a Cr target
            x-ray tube.

recent paper by Brown, Gilfrich, and M. C. Peckerar [2].

After Moseley's experiments x-ray spectra were measured throughout the world and were the basis for the discovery of Hf and Re. X-ray spectroscopy also contributed greatly to the emerging field of quantum mechanics in the 1920's. By 1930 analysts had succeeded in detecting parts per million of impurities in alloys, but accurate quantitiative analysis lagged because of the lack of stable x-ray tubes and suitable x-ray detectors (photographic film was difficult to interpret quantitatively).

From the early 1930's until the late 1940's there was a total lack of effort in x-ray spectrochemical analysis even though x-ray powder diffraction made great progress in the identification of crystalline compounds and the measurement of stress in alloys. Then in the late 1940's the use of fluorescent x-ray excitation outside sealed x-ray tubes and the use of a Geiger counter detector to measure x-ray intensity accurately led to a resurgence of interest.

With the improved instrumental capability came the realization that quantitative analysis requires accounting for matrix effects, i.e., the partial absorption of one element's radiation by other elements in the specimen, and the secondary fluorescence of some elements by characteristic radiation from other elements within the specimen. More is discussed about matrix effects in the section on Quantitative Analysis.

It should be pointed out that x-ray emission is an atomic process related to the inner-shell electrons. Figure 3 shows the atom schematically; in fluorescent excitation a primary x-ray photon ejects an electron from, say, the K shell in the atom. This vacancy

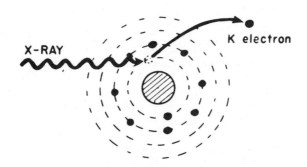

Figure 3.   Schematic of excitation of an atom by incident x-ray photon.

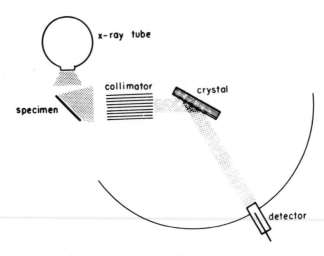

Figure 4.   Wavelength dispersion geometry.

is filled quickly by an electron from one of the outer shells and
this transition is what results in the emission of a characteristic
x-ray photon.   Thus the x-ray lines are essentially independent of
both the chemical combination or physical state of the sample.
The method is therefore equally as applicable to liquids or amorphous solids as it is to crystalline compounds or alloys.

SPECTRAL DISPERSION

     In order to identify the elements and measure their x-ray
intensities it is necessary to separate the characteristic lines
physically or electronically.   The two approaches used are called
wavelength dispersion and energy dispersion.

     Of the two approaches, wavelength dispersion is the classical
method and is illustrated in Figure 4.   In Figure 4 primary radiation from the x-ray tube excites the characteristic lines from the
sample elements.   This characteristic radiation emerges in all
directions and must be limited to a parallel beam by the collimator.
The polychromatic beam strikes the analyzer crystal which diffracts
only that wavelength which satisfies Bragg's law for the particular
angular setting, θ, of the crystal.   As the crystal is rotated each
wavelength is diffracted in turn and the intensity measured by the
detector which rotates at twice the angular velocity of the crystal.

     The second approach, energy dispersion, became practicable in
the mid 1960's with the advent of the solid-state Si(Li) detector.

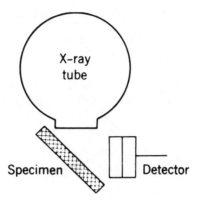

Figure 5. Energy dispersion geometry.

Figure 5 depicts the method. Again the characteristic lines are excited by the primary x-ray beam and emitted as before. But now, all of the characteristic radiation is detected simulatneously. With the Si(Li) detector each absorbed x-ray photon generates an electric pulse whose amplitude is proportional to the photon energy. Each pulse is stored in the proper energy bin in a multichannel analyzer, and the display of the stored pulses (number of pulses at each energy) is simply the energy spectrum of the sample. The solid-state detector does not have as good resolution as the crystal spectrometer and this results in more line interference between neighboring elements.

For a more detailed discussion of the relative merits of wave-length and energy dispersion the reader is referred to current x-ray literature [3].

QUANTITATIVE ANALYSIS

There are three common approaches for converting measured x-ray intensity to elemental concentration in the sample.

Calibration Curves

The most obvious approach to analysis is the use of calibration curves prepared from known standards. Figure 6 shows such curves for Ni in a variety of binary combinations. For systems of more than two elements families of curves result and become prohibitively time consuming and expensive to prepare. However, it is possible to

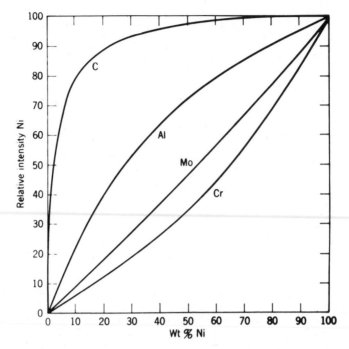

Figure 6.   Calibration curves for Ni in binary mistures.

use calibration standards for multi-element samples if only small changes in concentration from the standard are expected.

Regression Equations

Perhaps the most used approach for quantitative analysis involves the use of empirical coefficients, α_{ij}, to represent inter-element effects and correct for matrix absorption and secondary fluorescence (enhancement). These α coefficients are used in a family of homogeneous linear regression equations, one member of which is shown in Equation (2):

$$c_i/I_{ri} = \alpha_{io} + \alpha_{ii}c_i + \Sigma\, \alpha_{ij}c_j \qquad (2)$$

Here c_i is the concentration of element i and I_{ri} is its relative intensity (the intensity from the mixed matrix divided by the intensity from pure element i); α_{io} is the intercept of the tangent to the hypothetical calibration curve at the proper composition; α_{ii} is the effect of element i on its own radiation and is identically unity for the normalized I_{ri}; α_{ij} is the effect on element i due to the concentration c_j of element j. A similar equation

represents each element in the sample. The α values are determined
from a set of standards containing the same elements as are to be
analyzed in the unknowns. A small computer is needed to determine
the α's and then to calculate the c_i's.

It should be noted that the α coefficients must be determined
for standards similar to the unknowns (iron-base alloy standards
for iron-base unknown alloys, copper-base ore standards for copper-
base ores, etc.). It is also necessary that the same target x-ray
tube and the same operating voltage and detector be used when the
standards are measured and when the unknowns are measured. Because
of these limitations the regression-equation approach is most useful
when the same type of unknown is to be measured over and over. The
method is not versatile for random unknown samples because preparing
and measuring standards to determine the α's is an expensive and
laborious process.

Fundamental Parameters

The most versatile approach [4] to quantitative analysis makes
use of more elaborate mathematical expressions than Equation (2)
and relates x-ray intensity to concentration through the mass absorption coefficients and fluorescent yields which are known and
unchanging. It also requires knowledge of the primary spectral
distribution and can only be programmed for a large computer.
Within these requirements it can handle random unknown specimens
with up to 20 components and can generally carry out an analysis
at lower cost per element than the regression method (provided
access to a large computer is available by either time sharing or
batch processing).

For routine analysis the computer program would be used for
batch processing of data and the information required (x-ray
intensities, etc.) would be input on computer cards or tape.
However, the program is just as easy to use in an inter-active
mode where the input information is typed directly into the time-
sharing terminal and the answer is received immediately. Figure 7
illustrates the interactive mode and shows the information which
is input (each item preceded by a small star) and the results which
are output by the computer.

APPLICATIONS

The ability of x-ray spectroscopy to measure nearly all the
chemical elements in any physical state or chemical combination
allows a wide variety of applications. Table 1 is a generalized
listing but should not be considered as all inclusive. Of the many
possibilities the measurement of alloys, of minerals, and of
pollution represent the major applications.

```
(*    NORMALIZE                    (*    ANALYZE STEEL
(*    1 CR-STD
(*    2 MN-STD
(*    3 FE-STD                            LINE        RXI MEAS.
(*    4 NI-STD                       1 CR KA1        .2260E+00
(*                                   2 MN KA1        .6126E-02
#                                    3 FE KA1        .5445E+00
(*    COMPOSITION STEEL              4 NI KA1        .5735E-01
(*    CR ?
(*    MN ?                         COMPONENT      AMOUNT
(*    FE ?
(*    NI ?                         CR             17.0346
(*                                 MN              0.5948
#                                  FE             69.5891
(*    INTENSITIES STEEL            NI             12.0112
(*    1 230000 100 40 10 2         (TOTAL)        99.2296
(*    2 6700 100 45 10 2
(*    3 550000 100 50 10 2
(*    4 59000 100 55 10 2
(*
#
```

Figure 7.   Printout of interactive computer program for x-ray
            analysis.

TABLE 1

APPLICATIONS

1.   Solid samples - alloys,
     glass, ceramics, etc.

2.   Powders - ores, pigments,
     soils, etc.

3.   Biomedical - tissue, serum,
     bone, etc.

4.   Pollution - particulates by
     filtration, solubles by ion
     exchange

TABLE 2

1975 RESULTS USING FUNDAMENTAL PARAMETERS PROGRAM

	NBS 1151		NBS 1156		NBS 1160	
Element	Chem.	X-Ray	Chem.	X-Ray	Chem.	X-Ray
Fe	67.17	67.18	69.90	69.97	14.30	14.95
Ni	7.07	6.93	19.0	18.71	80.30	79.50
Cr	22.13	22.11	0.20	0.23	0.05	0.07
Cu	0.25	0.25	0.03	0.08	0.02	0.04
Si	0.37	0.33	0.18	0.19	0.37	0.35
Mo	0.76	0.72	3.10	2.95	4.35	4.35
P	0.011	0.008	0.01	0.06	0.003	0.07
Mn	2.17	2.35	0.21	0.25	0.55	0.63
S	0.004	0.034	0.012	0.026	0.001	N.D.

To illustrate typical results for alloys, Table 2 shows the results on three samples of iron-base and nickel-base samples. The specimens were taken from National Bureau of Standards certified standards but they were treated as unknowns in the x-ray measurements. The agreement between x-ray estimated and NBS chemical values is good with a few exceptions such as the low sulfur content where no attempt was made to correct for particle size effects, although the sulfur is probably present as precipitates.

Cost of Analysis

For a method to be practicable it must be competitive economically as well as capable scientifically. Compared to other chemical techniques the x-ray method is more costly in terms of capital investment but less costly on a total, per-sample basis because of the speed and automation which is possible. Table 3 compares the costs of the three most common types of x-ray equipment. The sequential instrument uses one analyzer crystal and measures one element at a time. It is a versatile instrument for routine analysis and still the most widely used of all the instruments. The energy dispersion instrument represents the popular solid-state-detector approach with convenient data treatment options. The multiple-crystal instrument with a fixed spectrometer for each of 20 to 30 elements is the kind of instrument installed in large-scale industrial plants where speed and accuracy are paramount and the elements to be measured can be specified in advance.

TABLE 3

COSTS FOR X-RAY ANALYSIS

	$70,000 Wavelength Sequential 20 samples/day	$50,000 Energy Dispersion 100 samples/day	$200,000 Wavelength Simultaneous 200 samples/day
Equipment*/yr	$14,000	$10,000	$ 40,000
Salaries/yr	15,000	30,000	30,000
Overhead/yr	15,000	30,000	30,000
	$44,000	$70,000	$100,000
Cost/sample	$ 11.00	$ 3.50	$ 2.50
Cost/element	$ 0.75	$ 0.25	$ 0.17

*Amortized over 5 years

Whichever x-ray approach is chosen the cost of analysis is low compared to traditional methods, and the accuracy is better than most indirect methods.

SUMMARY

Characteristic x-ray spectra and their simple relationship to atomic number were first demonstrated by Moseley in 1913 but routine quantitative analysis did not become practicable until the 1950's after better instrumentation had been developed for x-ray diffraction and then adapted for x-ray spectroscopy. Mathematical methods for data interpretation and refined instrumentation including the solid-state detector have brought x-ray spectroscopy to a very competitive position for quantitative analysis of alloys, minerals, biologicals, and pollution samples. Between 20 and 300 samples per day can be analyzed by a single x-ray instrument and the cost ranges from 20¢ per element to 75¢ per element depending on the instrument and the number of samples.

REFERENCES

1. Moseley, H.G.J., Phil. Mag., 26 (1913) 1024; 27 (1914) 703.

2. Brown, D.B., J.V. Gilfrich and M.C. Peckerar, J. Appl. Phys.,
    46 (1975) 4537.

3. Birks, L.S. and J.V. Gilfrich, Anal. Chem., 48 (1976) 273R.

4. Criss, J.W. and L.S. Birks, Anal. Chem., 40 (1968) 1080.

Chapter 9

REVIEW OF EUROPEAN ADVANCES IN NONDESTRUCTIVE XRD-TESTING

E. Macherauch

Institut für Werkstoffkunde I, Universität Karlsruhe (TH)
Karlsruhe, Germany

ABSTRACT

In practice, non-destructive x-ray diffraction testing is mainly applied to phase analysis and to stress analysis. Some short introductory remarks concerning the European situation in these two fields are made. Then a comprehensive review of recent European activities in the area of x-ray stress analysis is given. The principles underlying the $\sin^2\psi$ method are outlined. The progress achieved in x-ray elastic constants is discussed. In the field of the analysis of loading stresses recent results concerning stress states in cracked specimens are discussed. Some new investigations in the field of residual stresses are reviewed. Comments are given to the problem of milling respectively grinding residual stresses and fatigue. Recent contributions to the understanding of deformation stresses, machining stresses, and welding stresses are presented.

INTRODUCTION

With respect to the title of this chapter, there are two important fields of particular interest, especially as far as non-destructive XRD testing is concerned in connection with the mechanical behavior of materials: Quantitative Phase Analysis by Means of X-Rays, and Stress Analysis by Means of X-Rays. However, it is beyond the scope of this chapter to review both items in an appropriate manner indispensable for this purpose. Since the author's interests are more concentrated on the field of x-ray stress analysis, the following review is restricted mainly to the

Figure 1.   Places in Europe where study groups are active in the
            field of NDXRD-Testing.

field of stress meansurements by means of x-rays.

     At the beginning some short remarks should be made about the
actual situation in the field of applied non-destructive x-ray
diffraction in Europe. The study groups in Central Europe working
in this field are partially participating in a close exchange of
ideals and results with one another. In Figure 1 the places are
marked where active study groups exist. At present there are about
40 groups active in this field. Due to well-known reasons real and
active communication is possible only between the study groups in
Austria, France, Great Britain, Italy, the Netherlands, Switzerland,
and West Germany. Most of these have joined to form committees
which are attended by the West German "Association for Heat Treatment
and Materials Technology". These committees are named "Stress State
and Behavior of Materials" and "Measuring Technique". The members
of these committees assemble for two day meetings at half-year
intervals in order to discuss results obtained and problems encountered in a congenial and open manner. Contacts with East European
study groups are, unfortunately, not very intense. However, there
are communications to study groups in Bulgaria, Czechoslovakia,
Hungary, and Poland.

The cooperation of Central European experts in the field of applied x-ray diffraction in the committees mentioned above has recently led to two comprehensive publications. The first one was published three years ago and summarizes all substantial aspects of quantitative x-ray phase analysis in condensed form. This special issue of "Härtereitechnische Mitteilungen" [1], among others, contains contributions such as:

Physical fundamentals of the quantitiative x-ray phase analysis
Quantitative x-ray phase analysis on texturized materials
Quantitative x-ray phase analysis on multiphase materials with
        superposition of the diffraction lines
Experimental techniques in x-ray phase analysis
Uncertainties and errors in x-ray phase determinations

This year a second comprehensive publication appeared concerning the status of x-ray stress analysis [2]. This synopsis is comprised of contributions such as:

Definition of residual stresses
The $\sin^2\psi$ method of x-ray stress analysis
Evaluation of residual stress distributions
X-ray stress measurements with film and diffractometer techni-
        ques
X-ray elastic constants
Stress distributions in macroscopic inhomogeneous stressed
        materials
Residual stresses due to thermal treatments, phase transforma-
        tions, welding, plastic deformation, shot peening,
        turning, milling, and grinding
X-ray stress analysis on texturized materials
Recommendations to the practical application of x-ray stress
        measurements

In the following a selection of some European contributions to the field of x-ray stress analysis is presented which, of course, is determined by personal attitude. Nevertheless, the author hopes to give a feeling for the developments in this important field of applied x-ray diffraction which have been made in Central Europe in the last three to five years.

BASIC PRINCIPLES OF X-RAY STRESS ANALYSIS

X-ray stress analysis is based on the measurement of changes dD/D_0 of a special set of the lattice spacings D_0 of special oriented grains in the surface layers of polycrystals due to the influence of internal or external stresses. Such changes in the D-spacings are called lattice strains. The identification of these lattice strains dD/D_0 with the engineering strains ε calculated from the theory of isotropic elasticity is the main feature of x-ray stress analysis.

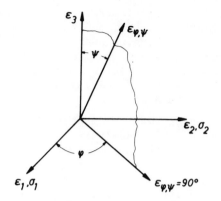

Figure 2.   System of co-ordinates.

To measure lattice strains a collimated x-ray beam of suitable
wavelength λ is used and high θ angle interference lines of the
specimens under investigation are measured using either back reflection film cameras or x-ray diffractometers. From the well-known
Bragg's law

$$n\lambda = 2 D_o \sin \theta_o \qquad (1)$$

the following equation can be obtained

$$\frac{dD}{D_o} = - \operatorname{ctn} \theta_o \, d\theta \qquad (2)$$

where θ_o is the Bragg angle of the stress-free specimen dD/D_o is the
lattice strain and $d\theta$ is the appropriate shift of the interference
line due to the lattice strain. Therefore x-ray diffraction lines
of stressed specimens undergo changes in their position. The determine lattice strains, line shifts have to be measured.

In the case of a biaxial state of stress (Figure 2) given by
the principal stresses σ_1 and σ_2 both parallel to the surface of a
specimen, the theory of elasticity yields for the engineering strain
in the direction ρ,ψ

$$\varepsilon_{\rho,\psi} = \frac{1}{2} s_2 \, (\sigma_1 \cos^2\rho + \sigma_2 \sin^2\rho) \sin^2\psi + s_1 \, (\sigma_1 + \sigma_2) \qquad (3)$$

where ρ is the azimuth angle with respect to σ_1 and ψ is the angle
between the normal L on the surface of the specimen and the direction
of strain. The quantities $1/2 \, s_2$ and s_1 are

$$\frac{1}{2} s_2 = \frac{\nu + 1}{E} \qquad (4)$$

and

$$s_1 = -\frac{\nu}{E} \tag{5}$$

E is Young's modulus and ν is Poisson's ratio. The first term in brackets in the above mentioned Equation (3) is equal to the surface stress component σ_ρ corresponding to the azimuth angle ρ. Then, assuming the strains derived from the theory of elasticity are identical with the lattice strains determined by x-rays, Equation (2) and Equation (3) yield

$$\varepsilon_{\rho,\psi} = (\frac{dD}{D_o})_{\rho,\psi} = \frac{1}{2} s_2 \quad \sin^2 + s_1 (_1 + _2) \tag{6}$$

This is the fundamental relation for every x-ray stress analysis. Important features of this equation are:

1. The lattice strains dD/D_o are linearly dependent on $\sin^2\psi$, independent of the azimuth angle ρ of the macroscopic plane given by L and ρ in which the lattice strains are measured.

2. The slope of the straight line through the lattice strains plotted as a function of $\sin^2\psi$ is determined by

$$\frac{d\varepsilon_{\rho,\psi}}{d \sin^2\psi} = \frac{1}{2} s_2 \sigma_\rho \tag{7}$$

    Therefore, the stress component σ_ρ can be obtained if an appropriate value of $1/2 \ s_2$ is available.

3. The intercept of the straight line with the ordinate axis, that is the lattice strain for $\psi = 0$, is given by

$$\varepsilon_{\rho,\psi=0} = s_1 (\sigma_1 + \sigma_2) \tag{8}$$

    Therefore, the sum of the principle stresses can be determined if an appropriate value of s_1 is available.

In determining stresses by means of x-rays, all scientists and engineers in Central Europe now use the so-called $\sin^2\psi$ method which the author proposed at the end of the 1950's [3]. The underlying idea is to measure four or more lattice strains in the azimuth of interest in different ψ-directions which are chosen such that the appertaining values of $\sin^2\psi$ assume equal intervals. Without knowledge of the directions of the principal stresses it is then possible to determine every surface stress component from the slope of the mean-square line obtained from the plot of the lattice strains versus $\sin^2\psi$ and the sum of the surface principle stresses from the intercept of this straight line with the ordinate axis.

In Europe, x-ray stress analysis is applied either for basic research or for solving problems of practical importance. Most work is done with the back-reflection diffractometer method, using so-called θ-diffractometers which work according to the classical Bragg-Brentano principle [4]. At present, however, six laboratories are already equipped with the Karlsruhe type diffractometer which is called ψ-diffractometer [5]. Some portable x-ray back reflection diffractometers are also in service [6,7]. Furthermore, in special cases the back-reflection film method is still used [8].

In the laboratories equipped with diffractometers frequently the diffracted x-ray intensities are automatically measured point by point using computer-controlled devices. The computer subsequently determines the net intensities, performs derived corrections, calculates centers of gravity as well as stress values from the slope of the lattice strain distribution over $\sin^2\psi$ in a given azimuth plane. On demand it also draws graphs. At present studies with position sensitive gas proportional detectors are in process at different places [9].

ELASTIC ANISOTROPY

Some European contributions very important to the field of x-ray stress analysis concern the influence of the elastic anisotropy. This influence occurs because the lattice strain measurements in different directions ψ_i are always done in the same crystallographic direction hkl of special oriented sets of grains of the polycrystal under investigation. Therefore, in the basic equation of x-ray stress analysis [Equation (6)], instead of $1/2\ s_2$ and s_1, the x-ray values of these quantities have to be used. They are named x-ray elastic constants and designated as $1/2\ s_2^{\text{x-ray}}$ and $s_1^{\text{x-ray}}$. As can be seen from Equation (6) with $\rho = 0$ and $\sigma_2 = 0$ the lattice strains $\varepsilon_{\rho=0,\psi} = \varepsilon_\psi$ are a linear function of $\sin^2\psi$ as well as of σ_1. For that reason, partial differentiations for these variables yield

$$\frac{\delta\ \varepsilon_\psi}{\delta\ \sin^2\psi} = \frac{1}{2}\ s_2^{\text{x-ray}}\ \sigma_1 \tag{9}$$

and

$$\frac{\delta 2\varepsilon_\psi}{\delta\ \sin^2\psi\ \delta\sigma_1} = \frac{1}{2}\ s_2^{\text{x-ray}} \tag{10}$$

resp.

$$\frac{\delta\ \varepsilon_{\psi=0}}{\delta\sigma_1} = s_1^{\text{x-ray}} \tag{11}$$

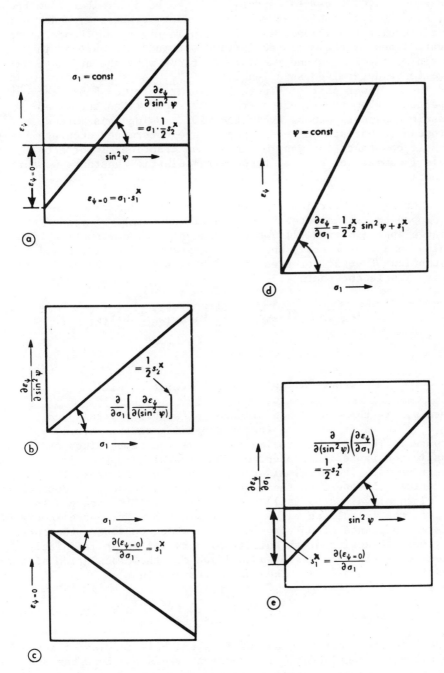

Figure 3.   Methods for the determination of x-ray elastic constants.

Consequently, $1/2\ s_2$x-ray and s_1x-ray can be determined from the ε_ψ, $\sin^2\psi$ relationships at several loading stresses σ_1 under uniaxial tensile conditions, as indicated in Figure 3. From the mean square lines of the measured lattice strain distributions in respect to $\sin^2\psi$, the slopes and the interceptions with the ordinate axis have to be determined as a function of σ_1. As shown in the middle and in the lower part of Figure 3, linear relationships are again expected. From the slopes of these straight lines the x-ray elastic constants can be evaluated [10]. Using this method, European scientists in recent years have determined accurate x-ray elastic constants for many materials of practical importance. Table 1 summarizes some of these x-ray elastic constants measured recently.

During the Thirties, the first theoretical methods were developed to estimate the x-ray elastic constants of polycrystals. Voigt's approximation [11] which assumes equal deformation in all grains in a polycrystal, and Reuss' approximation [12] based on the assumption of equal stresses of all grains in a polycrystal yield x-ray elastic constants [13-15]

$$(s_1)^{\text{x-ray}}_{\text{Voigt}} = f(s_{ik}) \qquad (\tfrac{1}{2}\ s_2)^{\text{x-ray}}_{\text{Voigt}} = g(s_{ik}) \qquad (12)$$

and

$$(s_1)^{\text{x-ray}}_{\text{Reuss}} = (s_{ik}, \Gamma) \qquad (\tfrac{1}{2}\ s_2)^{\text{x-ray}}_{\text{Voigt}} = (s_{ik}, \Gamma) \qquad (13)$$

where s_{ik} is the elastic compliances of single crystals and $\Gamma = \Gamma(h,k,l)$ is an orientation factor depending on the $\{h,k,l\}$ - planes measured. The Voigt approximation gives the same values of x-ray elastic constants independent of the crystallographic directions of measurements. The x-ray elastic constants according to the Reuss approximation depend on the $\{h,k,l\}$ - planes measured. Most of the experimental values of (s_1)x-ray and $(1/2\ s_2)$x-ray are placed between the calculated values [16]. Recently, Kroener's theory of elastic constants of polycrystals [17] which considers the correct elastic interaction of an anisotropic crystal in an infinitely extended matrix also has been modified for the claculation of x-ray elastic constants [18]. The theory results in

$$(s_1)^{\text{x-ray}}_{\text{Kröner}} = F(s_{ik}, \Gamma, G, K) \qquad (\tfrac{1}{2}\ s_2)^{\text{x-ray}}_{\text{Kröner}} = G(s_{ik}, \Gamma, G, K) \qquad (14)$$

where G is the shear modulus and K is the bulk modulus of the polycrystal under consideration. In recent years the European experimental work has given strong evidence that the x-ray elastic constants calculated according to Kröner's theory coincide well with the experimental findings [2,8]. The typical result of an extensive

TABLE 1

X-RAY ELASTIC CONSTANTS

Material	Radiation	Measured Planes	$\frac{1}{2} s_2^{\text{x-ray}}$ 10^{-6} mm^2/N	$s_1^{\text{x-ray}}$ 10^{-6} mm^2/N	Ref.
Ck 15	Mo –Kα	{800}	8.17	1.79	(a)
		{620}	6.79	1.53	
		{710/550/543}	6.15	1.40	
		{732/651}	5.85	1.15	
		{444}	5.58	1.27	
Ck 45	Cr – Kα	{211}	5.39	– 1.25	(b)
C 125	Cr – Kα	{211}	5.13	– 1.18	(b)
TiAl6V4	Co – Kα	{121}	14.7	– 3.00	(c)
TiAl6V6Sn2	Cu – Kα	{213}	11.59	– 2.94	(d)
Cu	Cu – Kα	{420}	11.00	– 3.10	(e)
CuNi 42	Cu – Kα	{420}	8.80	– 2.31	(e)
Zr	Co – Kα	{204}	10.50	– 2.55	(f)

(a) E. Macherauch, U. Wolfstieg, Proc. 25th Denver X-Ray Conference, 1976.
(b) W. Klein, Diss. Univ. Karlsruhe (1977).
(c) F. Bollenrath, W. Fröhlich, V. Hauk, H. Sesemann, Z. Metallkde. 63 (1972) 790.
(d) U. Wolfstieg, unpublished results.
(e) G. Faninger, Z. Metallkde. 60 (1969) 601.
(f) V. Hauk, W. Ott, H. Sesemann, Z. Metallkde. 63 (1972) 796.

study of x-ray elastic constants is shown in Figure 4 where the measured $(s_1)^{\text{x-ray}}$- and $(1/2 \ s_2)^{\text{x-ray}}$-values of copper-nickel alloys, including the pure metals, are given as a function of the nickel content of the alloys [19]. Measurements on {420} – planes of the polycrystals are compared with the theoretical values calculated according to Voigt, Reuss, and Kröner. As can be seen, there is an excellent correspondence with Kröner's theory. Also in the case of structural steels and plain carbon steels the estimates based on Kröner's theory are coinciding well with the

experimental findings of different European study groups.   In
Figure 5 the 1/s s_2^{x-ray} values of a structural steel (German grade
St 37) are shown in dependence of the orientation factor [20].   The
x-ray elastic constants according to Kröner's theory are given by
the straight line.   Again, a satisfying agreement exists between
theory and measurements.

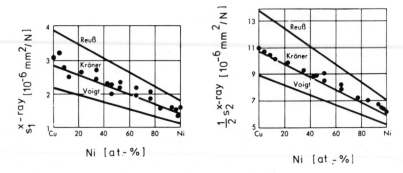

Figure 4.   X-ray elastic constants of CuNi-alloys.

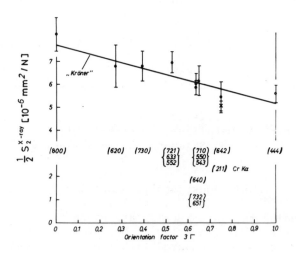

Figure 5. $1/2\ s_2^{x-ray}$-values of a structural steel in dependence
of the orientation factor.

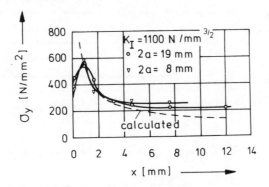

Figure 6. Measured and calculated stress distributions near the
tip of cracks.

LOADING STRESSES

The main objectives for recent x-ray investigations on mechanically loaded materials were a better understanding of loading (and
residual) stresses in notched and cracked specimens. An important
advantage of the x-ray method in the analysis of inhomogeneous
stress distributions is the possibility of measuring relatively
small specimen areas. This can be done either by reducing the primary x-ray beam or by covering the areas of the specimens which are
not of interest with suitable masks. Most of the following investigations were done with Cr-Kα radiation on a plain carbon steel with
a carbon content of 0.45 wt.-%. It is well known that the stress
field near the tip of an externally loaded cracked specimen which
follows the opening mode I of fracture mechanics, is described by
the Williams-Irwin equations [21] and is characterized by the
stress intensity factor

$$K_I = \sigma_n \sqrt{\pi a} \tag{15}$$

σ_n is the nominal tensile stress perpendicular to the crack surfaces
and 2a the crack length. Thus, in cracked specimens the same stress
distributions should exist for equal stress intensities.

In Figure 6 the full lines are showing the measured distributions of the longitudinal stresses $\sigma_y(x)$ in front of the crack tips
of center-cracked flat specimens with crack lengths of 2a = 8 mm
and 2a = 19 mm [22]. The specimens were loaded with the same
stress intensity K_I = 1100 N/mm^2. The stress distributions experimentally determined are in accordance with each other as well as
with the theoretical distribution sketched by the dashed line. In

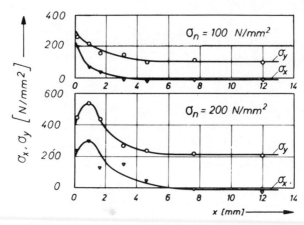

Figure 7. The distributions $\sigma_y(x)$ and $\sigma_x(x)$ in front of the tip of
           a cracked specimen under different nominal stresses.

the case of different crack lengths and a given K-value the
Williams-Irwin equations claim the same $\sigma_y(x)$-distribution inversely
proportional to the square root of the distance from the crack tips.
As can be seen from Figure 6, this claim is approximately valid
within a certain distance from the crack tip. A typical decrease
in stress appears in the plastic zone of the specimens in front of
the crack. The distribution of the transversal stresses $\sigma_x(x)$ in
the ligament of a cracked specimen also coincides well with the
theoretical expectation. As Figure 7 shows, the condition

$$\sigma_x(x) = \sigma_y(x) - \sigma_n \tag{16}$$

is nearly fulfilled for a flat specimen with a center crack of 19 mm
length at nominal stresses σ_n = 100 N/mm^2 as well as σ_n = 200 N/mm^2.

     Another problem of high actuality concerns the real stress dis-
tribution near the tip of a fatigue induced crack in a loaded
specimen. As a consequence of the inhomogeneous plastic deformations during the development of the crack, residual stresses are
not only to be expected in front of the tip of the crack, but also
at every side of the crack surfaces. Figure 8 shows the results
of corresponding residual stress measurements. In a fatigue cracked
specimen (2a = 9.5 mm, stress intensity range ΔK_I = 1460 N/mm$^{3/2}$)
longitudinal residual stress measurements were performed in front
of the crack tip as well as along the flanks of the crack. It is
worth noting that in front of the crack tip the compressive residual
stresses continue to the crack flanks beyond the crack tip. Only
at larger distances from the crack tip compressive residual stresses
change to tensile residual stresses. Thus, in fatigued specimens the

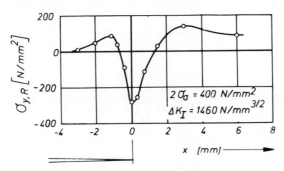

Figure 8.   Residual stress distribution along the ligament and the
             flanks of a fatigue crack.

crack flanks partly are charged with compressive residual stresses
in unloaded state [23].

    Unquestionably residual stresses caused by fatiguing a cracked
specimen influence the total stress state near the crack tip of a
specimen adjusted to a subsequent static loading. A superposition
of loading and residual stresses can be expected. Stress distributions as described by the Williams-Irwin equations do not occur any
longer. Stress distributions under different nominal stresses in a
cracked specimen (2a = 9.5 mm) with residual stresses due to fatigue
crack formation (0) are compared with those without residual stresses
(Δ) in Figure 9. The upper part of the picture shows the residual
stress distribution in front of the crack tip of the unloaded
specimen. As can be seen, under all loading conditions there are
marked differences between the stress distribution of the specimen
with and without residual stresses.

    As already mentioned, x-ray measurements of residual stress
distributions with high stress gradients demand small measuring
areas. Remarkable advances in this respect have recently been made.
It is of great practical importance to know the surface residual
stress state of CT-specimens produced according to the ASTM-standards
and used as standard specimens for the determination of K_{Ic}-values
[24]. Figure 10 shows the results of residual stress measurements
on a pre-loaded CT-specimen using measuring areas with a diameter
of 0.50, 0.22 and 0.11 mm [25]. The specimen was manufactured from
a fine grain alloyed steel with a yield strength of about 1000 N/mm^2.
After producing a fatigue crack, the specimen was loaded to a stress
intensity smaller than K_{Ic}, unloaded and measured. As can be seen
from Figure 10, the same tendency in residual stress distributions
was found applying different measured areas. However, the maximum
residual stresses near the crack tip depend on the extent of the
measured areas. Using an area of 0.5 mm diameter, the maximum
residual stress was about -750 N/mm^2, using an area of 0.11 mm
diameter the stress value increased to about -1000 N/mm^2 [25].

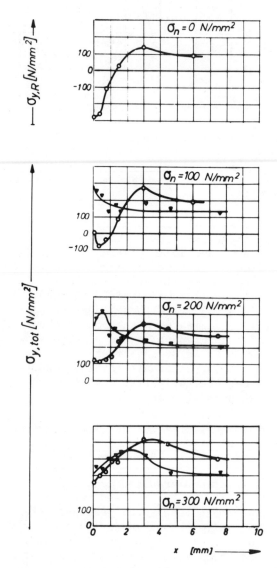

Figure 9.   Comparison of the total stress distribution in front of
            the crack tip of specimens, with and without residual
            stresses.

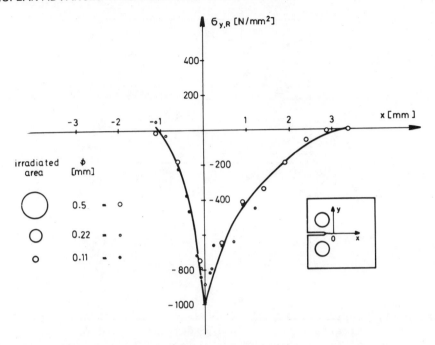

Figure 10.   Residual stress distribution near the crack tip of a
preloaded CT-specimen.

RESIDUAL STRESSES CAUSED BY UNIAXIAL RESP.
MULTIAXIAL PLASTIC DEFORMATION

The problems involved in the creation of residual stresses due
to uniaxial resp. multiaxial plastic deformation are of great importance basically as well as practically. For that reason, in recent
years strong European activities have been developed in this field.
Especially the creation of residual stresses in tensile deformed
homogeneous and heterogeneous materials has been studied.   Efforts
have also been made to determine the consequences of deformation-
textures on x-ray stress analysis.

The present state of knowledge in the first mentioned field
has been reviewed in two contributions to the 1976 Denver X-Ray
Conference [26,27]. For that reason a repeated presentation of
the problems involved does not seem to be necessary in this chapter.
Therefore, only the typical result of recent European cooperative
work concerning this part of x-ray stress analysis is presented.
The results of comparative residual stress measurements performed
by Austrian, Dutch, German and French scientists on plain carbon
steel specimens with different carbon contents are comprehended

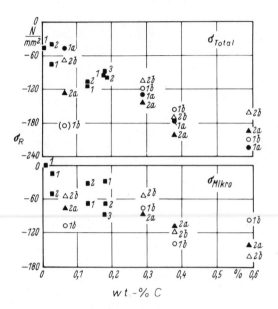

Figure 11.   Surface residual stresses of plain carbon steel specimens
             after a tensile plastic deformation of 8% as a function
             of the carbon content.   Top: Total residual stresses.
             Bottom: Micro residual stresses.   (1a, 2b, etc. mean
             different authors)

in Figure 11 [28].   All specimens were 8% plastically deformed by
an uniaxial tensile test.   In the upper part of the picture the total
residual stress values are shown as a function of the carbon content
of the specimens.   The lower part of the picture includes the portions of micro stresses on the total stress values.   With increasing
carbon content the same plastic deformation produces an increasing
amount of micro residual stresses.   But there is also evidence that
macro residual stresses are present in the uniaxially deformed
specimens.   These observations are in satisfactory agreement with
earlier studies on the formation of residual stresses during plastic
tensile deformation of plain steel specimens [29].   The second
problem mentioned about involves the influence of deformation textures on the x-ray stress analysis of heavily deformed materials.
It is well established today that the texture state influences the
residual lattice strain distributions of a plastically deformed
material [2,30].   As a consequence, non-linear dD/D_o, $\sin^2\psi$ resp.
$2\theta, \sin^2\psi$ distributions are performed if the material is loaded by
residual stresses.   For example, Austrian, German and Swedish
scientists have measured the $2\theta, \sin^2\psi$ distributions on similar
rolled steel sheets [31] and have found the results which are shown
in Figure 12.   Large systematic deviations from linear $2\theta, \sin^2\psi$

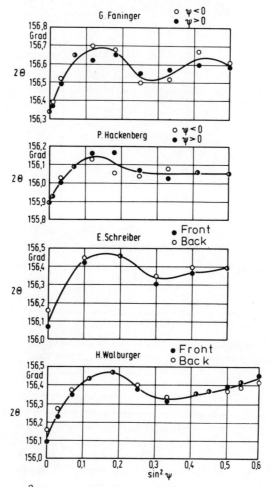

Figure 12. $2\theta, \sin^2\psi$ distributions measured on a texturized steel
specimen by different authors.

distributions occur.   As can be shown by supplementary experiments,
a strong correlation exists between the texture of the steel sheets
and the distribution of the lattice strains.   At the present, efforts
are being made to develop methods for a sufficiently accurate x-ray
stress analysis on texturized materials [32].

RESIDUAL STRESSES AND FATIGUE

Another European activity in the field of applied x-ray stress
analysis concerns the problem of the influence of residual stresses

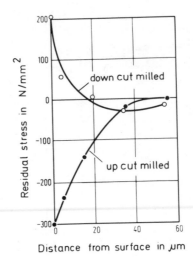

Figure 13.   Residual stress distributions near the surface of
downcut and upcut milled Ck 45 - specimens.

on the fatigue behavior of materials.   In this respect, the fatigue
behavior of machined metals is of great importance.   During
machining the surface of a material, in addition to heat, both
macroscopic and microscopic residual stresses are produced due to
inhomogeneous plastic deformation processes.   Also a characteristic
surface topography which is typical of the machining process is
formed.   If at all, in this context macroscopic compressive surface
residual stresses are generally thought to increase and tensile
surface residual stresses to reduce the fatigue strength of the
machined specimens [33].   However, a clearcut assessment of the
effect of residual macro stresses on the fatigue behavior can only
be undertaken by comparing the Woehler curves of machined specimens
with different residual macro stresses but identical residual micro
stresses and identical surface topography.   Such conditions have
successfully been produced in soft annealed plain carbon steel
specimens with a carbon content of 0.45 wt.-% by upcut milling and
downcut milling [34].   After both milling treatments in respect to
the surface topography the specimens were showing the same arith-
matical center line average heights R_a = 0.8 µm and the same surface
roughnesses, R_t = 6 µm.   Also, the half widths of x-ray interference
lines, as a measure of the micro stress state, agreed in both cases.
However, as indicated in Figure 13, the downcut milled specimens
revealed tensile surface residual stresses of +210 N/mm^2, whereas
the upcut milled specimens showed compressive surface residual
stresses of -300 N/mm^2.   The results of reversed bending fatigue
tests with both types of specimens are summarized in Figure 14.
Despite the difference of 500 N/mm^2 between the surface residual

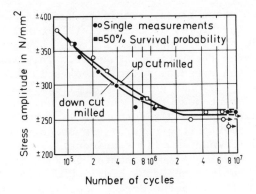

Figure 14.   Woehler curves of downcut and upcut milled Ck 45-specimens.

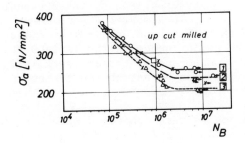

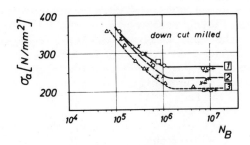

Series	State
1	milled
2	+ 2h 500 °C
3	+ 2h 700 °C

Figure 15.   Influence of different annealing treatments on the Woehler curves.

stresses, in both cases the appertaining Woehler curves do practically coincide. Therefore, it can be safely concluded that residual
macro stresses in the soft annealed state of a plain carbon steel
have no influence on the fatigue behavior. This statement permits
an assessment of the residual micro stresses in respect to the
fatigue strength. For that purpose the milled specimens have to be
subjected to different heat treatments and subsequently fatigued.
The results of correspondingly conducted experiments are shown in
Figure 15. The upper part of the figure belongs to upcut milled,
the lower part to downcut milled specimens. In both cases, Woehler
curves of milled specimens are compared to Woehler curves of milled
specimens subsequently annealed 2 h at 540°C resp. 2 h at 700°C.
In the upcut as well as in the downcut condition the Woehler curves
are shifted to lower stress amplitudes in a similar way with
increasing annealing temperature. The bending fatigue strength
decreases from about 260 N/mm^2 to 235 N/mm^2 and then to 205 N/mm^2.
As shown by supplementary experiments, the residual micro stresses
are reduced by annealing in almost the same manner as the macro-
stresses are. On the other hand, as indicated above, residual
macro stresses have no influence on the fatigue strength of the
milled specimens. Consequently, the entire gain in fatigue strength
due to milling (~ 30%) can be attributed to the creation of residual
micro stresses.

Similar experiments were conducted with hardened as well as
with quenched and tempered speciments of the same steel machined
by grinding [35]. For example, in the case of hardened specimens
with special grinding techniques, surface residual stresses
between –220 N/mm^2 and +890 N/mm^2 were produced. The whole influence
of the different grinding treatments on the bending fatigue behavior
is illustrated by the appertaining Woehler curves sketched in
Figure 16. Very pronounced differences appear in the fatigue
strengths. The specimens with compressive surface residual stresses
exhibit a 2.4 times larger fatigue strength than the specimens with
the highest tensile surface residual stresses. However, there
occurred differences in surface micro stresses and surface topography of the different ground specimens. These influences had to
be considered before a clearcut assessment of the consequences of
macro surface residual stresses on the bending fatigue strength
was possible [35]. Figure 17 summarizes the results of these considerations. Besides the results of the hardened specimens also
the results of quenched and tempered specimens are included. As
can be seen, in both cases compressive residual stresses in the
surface layers of the specimens improve the fatigue strength while
tensile residual stresses lower them. The harder the specimens are
the more pronounced is the influence of the residual stresses on
the bending fatigue strength.

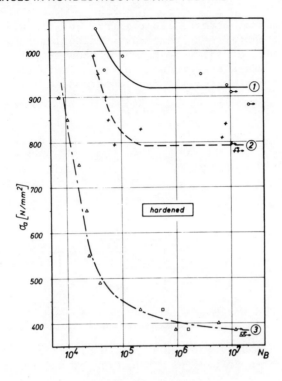

State	σ_R [N/mm²]	R_t [μm]	R_a [μm]	HV 0.1
1	- 220	5.0	0.47	765
2	+ 60	7.0	0.83	730
3	+ 890	15.5	1.83	650

Figure 16. Influence on grinding residual stresses on the bending
            fatigue behavior of hardened plain steel specimens with
            a carbon content of 0.45 wt.-%.

RESIDUAL STRESSES DUE TO WELDING

     The fusion welding process is of great practial importance.
An unavoidable and well-known consequence of this process is the
creation of residual stresses. Since the 1930's, in a large number
of publications the influence of residual stresses on the mechanical
behavior of welded joints has been discussed [36]. Nowadays it
becomes more and more important to know exactly the sign and magnitude of residual stresses in welded constructions. Earlier investigations [37] have mainly reported tensile residual stresses in the
middle part of the seam. In contrast to these, an increasing
number of recent experimental investigations [38] have shown

compressive residual stresses in this area of the weld. Therefore, the generally accepted statement that residual tensile stresses occur in that part of the weld reaching room temperature last, has to be modified. As a consequence, a reconsideration of the fusion welding process and especially the subsequent cooling process was

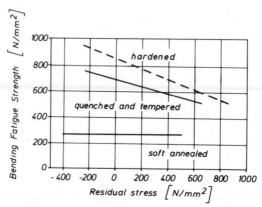

Figure 17.   The bending fatigue strength of different heat treated Ck 45-specimens as a function of the residual stress.

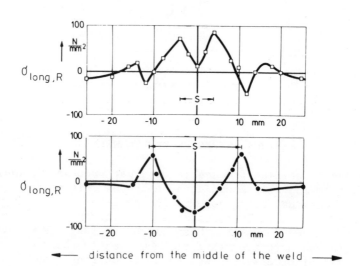

Figure 18.   Residual stress distribution transverse to the seam of a simulated weld. Top: Specimen thickness 2.5 mm. Bottom: Specimen thickness 10 mm.

necessary to understand the reasons for the unexpected sign and distribution of welding residual stresses. In recent years the x-ray
method has demonstrated its utility as an important tool in this
area of practical stress analysis. In the following a European
contribution to this problem is presented by some results of recent
experimental investigations on "simulated welds". At least three
different processes have to be considered regarding the complex
development of welding stresses: residual stresses due to the
shrinkage of the seam and the heat affected zones, residual stresses
due to transformation processes and residual stresses due to a surface quenching effect [39]. Figure 18 shows the residual stress
distributions transverse to the seam of simulated welds in 2.5 resp.
10 mm thick sheets of a structural steel of the German grade St 37
[40]. As a consequence of the surface quenching effect due to air-
cooling the thicker sheet exhibits higher compressive residual
stresses at the center line of the seam whereas the thinner sheet
shows tensile residual stresses. The width of the weld is designated
by s. The simulated welds were produced by means of an oxy-acetylen
torch which melted a well determined strip along the center line of
the steel sheets. If the quenching is more severe, in the case of
the 2.5 mm thick sheets compressive residual stresses do also occur.
This is shown in Figure 19. Localized water-quenching at the seam
develops high compressive residual stresses of about -200 N/mm^2.
Besides quenching stresses, transformation stresses are also involved
in this case. Similar experiments with modified cooling conditions
on different materials have made it obvious that, in real cases,
welding residual stresses arise not only due to the mostly considered
shrinkage process but also due to the surface quenching effect and
in some cases due to the transformation effect.

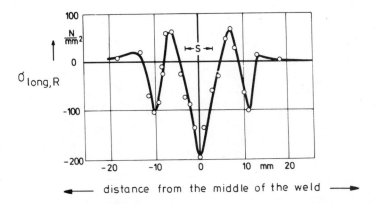

Figure 19. Development of compressive residual stresses in the
center of the seam in a simulated and water quenched
weld.

SUMMARY

    This chapter gives a short survey about some recent European
activities in the field of non-destructive XRD-testing. By way of
introduction, the European situation in the fields of X-Ray Phase
Analysis and X-Ray Stress Analysis is commented upon. Then,
restricting itself to recent advances in x-ray stress measurements,
some theoretical and experimental results are presented. These
have been achieved by some of the approximately 35 European study
groups active in this field. Some studies concerning x-ray elastic
constants are discussed. Examples are given for the utility of the
x-ray method in the analysis of loading stress states. Recent
investigations of inhomogeneous stress fields in cracked specimens
are considered. Typical results concerning the influence of uniaxial and multiaxial plastic deformation on the creation of residual
stresses are discussed. Some contributions to the consequences of
machining residual stresses on the fatigue behavior of different
heat-treated plain steel specimens are presented. Recent x-ray
investigations of residual stress distributions on simulated welds
are discussed.

REFERENCES

1. Härtereitechnische Mitteilunger (HTM), 27 (1974) 229-78.

2. HTM, Heft 1 + 2, 31 (1976) 1-124.

3. Macherauch, E., III. Int. Coll. Hochsch. Elekt. Illmenau
    (1958) 360; Macherauch, E. and P. Müller, Z. Angew. Phys., 13
    (1961) 305.

4. Trost, A., R. Lindemann, Z. Angew Phys., 115 (1940) 171;
    Bragg, W.H., Proc. Phys. Soc., London, 33 (1921) 222;
    Brentano, J., Proc. Phys. Soc., London, 37 (1925) 184.

5. Macherauch, E. and U. Wolfstieg, Proc. 25th Annual Denver X-Ray
    Conference, 1976, in press.

6. Kolb, K. and E. Macherauch, J. Soc. Mat. Sci. Jap., 13 (1964)
    14; Kolb, K., Z. Angew. Phys., 19 (1965) 360.

7. Lange, H., VDI-Bericht, 102 (1966).

8. Macherauch, E., Handbuch zerst. Werkstoffprüf., T 4 (1975).

9. Voskamp, A.P., private communcation.

10. Macherauch, E. and P. Müller, Arch. Eisenhüttenwes., 29 (1958)
    257.

11. Voigt, W., Lehrbuch der Kristallphysik, Teubner-Verlag Leipzig/
    Berlin (1928).

12. Reuss, ZAMM 9 (1929) 49.

13. Glocker, R., Z. techn. Phys., 19 (1938) 289.

14. Schiebold, E., Berg-u Hüttenmännische Monatshefte, 86 (1938)
    287.

15. Möller, H. and G. Martin, Mitt. KWJ-Eisenforschg., 21 (1939)
    261.

16. Neerfeld, H., Mitt. KWJ-Eisenforschg., 24 (1942) 61.

17. Kröner, E., Z. Phys., 151 (1958) 504.

18. Bollenrath, F., V. Hauk and E.H. Müller, Z. Metallkde., 58
    (1967) 76.

19. Ganinger, G., Z. Metallkde., 60 (1969) 601.

20. Wolfstieg, U. and E. Macherauch, unpublished results.

21. Williams, M.L., Trans. Am. Soc. Mech. Engrs., J. Appl. Mech.,
    24 (1957) 109; Irwin, G.L., Trans. Am. Soc. Mech. Engrs., J.
    Appl. Mech., 24 (1957) 361.

22. Hellwig, G. and E. Macherauch, Z. Metallkde., 65 (1974) 75.

23. Hellwig, G. and E. Macherauch, Z. Metallkde., 65 (1974) 496.

24. ASTM STP E 399-72

25. Wolfstieg, U. and E. Macherauch, unpublished results.

26. Cullity, B.D., Proc. 25th Annual Denver X-Ray Conference, 1976,
     in press.

27. Macherauch, E. and U. Wolfstieg, Proc. 25th Annual Denver
     X-Ray Conference, 1976, in press.

28. Faninger, G. and V. Hauk, HTM, 31 (1976) 72.

29. Kolb, K. and E. Macherauch, Arch. Eisenhüttenwes., 36 (1965) 9.

30. Marion, R.H. and J.B. Cohen, Advances in X-Ray Analysis, 18
     (1974) 466.

31. Faninger, G. and V. Hauk, HTM, 31 (1976) 98.

32. Hauk, V., private communication; Evenschor, P.D. and V. Hauk,
     Z. Metallkde., 66 (1975) 167; Hauk, V., D. Herlach and
     H. Sesemann, Z. Metallkde., 66 (1975) 734.

33. Frost, N.E., K.J. Marsh and L.P. Pook, Metal Fatigue,
     Clarendon Press, Oxford (1974).

34. Syren, B., H. Wohlfahrt and E. Macherauch, Arch. Eisenhüttenwes.,
     24 (1975) 735.

35. Syren, B., Diss., Univ. Karlsruhe (1975).

36. Wohlfahrt, H., HTM, 31 (1976) 56.

37. e.g., Tall, L., Weld. J. Res. Suppl., 43 (1964) 10;
          Wellinger, K., Schweissen u. Schneiden, 5 (1953) 157.

38. e.g., Bratt, J.F. and N. Berne, Brit. Weld. J., 13 (1966) 707;
          Lusemoore, A., E.K.Y. Dan and G.R. Egan, Met. Constr.
          and Brit. Weld J. (1970) 229.

39. Wohlfahrt, H. and E. Macherauch, IX. Schweisstechn.
     Hochschulkoll., Essen (1975) 45.

40. Hickel, H., H. Wohlfahrt and E. Macherauch, DVS-Bericht
     Strahltechnik VI, 26 (1973) 49.

Chapter 10

POLYCHROMATIC X-RAY STRESS ANALYSIS AND ITS APPLICATION
(NONDESTRUCTIVE DISTRIBUTION MEASUREMENT OF SUB-
SURFACE PHYSICAL QUANTITIES)

M. Nagao and S. Kusumoto

Hitachi Research Laboratory, Hitachi Ltd., Japan

ABSTRACT

The basic equations for a method of polychromatic x-ray stress analysis that uses a semiconductor detector were introduced and the accuracy of the experimental apparatus and the measuring process were confirmed.

The characteristics of this technique are as follows:

1. No mechanical scanning is necessary, more than one diffraction line is measurable at the same time, and accuracy is independent of diffraction angle.

2. Wide parallel beam slits allow a rather large sample missetting and rapid measurement. No accurate diffraction angle nor absolute peak energy value is required for stress measurement.

3. These characteristics permit a wide variety of applications, such as distribution measurements of sub-surface physical quantities across the thickness of thin samples, stress measurement of complex machine parts, etc.

Various applications are possible. Different diffraction peaks appear at different energy levels. The effective x-ray penetration depth (t_e) is quite different at different x-ray energy levels, because the x-ray absorption coefficient ($\delta\mu$) is quite different at different levels. Using this characteristic, non-destructive distribution measurement of physical quantities was demonstrated.

First, residual stress distribution in lathe-machined stainless

steel (Type 304L) and sandpaper polished samples were measured.
These results were compared with those of monochromatic x-ray stress
measurements made with an electro-polishing process. The results
were in good agreement with one another.

Second, the sub-surface peak intensity (I_f/I_a) distributions
of fatigue fracture samples fractured at various stress amplitudes
were measured. True surface peak intensity, intensity gradient and
plastic zone size were observed by this method. These quantities
were closely related to the stress intensity that caused the fatigue
crack propagation. These results were confirmed even in cases where
the fracture surface was slightly corroded or was cracked along the
grain boundary.

X-ray broadenings of both monochromatic and polychromatic x-ray
techniques were also found to good parameters for fracture sub-surface analysis. The quantitative x-ray failure analysis method
demonstrated here gives more failure analysis information than does
photographic fractography. We call this new method of analysis
x-ray fractography.

INTRODUCTION

The semiconductor radiation detector has a high resolving
power. It has been used in radiation detector, fluorescence x-ray
analyzers, x-ray micro-analyzers, and scanning electron microscopes
as a non-diversive radiation detector. A polychromatic x-ray diffraction technique has been used as a rapid and versatile technique
in studies which would require time consuming and difficult work
if a monochromatic x-ray method were used, since the first report
of Geissen and Gordon [1]. Albritton and Margrave [2] measured
the lattice parameter "a" of metals under high temperature and high
pressure conditions in only 3 seconds reasonably accurately. It
is said today that an accuracy of 10^{-4} should be possible in careful
measurements [3]. The main characteristics of this polychromatic
x-ray technique are as follows:

1. The relative angle between the primary and the secondary beams
    is constant, i.e., no mechanical scanning of the detector is
    necessary. Hence, measurement under high pressure, high tem-
    perature and very low temperature conditions is possible even
    through a small aperture. Periodical measurement is also
    possible.

2. More than one diffraction peak can be measured at the same time.
    Of course, quantitative phase analysis of alloys is possible.
    As is mentioned later, the distribution measurement of sub-
    surface physical quantities is possible utilizing this
    characteristic.

3. Accuracy is independent of the diffraction angle. It depends
    on the energy level itself. Thus low angle diffraction analysis
    is possible.

     In this chapter, the equations for a polychromatic x-ray stress
analysis method are introduced and practical measuring systems are
examined.

     Next, the sub-surface residual stress distribution of machined
austenite stainless steel (AISI type 304L) and the peak intensity
distribution of fatigue-fracture sub-surfaces were measured. The
latter will be demonstrated to be effective failure analysis, through
the concept of fracture mechanics.

BASES FOR POLYCHROMATIC X-RAY STRESS ANALYSIS

     Table 1 shows the basic equations for polychromatic x-ray
stress analysis. Equation (1) is Bragg's diffraction condition.
In this equation d is diffraction plane spacing, θ is the diffraction angle, h is Plank's constant, c is the speed of light, and
En is x-ray energy.

     Differentiation of Equation (1) gives Equation (2), where θ is
constant for the polychromatic x-ray method. From Equation (2), it
is understood that measureing accuracy is independent of θ and is
dependent only on the energy level.

     On the other hand, the theory of elasticity gives the stress
and strain relationhip of Equation (3), with the assumption that
the normal stress of the sample surface σ_z is zero, where σ_x is the
stress along the X-axis on the surface, E and ν are elastic constants, ψ is angle inclined to the X-axis from the specimen normal;
and ε_ψ is the strain in ψ direction as expressed in Equation (4).
Thus, the stress σ_x is calculated by measuring more than one ε_ψ of
different ψ angles.

     The combination of Equation (1) with Equation (3) yields
Equation (5), which can then be modified to Equation (6). It is
shown in Equation (5) that the degree of accuracy obtained in
measuring stress σ_x is not so much influenced by the accuracy of
the angle θ setting and that the absolute value of peak energy is
not necessarily required but that the accuracy in measuring the
relative peak energy, i.e., peak energy shift, ΔEn in different ψ
angles is required.

     Photo 1 shows a photo-geometric system of polychromatic x-ray
stress measurement. The x-ray source is on the left hand side.
The primary beams are directed through the parallel beam slit on

TABLE 1

Basic Equations for Polychromatic X-Ray Stress Analysis

Bragg's eq.	$2d \sin\theta = \dfrac{hc}{En} = \dfrac{12.398}{En}$ ----- (1) where d in $\overset{\circ}{A}$, En in k.eV
Differentiation of eq. (1)	$\dfrac{\Delta d}{d} = -\dfrac{\Delta En}{En}$ -------------------- (2) . where θ is constant
Stress-strain relationship	$\sigma_x = \dfrac{E}{1+\nu} \cdot \dfrac{\partial \mathcal{E}_\psi}{\partial \sin^2\psi}$ --------- (3) where E and ν are elastic constants, $\mathcal{E}_\psi$: strain in ψ direction $\left(= \dfrac{d\psi - do}{do}\right)$ ψ: angle inclined to the measuring direction from the specimen normal $\mathcal{E}_\psi = \dfrac{d\psi - do}{do}$ -------------------- (4)
Combination of eqs. (1) and (3)	$\sigma_x = -\dfrac{E}{1+\nu} \cdot \dfrac{1}{En} \cdot \dfrac{\partial En}{\partial \sin^2\psi}$ --(5) . where En : peak energy of stress free state. Thus, $\sigma_x = K \cdot \dfrac{En_2 - En_1}{\sin^2\psi_2}$ -------------- (6) . where $K \equiv -\dfrac{E}{1+\nu} \cdot \dfrac{1}{En}$ $\psi_1 = 0, \ \psi_2 \neq 0$

to the specimen which is placed on the sample inclination holder. As the diffraction angle θ is small, the sample is inclined three dimensionally as in Figure 2.

The secondary beams right hand pass through the parallel beam receiving slit into the semiconductor detector. Photo 2 shows an

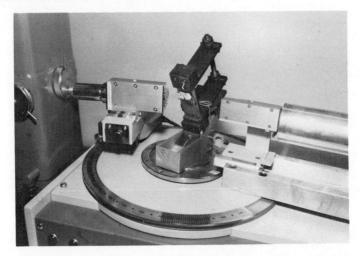

Photo 1. Diffractometer assembly (parallel beam-slits, three
dimensional sample-inclination method, x-ray generator
and a semiconductor detector).

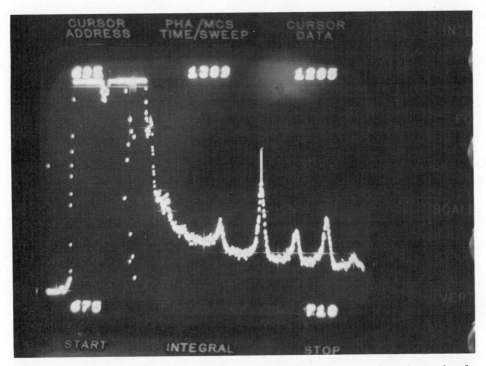

Photo 2. Oscilloscope display of intensity vs. energy channel of
diffraction and fluorescent x-rays.

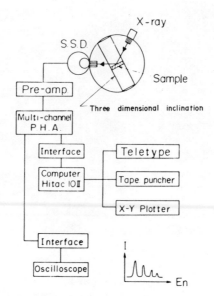

Figure 1.   Basic system for polychromatic x-ray stress measurement.

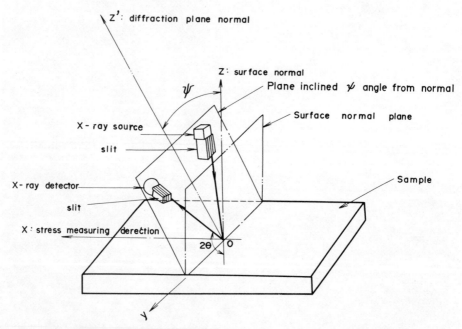

Figure 2.   Three dimensional inclination method.   The plane of the
             primary and the secondary x-ray beams is inclined by ψ
             from the specimen normal to the stress measuring direction.

oscilloscope display of fluorescent and diffracted x-rays, on the
left and right hand sides respectively, obtained with a multi-
channel analyzer. Figure 1 shows a block diagram of the apparatus.
The multi-channel analyzer has a stability of more than 10^{-4}.
Figure 2 explains the sample inclination method used in our experiments. The x-ray source and the detector are inclined by ψ angle
from the specimen normal, as shown in the figure. As this can be
explained by a three dimensional diagram, we call this "the three
dimensional inclination method". (The method is sometimes called
the "Side Inclination" or "ψ Method".) While sample inclination
methods which can be explained by two dimensional diagrams are
called "two dimensional inclination methods". The three dimensional
inclination method should be employed in order to obtain a large
ψ angle, because of the low diffraction angle θ.

     In such a case, the parallel beam slits should have parallel
slits in both the X and Y directions as explained in Figure 3.   The
divergent angles for θ and for ψ are 0.42 and 5 degrees, respectively.
The divergent angle β is determined only by the parallel beam slit
as shown in Figure 4 and is independent of the forcus size of the
x-ray source and the size of the x-ray exposed area.   Thus, a large
x-ray source and a large x-ray exposed area allow rapid measurement,
without losing photogeometric accuracy as shown in Figure 7.
Figure 5 shows the peak shift of diffraction profiles of the specimen with different ψ angles. The peak shift (En_2-En_1) is directly
proportional to the measured stress, according to Equation (5).

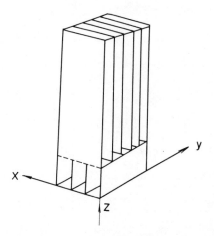

Figure 3.   A parallel beam-slit example for three dimensional in-
            clination method.   Parallel slits are inserted both in
            X.Z plane and in Y.Z plane.

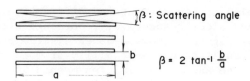

$$\beta : \text{Scattering angle}$$

$$\beta = 2 \tan^{-1} \frac{b}{a}$$

Figure 4.   The scattering angle β in relation to slit geometry.

The peak position was obtained by the half-value center method, which had proved to be the most reliable method [5].   Figure 6 shows the peak shift due to specimen misalignment, Δl for $\psi = 0$ and $\psi = 45°$, with and without external load ($\varepsilon = 400 \times 10^{-6}$).   Because of compressive residual stress, the peak intensity channel number (i.e., energy) is different for $\psi = 0$ and $\psi = 45°$.   An application of tensile strain caused the peak energy of $\psi = 45°$ to decrease. A plateau was observed in the sample mis-setting between $\pm$ 0.5mm. The slit width in this case was 5mm.   Figure 7 shows the relation-ship between x-ray path difference Δy and mis-exposure length Δx, and mis-setting Δl for the parallel beam polychromatic x-ray method. Table 2 shows the calculated path difference Δy caused by sample mis-setting Δl for various diffraction angle θ.   The area inside the bold line indicates that Δy is less than 3mm.   Because the detector diameter is 30mm, a path difference Δy of up to one half of 30mm caused by a sample mis-setting is allowed if the width of the receiving slit is 30mm.

Figure 8 shows peak energy versus $\sin^2\psi$.   The data points are

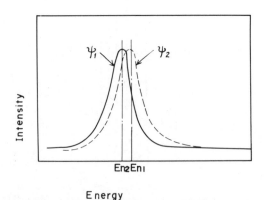

Energy

Figure 5.   Comparison of diffraction patterns from the specimen at different ψ angles.

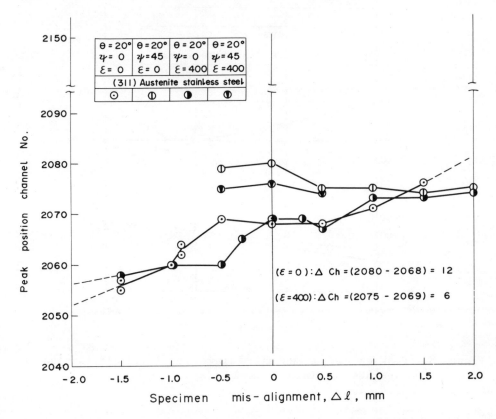

Figure 6.   Peak shifts due to specimen mis-alignment.

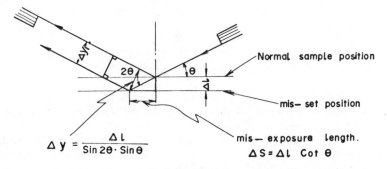

$$\Delta y = \frac{\Delta l}{\sin 2\theta \cdot \sin \theta}$$

mis- exposure length.
$$\Delta S = \Delta l \, \cot \theta$$

Figure 7.   X-ray path difference Δy and mis-exposure length Δs
caused by mis-setting Δl in parallel beam polychromatic
x-ray method.

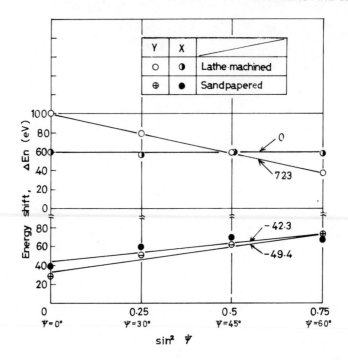

Figure 8.   Peak energy versus $\sin^2\psi$ diagram of {311} diffraction of
type 304L stainless steel.

{311} diffraction peaks of lathe-machined and sandpaper polished
type 304L stainless steel. Y indicates circumferencial and X
longitudinal direction. Using this diagram, stresses can be calculated with Equation (5). The energy resolution (i.e., repeatability
of the peak energy) was ± 2 eV. Thus, the polychromatic x-ray
stress analysis method was experimentally confirmed.

NON-DESTRUCTIVE DISTRIBUTION MEASUREMENT OF
SUB-SURFACE PHYSICAL QUANTITIES

Tables 3 and 4 show polychromatic and monochromatic x-ray
conditions, respectively. Table 5 shows the mechanical and the
chemical compositions of the measured materials. Notice that the
measured residual stress of stainless steel (Figure 8) exceeds
tensile strength (Table 5). Table 6 shows the surface finishing
conditions of the ype 304L stainless steel.

TABLE 2

Mis-Exposure Length Δs and X-Ray Path Difference Δy
Caused by Sample Mis-Setting Δl for Various Diffraction
Angles θ. The Area Inside Bold Line Indicates That Δy
Is Less Than 3mm.

θ	$\Delta = \Delta l \times \left(\frac{1}{cot\theta}\right)$	$\frac{1}{Sin2\theta \cdot Sin\theta}$	$\Delta y = \left(\frac{\Delta l}{Sin2\theta}, Sin\theta\right)$							
			ΔL=0.1	0.2	0.5	1.0	1.5	2.0	3	5
5	11.43	66.074	6.6074	13.215	33.037	66.074	99.112	132.150	198.223	330.372
10	5.67	16.837	1.6837	3.3675	8.4187	16.837	25.256	33.675	50.512	84.187
15	3.73	7.7274	0.7727	1.5455	3.8637	7.7274	11.591	15.455	23.182	38.637
20	2.75	4.5486	0.4548	0.9097	2.2743	4.5486	6.8230	9.0972	13.648	22.743
25	2.14	3.0888	0.3088	0.6176	1.5444	3.0888	4.6332	6.1776	9.2664	15.444
30	1.73	2.3094	0.2310	0.4620	1.1547	2.3094	3.4641	4.6188	6.9282	11.547
35	1.43	1.8553	0.1855	0.3710	0.9276	1.8553	2.7830	3.7106	5.5660	9.2765
40	1.19	1.5797	0.1580	0.3160	0.7900	1.5797	2.3950	3.1594	4.7391	7.8985
45	1.00	1.4142	0.1414	0.2830	0.7071	1.4142	2.1213	2.8284	4.2426	7.0710
50	0.84	1.3255	0.1325	0.2651	0.6627	1.3255	1.9882	2.6510	3.9765	6.6275
55	0.70	1.2991	0.1300	0.2600	0.6495	1.2991	1.9486	2.5982	3.8973	6.4955
60	0.58	1.3333	0.1335	0.2666	0.6666	1.3333	2.0000	2.6666	4.0000	6.6665
65	0.47	1.4403	0.1440	0.2880	0.7201	1.4403	2.1604	2.8806	4.3210	7.2015
70	0.36	1.6555	0.1655	0.3311	0.8277	1.6555	2.4832	3.3110	4.9665	8.2775
75	0.27	2.0705	0.2070	0.4141	1.0352	2.0705	3.1057	4.1410	6.2115	10.352
78	0.21	2.5135	0.2513	0.5027	1.2567	2.5135	3.7702	5.0270	7.5405	12.567
85	0.087	5.7807	0.5781	1.1561	2.8903	5.7807	8.6710	11.5614	17.3421	28.903

Bases

Table 7 shows equations for the non-destructive distribution
measurement of physical quantities under the surface. Equations
(7) to (9) show the intensity ratio I_t/I_∞ of cases (a) to (c) which
are explained in Figure 9, where I_t is the diffraction x-ray intensity of the material from the surface to the depth "t" and I_∞ is
total x-ray diffraction intensity.

By defining the effective penetration depth "t_e" as shown in
Equation (10), the intensity ratios of Equations (7) to (9) are
expressed as Equation (11). Differentiation of Equation (11) yields

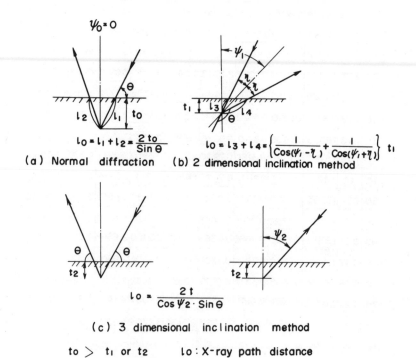

$$\psi_0 = 0$$

(a) Normal diffraction

$$l_0 = l_1 + l_2 = \frac{2 \, t_0}{\sin \theta}$$

(b) 2 dimensional inclination method

$$l_0 = l_3 + l_4 = \left\{ \frac{1}{\cos(\psi_1 - \eta)} + \frac{1}{\cos(\psi_1 + \eta)} \right\} t_1$$

(c) 3 dimensional inclination method

$$l_0 = \frac{2 \, t}{\cos \psi_2 \cdot \sin \theta}$$

$t_0 > t_1$ or t_2 l_0 : X-ray path distance

Figure 9. Effective x-ray penetration depth differences
of the various x-ray diffraction methods.

TABLE 3

Polychromatic X-Ray Conditions

Bragg angle (deg.)	20°
Divergent angle of slits	0.42° with 5x10m² window
Diffraction planes (h, k, l)	γ-Fe: 111 , 200 , 311 , 222 , 220 400 , 420 , 422 , 511 , 331 440 , 531 , 600 , 620 α-Fe: 110 , 200 , 211 , 220 , 310 222 , 321 , 400 , 411 , 420 332 , 422 , 510 , 521
Peak position determination	half-value center method
Energy step/channel	10.00 eV
X-ray exposed area	2 x 5 mm²
Counting time (sec.)	30 ~ 1,200
Inclination	3 dimension inclination ψ = 0°, 30°, 45°, 60°
Elastic constants (SUS 304L)	E = 19,000 kg/mm², ν = 0.28
X-ray source	50kV, 100mA
Detector	Hp Ge Detector
System resolution	155eV,FWHM at 5.9 keV 610eV,FWHM at 122 keV

TABLE 4

Monochromatic X-Ray Conditions

Measurements	Stress	Broadening	
Crystal	(SUS304L) Austenite	(SUS304L) Austenite	(13Cr Steel) Martensite
X-ray	CrKβ	CrKβ	MnKα
Diffraction plane	$\{311\}$	$\{311\}$	$\{211\}$
Tube voltage	30 kV	30 kV	30 kV
Tube current	10 mA	10 mmA	15 mmA
X-ray exposed area	4 x 10 mm^2	2 x 5 mm^2	2 x 5 mm^2
Slit	Parallel beam type slite with 0.42 divergent angle		
Method	2 dimension inclination $\psi=\eta, \eta+15°, \eta+30°,$ $\eta+45°$	$\psi = 0$	$\psi = 0$
Elastic constants	E = 19,000 Kg/mm^2 ν= 0.28		
Counter	S.C.	S.C.	S.C.
Counting rate	4K C.P.S	4K C.P.S	2K C.P.S.
Time constant	16 sec.	16 sec	8 sec.
Scanning speed	2 deg./min.	2 deg./min.	2 deg./min.

Equation (12). Let us define A(t) as shown in Equation (12). A(t) is the intensity ratio at any depth from "t" to total diffraction. Hence, A(t) can show how much information from "t" contributes to the sum of information obtained from x-ray diffraction measurement. Let M(t) be a polynomial Equation (14), Equation (13), can be

TABLE 5

Mechanical Properties and Chemical Compositions of Tested Materials

Materials	Chemical Compositions %							Mechanical properties kg/mm^2, %				
	C	Si	Mn	P	S	Ni	Cr	0.2% Yield Stress	Tensile Strength	Elonga-tion	Reduction of area	Hardness H_B
13Cr Steel 1" Bolt	0.013	0.35	0.56	0.021	0.014	0.09	12.30	56	74	28	71	207
SUS 304L Stainless Steel	0.022	0.53	1.80	0.036	0.023	10.87	19.32	-	57	-	56	-

TABLE 6

Machining Conditions of Type 304L Stainless Steel

Lathe-machined	High speed steel bit	
(Roughness=5μ)	Cutting depth (mm)	0.5 and 0.2
	Cutting angle (deg.)	5 - 12
	Cutting speed (m/min.)	120 - 140
	Feed (mm/rev.)	0.1
Sandpaper (Roughness = 1μ)	Emery No. 800	
Wire peening	Wire : 2 mm in dia. hardened steel	
	Pressure	7 at$_g$.
	Time	3 sec.

integrated as shown in Equation (15), where a_n is unknown constant.

Thus, whenever there are more than (m+1) measured values (Mj), the true distribution of physical quantities M(t) can be obtained. Assuming that M(t) is a linear function, Mj is expressed by Equation (16), which is the same as Equation (14).

Residual Stress Distribution of Machined Type 304L Stainless Steel

Table 8 shows the measured residual stress of lathe-machined and sandpaper polished stainless steel (SUS304L). (hkl) stands for

TABLE 7

Equations for Non-Destructive Distribution Measurement
of Sub-Surface Physical Quantities

Intensity ratio	Case (a) $\dfrac{I_t}{I_\infty} = 1 - \exp\left(-\beta\mu \cdot \dfrac{2t}{\sin\theta}\right)$ --- (7) Case (b) $\dfrac{I_t}{I_\infty} = 1-\exp\left[-\beta\mu\left\{\dfrac{1}{\cos(\psi-\eta)}+\dfrac{1}{\cos(\psi+\eta)}\right\}t\right]$ ----- (8) Case (c) $\dfrac{I_t}{I_\infty} = 1-\exp\left(-\beta\mu\dfrac{2t}{\sin\theta\cos\psi}\right)$ ---- (9)
Definition of effective penetration depth, t_e	Let "t_e" be the depth where the intensity ratio is eq. (10). $\dfrac{I_t}{I_\infty} = 1-e^{-1} = 0.63$ ------------------ (10) using t_e, eqs. (7) ~ (9) can be expressed as eq. (11) $\dfrac{I_t}{I_\infty} = 1-\exp\left(-\dfrac{t}{t_e}\right)$ -------------- (11)
Differentiate eq.(11), i.e. intensity ratio, from the depth "t" to the total diffraction. Distribution of physical quantities	$\dfrac{\Delta I_t}{I_\infty} = \dfrac{1}{t_e}\exp\left(-\dfrac{t}{t_e}\right)\cdot\Delta t \equiv A(t)\cdot\Delta t$ -- (12) Let $M(t)$ be the distribution function of physical quantities. Measured value Mj can be expressed by eq.(13). $Mj = \displaystyle\int_0^\infty A(t)\,M(t)\,dt$ -------------------- (13) Let $M(t)$ be the polynomial $M(t) = \displaystyle\sum_{n=0}^{m} a_n \cdot t^n$ -------------------- (14) where $n=1,2,3,\ldots,m$, a_n: constants. $Mj = \displaystyle\sum_{n=0}^{m} n!\, a_n\, (t_{ej})^n$ -------------- (15) when $n=1$ $Mj = a_0 + a_1 t_{ej}$ ---------------------- (16)

TABLE 8

Measured Residual Stress of Lathe Machined and Sandpaper
Polished Stainless Steel (SUS 304L)

(kg/mm^2)

(h k l)	$\overline{t}_e$, (μ) ($\psi = 30°$)	Lathe-Machined		Sandpaper		Const., K** $(kg/mm^2/eV)$
		Y *	X *	Y	X	
1 1 1	0.8	–	–	0 ± 10	–	1.689
2 0 0	1.1	87.7	–22.6	0 ± 10	0 ± 10	1.462
2 2 0	2.9	118.6	–14.6	–33.1	–20.7	1.034
3 1 1	4.6	72.3	0	–49.4	–42.3	0.882
2 2 2	5.3	82.7	9	–44.5	–58.3	0.844
4 0 0	7.4	72.6	22.2	–28.7	–45.8	0.731
3 3 1	9.8	66.6	26.3	–21.5	–38.3	0.671
4 2 0	10.9	66.7	33.8	–13.8	–26.8	0.654
4 2 2	13.4	61.9	11.3	–5.6	–11.9	0.597
5 1 1	15.3	49.6	9	7.4	0	0.563
4 4 0	20.4	50.1	–	16.5	–	0.517
5 3 1	23.0	33.8	–	18.8	–8.9	0.494
6 0 0	23.7	32.2	–	–	–	0.487
6 2 0	27.3	50.7	–	20.4	–	0.462
CrKβ 311	5.8	94	0	–48	–46	

* Y : Circumferential direction

* X : Longitudinal direction

** $K = \dfrac{E}{|t|\nu} \cdot \dfrac{1}{E_n}$

diffraction plane. The effective penetration depth "t_e" was calculated from Equation (9), assuming that $\psi = 30°$. As a reference,
the elastic constant K's in Equation (6) for various diffraction
planes are also listed.

Y stands for circumferential and X for longitudinal direction.

Figure 8 shows the measured residual stress as a function of effective penetration depth "t_e". In the figure, the measured stress
obtained with the monochromatic x-ray method (Crk β, {311} diffraction) with an electro-polishing porcess is also listed for comparison.
It can be seen that the results of both the polychromatic and the
monochromatic x-ray methods coinside rather well. However, a more
detailed distribution can be seen with the polychromatic method.

The residual stress in a circumferential direction in the
lathe-machined sample is a very large tensile stress. The values
exceed the tensile strength of this material from the surface to a
depth of 10 microns. This large stress may be explained by the
phenomena of work hardening and true tensile stress instead of
apparent tensile strength. The tensile stress region extended from
the surface to a depth of about 30 microns. However, even though
the tensile stress region is only 30 microns in depth, the stress
present there may accelerate stress corrosion crack initiation.
The residual stress in the sandpaper polished sample showed almost
the same compressive stress in both Y and X directions from the
surface, a depth of 15 microns. Thus, surface finishing condition
is very important, especially in the case of austenite stainless
steel.

Surface Work Hardened Layer

Peak intensity, which should be inversely proportional to
diffraction line broadening, is quite different in different work
hardening conditions. Figure 11 shows the profiles of the fatigue
fracture and final fracture surfaces of plain carbon steel (SF55:
JIS). Figure 12 shows the polychromatic x-ray diffraction patterns
of the fatigue fracture surface compared with those of electro-
polished plane carbon steel (SF55LJIS). The horizontal axis is
x-ray energy. It can be easily seen that the difference in the
diffraction peak intensity of steel fatigue fracture I_f and the
electro-polished I_a in the lower energy range is quite large, but
that the difference becomes smaller as x-ray energy increases.
That is, when the effective x-ray penetration depth "t_e" is small
the ratio (I_f/I_a) is very small, and when "t_e" increases the ratio
(I_f/I_a) increases. Figure 13 shows the intensity ratio (I_f/I_a)
versus effective penetration depth "t_e". The following factors
were obtained by the linear approximation of the distribution. The
plastic zone (or, thickness of work hardened layer) r_p = 16 microns,
the true surface peak intensity ratio I_f/I_a = 0.3, and the gradient
of the peak intensity ratio is 0.43/micron. The peak intensity
ratio I_f/I_a is closely related to engineering factors such as
effective strain, effective stress and hardness. When the measured
surface is a fracture surface, the ratio should be closely related
to stress intensity and fracture strain, because a plastic zone will
be induced at the crack tip. Figure 14 shows the intensity ratio

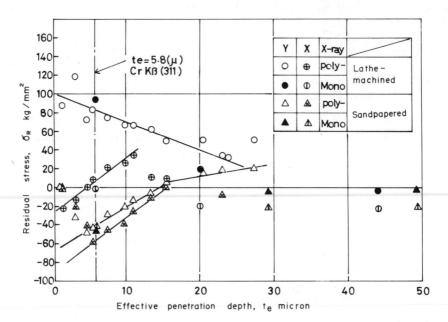

Figure 10. Sub-surface residual stress distribution in type 304L stainless steel.

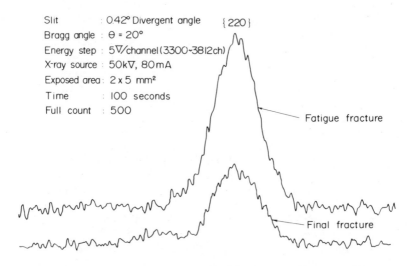

Figure 11. Polychromatic x-ray diffraction patterns for fatigued and final-brittle fracture surfaces (SF 55).

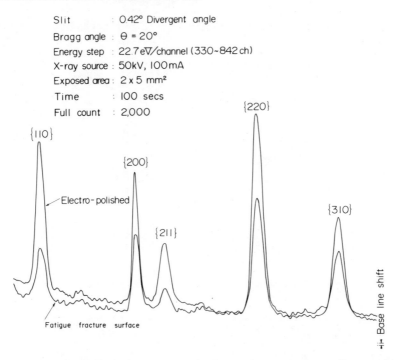

Slit   : 0.42° Divergent angle
Bragg angle : $\theta = 20°$
Energy step : 22.7 eV/channel (330~842 ch)
X-ray source : 50kV, 100mA
Exposed area : 2 x 5 mm²
Time   : 100 secs
Full count : 2,000

{110}

{200}

{220}

{211}

{310}

Electro-polished

Fatigue fracture surface

Base line shift

Figure 12. Polychromatic x-ray diffraction patterns for carbon
steel (SF 55) fatigue fracture surface and electro-
polished surface.

I_f/I_a of various machined austenite stainless steel samples as a
function of effective penetration depth "t_e". In the figure the
half-value breadth distributions "b" of the monochromatic x-ray
method measured by an electro-polishing procedure is also listed.
The thickness of work hardened layers induced by various machining
conditions can be obtained non-destructively with the linear distribution assumption. The comparison of I_f/I_a distribution with
the "b" distribution is not clear, because the surface of the
sample was removed to too great a depth by electro-polishing. In
any case detailed hardness distribution may be obtained with the
polychromatic method for the thin surface layers induced by lathe-
machining and sandpapering.

X-RAY FRACTOGRAPHY, QUANTITATIVE FAILURE ANALYSIS

Photographic fractography, e.g., observations by scanning
electron microscope, transmission electron microscope, etc., shows
only the topography of the fracture surface. Hence, analysis is
limited to qualitative fracture analysis. In the case of fatigue
fracture surfaces, striations may occasionally be observed. In

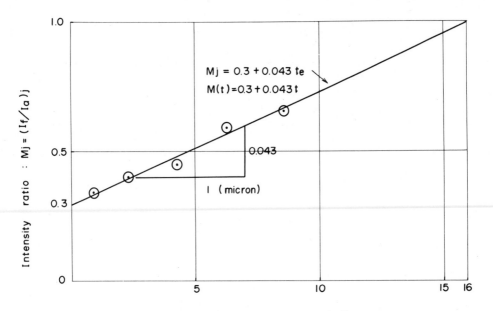

Figure 13.   Intensity ratio: I_f/I_a versus effective penetration
             depth: t_e.   Plastic zone size of 16 microns was
             obtained.

general, striation spacing can be measured in regions where the
crack propagation rate exceeds 10^{-4} mm/cycle (i.e., low cycle
fatigue region).   On the other hand, x-ray diffraction techniques
reveal the micro-structure of the fracture sub-surface.   More information should be expected from x-ray observation of fracture sub-
surfaces rather than from photographic fractography.   But, until
now, x-ray observations of [6,7] fracture surfaces have been qualitative rather than quantitative.   In this chapter, a quantitative
fatigue failure analysis method using non-destructive sub-surface
distribution measurement techniques will be demonstrated.

    Because of the roughness of the fracture surface, a parallel
beam method was employed in both the polychromatic and the monochromatic x-ray techniques.

Plastic Zone Size r_p of the Corrosion Fatigue Fracture Surface

    Thirteen chromium steel bolt specimens (one inch in diameter)
were fatigued in the water under zero-tension and tension-tension
load cycles.

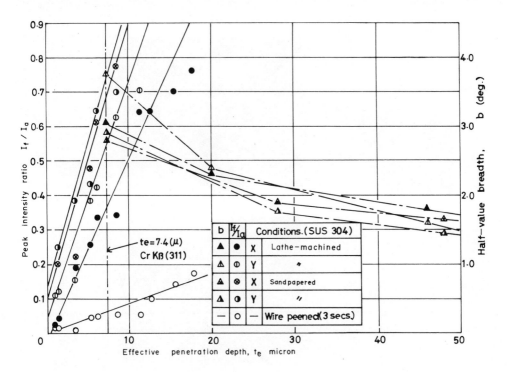

Figure 14.   Sub-surface distributions for peak intensity ratio
I_f/I_a and half-value breadth b (I_f/I_a by polychromatic
method and b by Crk β, {311} diffraction).

        Micro- and macro-fractographs of the specimens are shown in
Photo 3. Macro-fractography of the program loaded sample [Photo
3(e)] shows shell pattern while micro-fractography shows only a thin
corroded film [Photo 3(a)].   The shell pattern was observed only on
the program loaded sample, but the thin corroded film was observed
on most of the samples. In a few cases [Photo 3(b) to (d)] samples
were seen clearly without a thin corroded film, but in no case were
striations observed. Photo 3(b) is a dimple shaped fatigue fracture
surface, and (c) and (d) show grain boundary cracking. Polychromatic x-ray peak intensity measurement was carried out on these
fracture surfaces. The x-ray exposed area was 2 x 5 mm^2, including
the crack initiation part. Parts of the sample other than the
fatigue fracture surface were covered with vinyl tape so that only
the diffraction x-rays from the exposed area would be detected.
The measured peak intensity ratios are listed in Table 9.

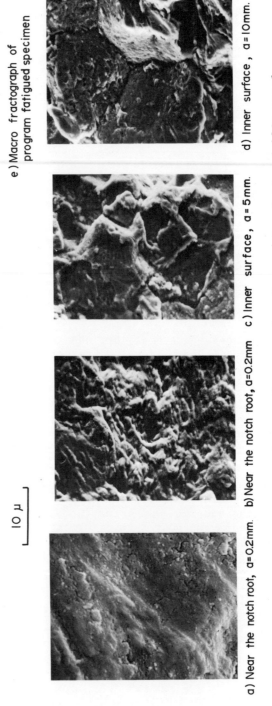

e) Macro fractograph of program fatigued specimen

d) Inner surface, a=10mm.   c) Inner surface, a=5mm.   b) Near the notch root, a=0.2mm.   a) Near the notch root, a=0.2mm.

10 μ

Photo 3: Micro- and macro-fractography of corrosion fatigued 13Cr steel specimens. Magnification: X2,000 (Reduced 20% for purposes of reproduction.).

TABLE 9

Peak Intensity Ratio (I_f/I_a) of Fatigue Fracture Surface

Stress range σ (kg/mm²)		20	25	35	49	20 σ_m =27.5	49 σ_m=27.5	49 & 25 Program
$(h\ k\ l)$	t_e (μ)	Peak	intensity	ratio ,	I_f/I_a			
1 1 0	1.2	0.705	0.61	0.39	0.18	0.675	0.10	0.33
2 0 0	2.3	0.715	0.63	0.46	0.25	0.70	0.20	0.40
2 1 1	4.3	0.75	0.69	0.57	0.275	0.76	0.28	0.49
2 2 0	6.3	0.83	0.71	0.58	0.30	0.80	0.29	0.53
3 1 0	8.5	0.875	0.72	0.66	0.31	0.815	0.31	0.54
2 2 2	11.3	0.925	0.76	0.665	0.34	0.85	0.315	0.57
3 2 1	14.2	0.95	0.76	0.68	0.38	0.88	0.36	0.59
4 0 0	16.2	0.955	0.78	0.71	0.40	0.89	0.38	0.59
4 1 1	19.8	0.975	0.80	0.73	0.43	0.90	0.39	0.67
4 2 0	21.8	0.96	0.84	0.74	0.45	0.925	0.42	0.68
3 3 2	25.8	0.98	0.88	0.755	0.47	0.95	0.45	0.72
4 2 2	28.8	0.99	0.90	0.78	0.48	0.95	0.50	0.74
5 1 0	32.7	—	0.98	0.83	0.56	—	0.48	—
5 2 1	40.5	—	0.99	0.88	—	—	—	—
Surface peak intensity ratio, I_f/I_a		0.68	0.64	0.56	0.24	0.66	0.24	0.47
Plastic zone size r_P (μ)		14 (30)	40	54	80	16.5 (34)	80	57.5

The results are shown in Figure 15 for various stress amplitudes, with and without mean tensile stress. The peak intensity
ratio I_f/I_a distribution is clearly dependent of the stress range.
The plastic zone size r_p can be obtained as an intersection of
$I_f/I_a = 1$ and the I_f/I_a distribution line with the linear intensity ratio distribution assumption. The larger the stress range
of the specimens, the larger the plastic zone, and the smaller the
peak intensity ratio becomes. The triangular simboles indicate
the program fatigued specimen, which had stress ranges Δσ, 25 and

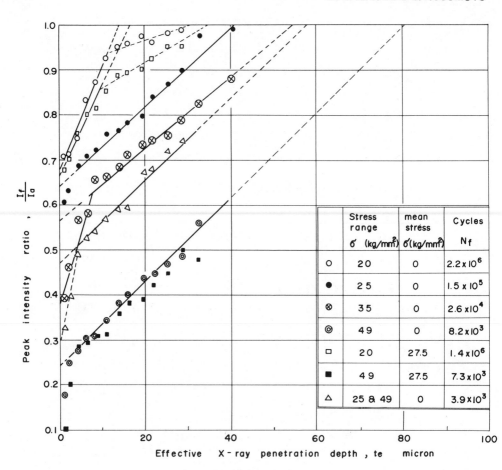

Figure 15. Sub-surface peak intensity distribution for 13Cr steel
            bolt specimens fatigued in the water.

49 kg/mm^2. The cycle ratios of both stress range are equal, i.e.,
σ_1, = 49 N$_1$ = 100 and σ_2 = 25 N$_2$ = 3,200 total cycles N$_f$ = 39,000
and N$_1$/N$_{f1}$ = N$_2$/N$_{f2}$.

     It is known from the number of shell patterns, shown in Photo
3(e), that the crack was initiated at a very early stage of fatigue
life (cycles to crack initiation is less than 7 percent of fatigue
life). The peak intensity ratio I$_f$/I$_a$ of the program fatigued
specimen was intermediate between those for the stress range $\Delta\sigma$,
49 kg/mm^2 and 25 kg/mm^2, except near the surface at a depth of less
than 3 microns where I$_f$/I$_a$ was closer to the I$_f$/I$_a$ of the larger
stress range, $\Delta\sigma$, 49 kg/mm^2. In Figure 16, plastic zone size r$_p$
and surface peak intensity ratio I$_f$/I$_a$ are expressed as a function
of stress range.

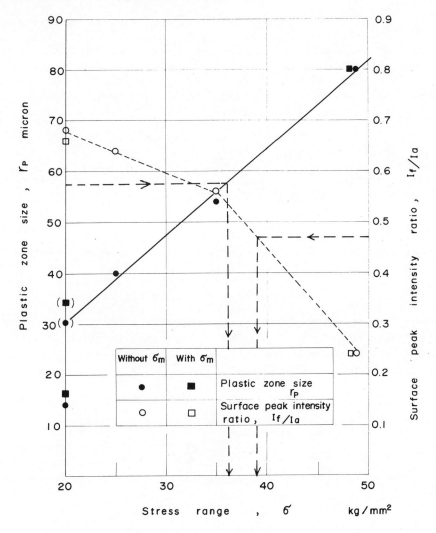

Figure 16. Plastic zone size and surface peak intensity ratio as
a function of stress range $\Delta\sigma$.

The dashed lines indicate plastic zone size r_p and surface
peak intensity ratio I_f/I_a. Following the arrows, the stress range
$\Delta\sigma$ of the program fatigued specimen was estimated to be 36 to 38
kg/mm^2. Thus, quantitative failure analysis is possible. An
interesting point is that the peak intensity ratio distribution
and the plastic zone size were expressed as a unique function of
stress range and were not so much influenced by the existence of
mean stress (i.e., pre-set mean load). Figure 17 shows the S-N

curves for one inch bolt specimens fatigued in water under various
load conditions. The solid line stands for zero-tension load
cycle fatigue and the dashed line for tension-tension load cycle
fatigue where the minimum stress is 27.5 kg/mm^2. But the fatigue
strengths of the specimens under these various conditions are not so
different even though the maximum stresses are quite different in
the zero-tension load cycles and the tension-tension load cycles.

Also, more than 93% of fatigue life consists of the fatigue
crack propagation process. Hence, it can be said that the mean
stress (or maximum stress) does not so much influence the fatigue
crack propagation rate. Thus, the measured plastic zone size, r_p,
and the peak intensity ratio I_f/I_a are dependent rather on fatigue
life (i.e., fatigue crack propagation rate), rather than on maximum
stress. This phenomenon can be explained in three ways.

First, plastic low cycle fatigue damage would be expected in
the region of the crack tip. In low cycle fatigue process, x-ray
broadening, which is considered to be a function of micro-stress
and particle size, is expressed, after a cyclic strain hardening
and softening process, as a unique function of strain range irre-
spective of the initial pre-strain conditions. Hence, the microstructure at the crack tip may become a unique function of strain
range (i.e., gross stress range), rather than of maximum stress.

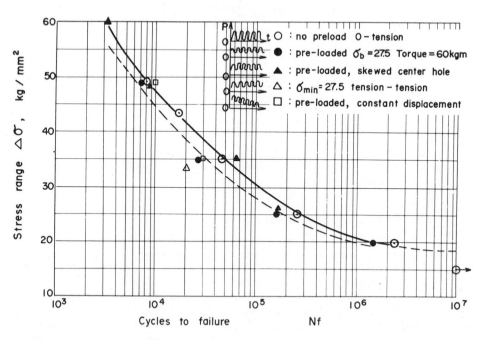

Figure 17.   S-N curves for water corrosion fatigue of 13Cr steel
             bolt specimens.

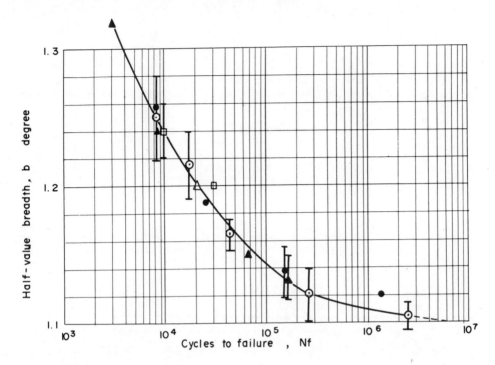

Figure 18.   X-ray broadening as a function of cycles to failure.
(13Cr steel bolt specimens)

Second, the cyclic plastic ragion ahead of the crack tip may
be dependent on the stress range [8], even though the plastic zone
created by the maximum stress may be very large.   This small cyclic
plastic zone is active and influences the crack propagation rate.
After crack propagation, this active cyclic plastic zone always
remains on fracture surface and can be observed with x-rays.

Third, the stress-strain hysterisis energy accumulated by
stress cycles is much larger than the energy expended by a monotonic
stress load.   Because of this energy difference, the micro-structures
near the fatigue fracture surface were dependent on cyclic load.
Fatigue strength was also influenced more by cyclic load than by
maximum load.

Mono-Chromatic X-Ray Line Broadening of Fatigue
Fracture Surface

The 13Cr steel bolt fatigue fracture specimens that were
mentioned above were also observed by Mn Kα radiation with {211}

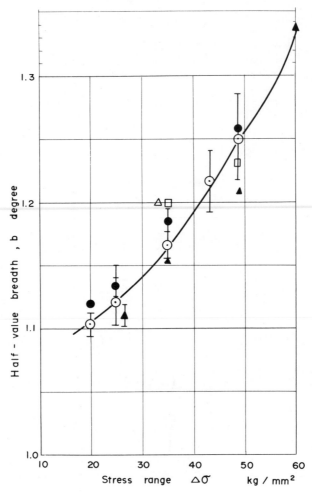

Figure 19. X-ray broadening at crack initiation area of fracture
            surface as a function of stress range. (13Cr steel
            bolt specimens)

diffraction. Figure 18 shows the half-value breadth "b" of the
fatigue fracture surface versus cycles to failure. Figure 19
shows the half-value breadth versus stress range. The symbols in
Figure 18 and 19 are the same as those in Figure 17. Data scatter-
ings are represented by the vertical lines drawn through the
measured points. All the half-value breadth "b" data can apparently
be expressed in a small scatter band as a unique function of cycles
to failure and stress range, whether or not mean stress is present

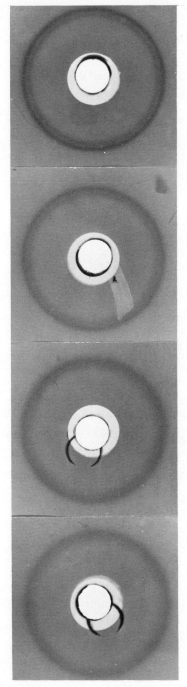

$\Delta\sigma=49$, Nf=7.26 x 10^3 $\Delta\sigma=35$, Nf=2.61 x 10^4 $\Delta\sigma=33.5$, Nf=1.94 x 10^4 $\Delta\sigma=25$, Nf=1.5 x 10^5

(a) Fatigue fracture surface at crack initiation part (preload: σ_{min} = 27.5, tension-
     tension test)

Photo 4. X-ray diffraction patterns from fracture surfaces of 13Cr steel bolt specimens,
          corrosion fatigue.

Final fracture   Crack length = 12mm   Crack length = 8mm   Crack initiation part

Crack length = 20mm   fatigue   fatigue   fatigue

Photo 4(b). Fatigue by program loading   ($\sigma_1 = 49$, $N_1 = 100$

                                        ($\sigma_2 = 25$, $N_2 = 3,200$, total $Nf = 39,000$

      Co$K\alpha$ radiation, {3 10} diffraction

      Film - sample distance: 70mm

      50kV, 80mA, 10 min

      1mm double pin-hole collimator with Fe filter

or screw holes are skew. These results should be considered identical to the results obtained with polychromatic x-ray analysis (i.e., Figures 15 and 16). Photo 4 shows $CoK\alpha$ back reflection diffraction patterns from the fatigue fracture surfaces of 13Cr-steel bolt specimens. Photo 4(a) shows diffraction patterns of fatigue fracture surfaces under various stresses at crack initiation. The Debye ring doublet can be observed in the photographs of the lower stress ranges, e.g., $\Delta\sigma$ = 25 and 33.5 kg/mm^2. It is clear that the patterns became diffused as the stress range increases in size. Photo 4(b) shows diffraction patterns of different crack length. The specimen was fatigued by program loading. The longer the crack length, the greater ring diffusion becomes. Figure 20 shows the x-ray broadening distribution along crack depth for fatigued specimens under various loading conditions. In the figure, "*" indicates the bolt specimen fatigued with a 0.5mm skew hole and "**" indicates the bolt specimen under constant displacement. All other specimens were fatigued under a constant load.

In the case of the constant load fatigue fracture surface, half-value breadth increased rapidly along the crack length. While, in the case of the constant stroke fatigue fracture surface, the rate of increase of half-value breadth was not as high as that of constant load fatigue. Thus, it can be concluded that x-ray broadening of fracture surfaces is a function of stress intensity. This relation should prove to be a useful tool in the quantitative failure analysis.

CONCLUSIONS

The foundations for a polychromatic x-ray stress analysis that uses a semiconductor detector were presented. A pure germanium detector that has a high energy resolution power and high detecting efficiency at high x-ray energy levels was employed in the experiments. Using parallel beam-slits proto-geometry and a three dimensional sample-inclination method, the usability of this technique was confirmed.

The characteristics and applications of this technique were as follows:

1. No mechanical scanning is necessary, more than one diffraction line is measurable at the same time, and accuracy is independent of the diffraction angle.

2. Wide, parallel beam-slits allow a rather large, sample mis-setting. This will simplify the x-ray stress measurement of real machine parts. Because the area exposed to x-ray is a large one, x-ray intensity is high enough for rapid measurements. No precise diffraction angle setting or absolute peak energy value is required for the stress measurement. However, a

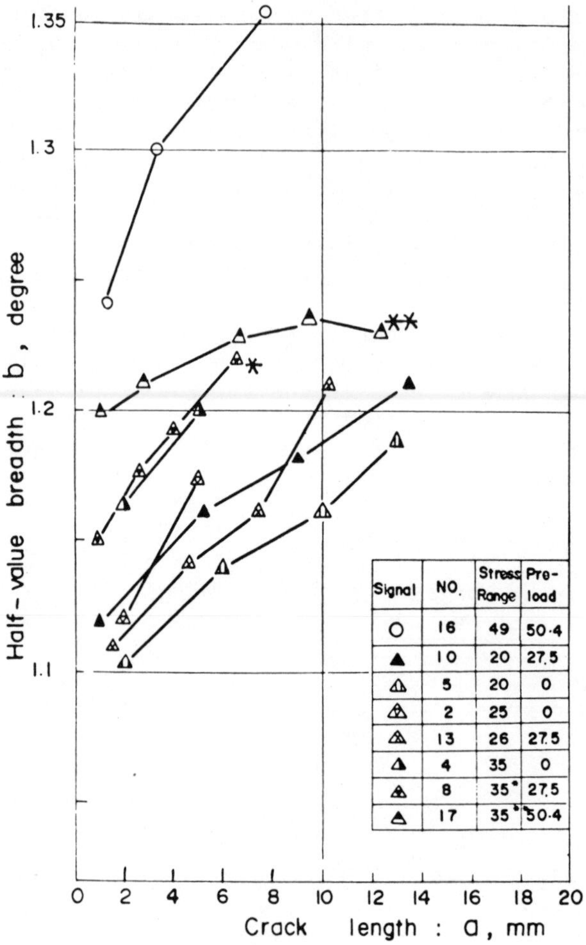
1.35
1.3
1.2
1.1
Half-value breadth : b , degree
0 2 4 6 8 10 12 14 16 18 20
Crack length : a , mm
*
**
Signal | NO. | Stress Range | Pre-load
○ | 16 | 49 | 50·4
▲ | 10 | 20 | 27.5
◭ | 5 | 20 | 0
◬ | 2 | 25 | 0
◬ | 13 | 26 | 27.5
◭ | 4 | 35 | 0
◬ | 8 | 35* | 27.5
△ | 17 | 35** | 50·4

Figure 20. X-ray broadening distribution along crack depth. (fatigued in water, 13Cr steel bolt specimens)

relative peak energy difference is required.

3. These characteristics allow a wide variety of applications, especially in those fields where monochromatic x-ray methods could be used only with difficulty.

4. As an application of this technique, non-destructive distribution measurement of physical quantities was demonstrated by utilizing the differences in effective penetration depth at different energy levels. First, the residual stress distribution of lathe-machined stainless steel (type 304L) and

sandpaper polished samples was measured. These results were
compared with those of monochromatic x-ray (CrKα) stress
measurements made with an electro-polishing process. The
results agreed well with one another.

5.   Second, the sub-surface peak intensity ratio (I_f/I_a) distribu-
     tions of fatigue fracture surfaces fractured at various stress
     amplitudes were measured. The plastic zone size and the surface
     peak intensity obtained with this method were closely related
     to the cyclic stress intensity that caused the fatigue crack
     propagation.

     These results were confirmed even in cases where the fracture
surface was slightly corroded or was cracked along the grain
boundary. X-ray broadening of monochromatic x-rays (MnKα, {211})
was also found to be a good parameter for fracture sub-surface
analysis. The obtained micro-structure parameters (i.e., plastic
zone size and x-ray broadening) were related more to the stress
amplitude rather than to the maximum stress in the case of fatigue
under mean tensile stress. On the other hand, the fatigue strength
or fatigue crack propagation rate, was influenced more by stress
amplitude than by maximum stress under tension-tension cyclic
loads. Hence, the x-ray parameters (i.e., plastic zone size and
x-ray broadening) of the fracture surface are more closely related
to the fatigue strength itself. This phenomenon was explained by
the strain amplitude dependency of the micro-structure during the
low cycle fatigue process, and after a cyclic strain hardening and
softening process. Thus, it was found that the micro-structure
remaining on the fracture sub-surface were closely related to the
resultant macro-scopic phenomena and mechanical factors.

     The rapid method of quantitative x-ray failure analysis demon-
strated here gives more failure analysis information than does
photographic fractography. We have called this new method of
analysis x-ray fractography.

REFERENCES

1. Geissen, B.C. and G.E. Gordon, Science, 159 (1968) 973.

2. Albritton, L.M. and J.L. Margrave, "Poly-chromatic X-ray Diffraction, a Rapid and Versatile Technique for the Study of Solid Under High Pressure and High Temperature", Rice Univ. (1972).

3. Fukamachi, T. and S. Hosoya, Japan Society for Physics, Vol. 28, No. 5 (1973.5) 387-402.

4. Nagao, M. and V. Weiss, ASME, NDE Conf. (1976) 76-WA/Mat-10.

5. Hayama, T. and S. Hashimoto, "Automation of X-ray Stress Measurement by Small Digital Computer", The 1st Conference on Mechanical Behavior of Materials (August 1973).

6. Orowan, E., Weld J. Res. Suppl. (March 1955) 157; Holden, J., Phil. Mag., 6 (1961) 547.

7. Taira, S. and K. Hayashi, Trans. Iron and Steel Institute of Japan, Vol. 8 (1968) 220-29.

8. Liu, H.W. and T.S. Kang, Int. J. Fracture, 10 (June 1974) 201-22.

9. Feltner, C.E. and C. Laird, Acta Met., 15 (1967) 1621, 1633.

Chapter 11

OVERVIEW OF ULTRASONIC NDE RESEARCH

R. B. Thompson

Science Center, Rockwell International

Thousand Oaks, California

ABSTRACT

One of the shortcomings of current ultrasonic NDE technology
is the inability to obtain quantitative measurements of defect
parameters for failure prediction. Flaws near criticality may produce smaller indications in standard testing procedures than more
benign flaws, and parts may be unnecessarily reworked or scrapped
at great expense. There is, consequently, a large need to broaden
our ultrasonic capabilities so that the size, shape, and orientation
or defects can be determined. This chapter describes a research
program jointly undertaken by a number of university and industrial
scientists which is making significant progress towards this goal.
Included are discussions of disciplinary research results such as
improvements in ultrasonic transducers and signal processing using
modern electrical engineering techniques, and advances in our understanding of the flaw-ultrasound interaction based on the physics
of elastic wave scattering. Interdisciplinary interpretative
approaches, such as imaging and adaptive learning procedures, to
combine these capabilities to identify defects are also presented.
Applications of the results of this research to ultrasonic standards
and the inspection of ceramic components are summarized.

INTRODUCTION

The previous chapters have discussed, in depth, both the
history and modern uses of x-rays, probably the first form of penetrating radiation used in the characterization of materials. Nondestructive testing with ultrasonics has also had a long and
successful history, dating back to the early work of Firestone [1].

257

Through the years, the vast majority of applications have been
directed towards the location of defects within materials. However,
recent changes in the philosophy of structural design have presented
new challenges [2]. In order to efficiently apply fracture mechanics
or other failure prediction disciplines in damage tolerant design
procedures [3], it is now necessary to nondestructively measure such
properties as the size, shape, and orientation of defects to determine whether they can be considered to be benign or potentially
catastrophic under the expected life loading cycle of the component.
The ultrasonic energy reflected from a defect has long been known
to contain considerable information about the nature of the scattering object that should be quite useful in making such an identification. This information is not presently used in standard industrial
test procedures. However, in response to the needs of damage
tolerant design, a vigorous research effort has developed to provide
the necessary understanding and technology so that we may do so in
the future.

In this chapter, the most recent results of a unique program
in which much of this work is taking place are summarized. The
Interdisciplinary Program for Quantitative Flaw Definition, sponsored
jointly by the Advanced Research Projects Agency and the Air Force
Materials Laboratory, consists of a set of cooperative technical
tasks performed by industrial and university scientists [4]. The
Science Center, Rockwell International, acts as a focal point by
both performing in-house research tasks and by coordinating and
integrating the results of all tasks so that they remain directed
at the overall technical problems of concern to the Department of
Defense.

The Interdisciplinary Program for Quantitiative Flaw Definition
is divided into two projects: Flaw Characterization by Ultrasonic
Techniques, and Measurement of Strength Related Properties. This
chapter will be restricted in scope to the work done as part of the
first project. Other closely related research results performed
as part of independent research programs are described in other
chapters in this volume. In addition, Buck [5] discusses some of
the results obtained in the second project of the ARPA/AFML
Program.

THE PROBLEM: QUANTITATIVE FLAW DEFINITION

It is well accepted that ultrasonic inspection is one of the
most useful NDE techniques. Ultrasonic waves are one of the few
forms of energy that can penetrate deep into materials and detect
interior flaws. Furthermore, they are relatively easy to use and
appear to provide considerable information about the geometry of
small defects. An intensive research effort has been established
to improve our ability to exploit this potential.

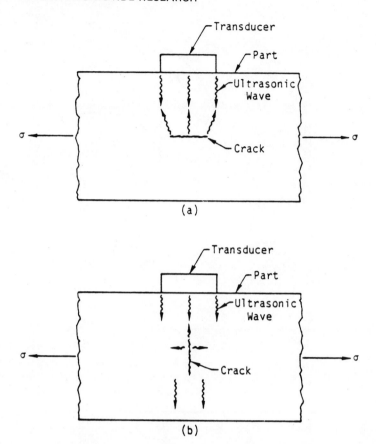

Figure 1. Comparison of ultrasonic reflections from two crack
           orientations. (a) Crack favorably oriented. (b) Crack
           unfavorably oriented.

     Figure 1 schematically illustrates one problem of standard
ultrasonics when quantitative size information is needed. In the
most common configuration, a single transducer is moved over the
surface of a part, and the amplitude of the ultrasonic echo is used
as a measure of the severity of a flaw. Consider the case of a
thin, penny-shaped crack. When this is oriented parallel to the
surface of the part, one can expect a rather large ultrasonic echo.
If it is oriented perpendicular to the surface of the part, the
echo will be relatively small. Under the tensile load shown, the
second crack is the one that would be most likely to produce failure
despite the fact that it would produce the smallest ultrasonic echo.

     Some of the consequences of this poor correlation between
defect size and echo strength are shown in Figure 2, where

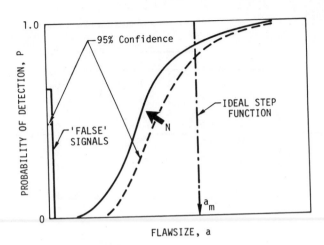

Figure 2.   Probability of flaw detection, for specific sensitivity
             setting, as function of flaw size.

probability of detection is plotted as a function of flaw size. In
the ideal case, the probability is either zero or unity, and a test
could be devised to reject all flaws whose sizes exceed a predeter-
mined value. However, in practice, for reasons including the one
illustrated in Figure 1, there is a continuous broad distribution
of probabilities. Accept/reject criteria must be a compromise
between Type I errors (missing bad parts) and Type II errors (rejecting good parts). The economic impact of approaching the ideal
probability of detection curve and minimizing this compromise can
be great [6].

   As the situation is approached where defect size can be unam-
biguously determined from ultrasonic measurements, it should be
possible to directly incorporate the principles of fracture mechanics
or other failure prediction disciplines in the formulation of accept/
reject criteria. For example [7], when a static load σ_a is applied
to a material in which failure is controlled by crack propagation,
the remaining lifetime t_r is expected to be

$$t_r = \frac{2}{v_o(n-2)} \left(\frac{K_o}{\sigma_a Y}\right)^n \left[\frac{1}{a^{(n-2)/2}} - \left(\frac{\sigma_a Y}{K_c}\right)^{n-2}\right] \qquad (1)$$

where 2a is the flaw diameter in the plane or propagation, Y is the
flaw profile, and v_o, n, K_o, K_c are crack propagation resistance
parameters. Use of ultrasonic data to determine defect size to be

TABLE I

Order of Magnitude Estimates of Critical Flaw Sizes In
Some Metal and Ceramic Systems

Materials		Flaw Size (mm)	Frequency for $\lambda_\ell = 2a_c$ (MHz)
Steels	4340	1.5	2.0
	D6AC	1.0	3.0
	Marage 250	5.0	0.59
	9N14Co 20C	18.0	0.16
Aluminum	2014-T651	4.5	0.71
Alloys	2024-T3511	25.0	0.26
Titanium	6Al-4V	2.5	1.2
Alloys	8Al-1Mo-IV(β)	14.5	0.21
Silicon	Hot Pressed	0.05	100
Nitrides	Reaction Sintered	0.02	250
Glasses	Soda Lime	0.001	2,500
	Silica	0.003	830

used in such a formula is one example of the ultimate goal of
quantitative ultrasonics; to predetermine the in-service failure
probability of a structural component with the best possible
confidence.

Catastrophic failure is expected when Equation (1) equals zero.
Table I gives estimates of this critical flaw size under typical
loading conditions for a variety of materials. Note that the critical flaw size ranges from 25 mm in 2024-T3511 aluminum down to
0.02 mm in reaction sintered silicon nitride and ever smaller in
some glasses. A useful yardstick is the frequency at which the
ultrasonic wavelength is equal to the defect diameter, and this
ranges from 260 kHz to 1.25 GHz. There are two important conclusions to be drawn from this information. First, it is clear that
no universal ultrasonic technique or instrument will emerge that
is going to be applicable to such a wide range of diverse materials.
However, it is equally clear that there must exist a sound foundation of basic information and design data so that specific instruments can be designed for specific problems.

RESEARCH RESULTS

Figure 3 illustrates the general components and functions that must be incorporated in a defect characterization system. For simplicity of coupling, a water bath system is shown. Since it is necessary to obtain large amounts of information as quickly as possible, an array of transducers would be desirable. This could be used to illuminate the part with a plane wave, or alternatively, with a focussed beam. When the elastic energy strikes the defect some will be reflected. Owing to the elastic nature of the medium, mode conversion will also take place and both longitudinal waves and mode converted shear waves will return to the surface of the part. There, both will be partially transmitted into the water and return to the array for detection. They can then be differentiated on the basis of their arrival times or the apparent position of their source. It is necessary, next, to perform signal processing to ensure that the ultrasonic information is obtained with the highest signal-to-noise ratio and bandwidth possible. Finally, and crucially, an interpretation criteria must be applied so that the desired characterisitics of the object can be determined.

Until recently, a number of gaps have existed in our understanding of the basic elastic wave phenomena and in our mastery of the

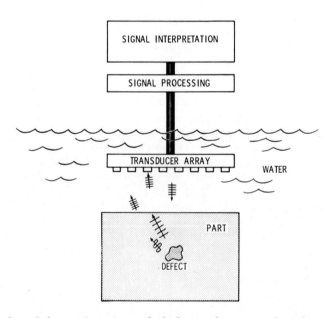

Figure 3.   Schematic view of defect characterization system.

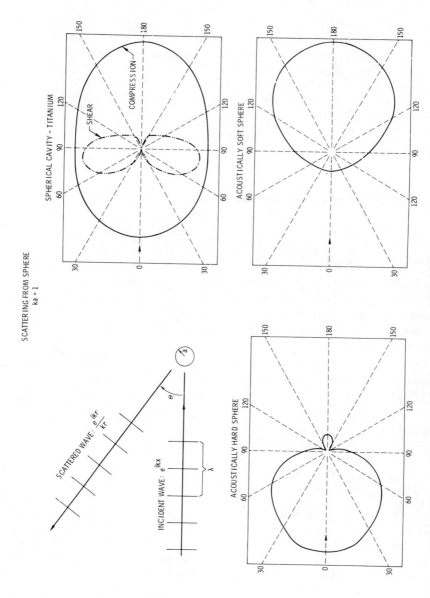

Figure 4. Comparison of fluid and solid scattering when ka = 1.

relevant technology which prevented us from constructing such a
system. The major purpose of the present program has been to
rectify this situation, and develop that information necessary so
that such a system can be designed. A number of the results that
have been obtained are reviewed below.

An understanding of the interaction of ultrasound with flaws
is essential. Figure 4 illustrates that this is nontrivial by
showing that experience gained in fluid acoustics cannot be simply
generalized to the solid case. These polar plots show the scattered
ultrasonic power as a function of angle for a spherical object when
ka = 1. The lower two scattering plots are for a rigid and a soft
sphere in a fluid, while the upper right-hand plot is for a cavity
in a solid. In the solid, not only is there a scattered shear wave
that does not exist in a fluid, but the longitudinal scattering has
a different angular dependence as well.

To generate the necessary understanding of elastic wave
scattering in a solid, a coordinated experimental and theoretical
effort has been established. The diffusion bonding technique has
been used to fabricate a set of samples, each having a spheroidal
cavity embedded in an otherwide homogeneous Ti-6%Al-4%V alloy.
These cavities, as shown in Figure 5, vary from oblate spheroids
(pancake shaped) with an aspect ratio of 4:1 to prolate spheroids
(cigar shaped) with an aspect ratio of 1:4. Also included are some
circular and elliptical cylindrical cavities of very short height
which simulate cracks. The choice of such simple shapes has been
made for two reasons. First, they are shapes which are amenable
to analysis and, hence, will allow the direct comparison of theory
and experiment necessary to gaoin a fundamental understanding of
the scattering process. Second, they form a set which model a
variety of real defect situations; e.g., both the oblate spheroids
and circular cylinders are similar to a penny-shaped crack.

Theoretical models for scattering that have been studied
include exact calculations for special geometries, various approxi-
mations to the solution of the scattering problem when formulated
as an integral equation including the Born Approximation, and a
modification of Keller's geometrical theory of scattering. These
have been compared to experiment in both contact and immersion
geometries. Figure 6 is a direct comparison of exact calculations
by Cohen [9], the Born Approximation as formulated by Gubernatis,
et al. [10], and experimental data of Tittmann [9] for the case of
a spherical cavity. By comparisons such as these, it has been
established that the Born Approximation provides useful results,
not only for the case of scattering from an inclusion whose properties are similar to those of the host medium, but also for a cavity
in the backscattering directions when the wavelength is comparable
to the dimensions of the object, ka $\overset{<}{\sim}$ 1. This approximation is
particularly attractive because of its analytic simplicity.

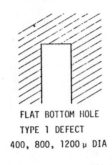

FLAT BOTTOM HOLE
TYPE 1 DEFECT
400, 800, 1200 μ DIA

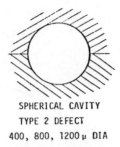

SPHERICAL CAVITY
TYPE 2 DEFECT
400, 800, 1200 μ DIA

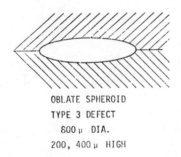

OBLATE SPHEROID
TYPE 3 DEFECT
800 μ DIA.
200, 400 μ HIGH

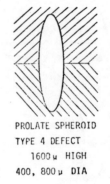

PROLATE SPHEROID
TYPE 4 DEFECT
1600 μ HIGH
400, 800 μ DIA

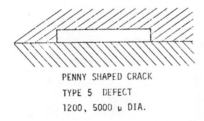

PENNY SHAPED CRACK
TYPE 5 DEFECT
1200, 5000 μ DIA.

Figure 5.   Defect shapes placed in titanium by diffusion bonding.

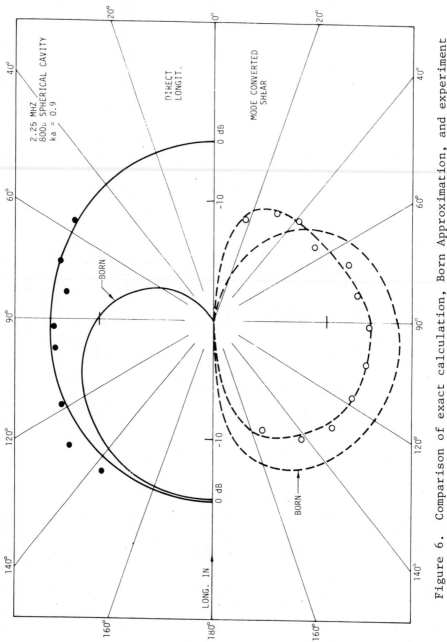

Figure 6.   Comparison of exact calculation, Born Approximation, and experiment
for longitudinal wave scattering from spherical cavity when ka = 1.

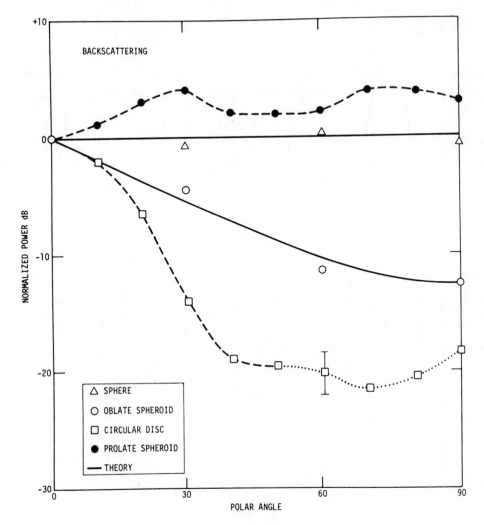

Figure 7.   Angular dependence of pulse-echo intensity from various
            defect shapes.

       The establishement of approximate techniques are particularly
important since they provide a theoretical basis for more complex
geometries than the sphere which are not amenable to exact analysis.
Figure 7 is a comparison of theoretical predictions and measured
ultrasonic backscattering (single transducer, pulse echo) as a
function of angle for spherical, oblate spheroidal, prolate
spheroidal, and penny-shaped cavities [11]. For the spherical and
oblate spheroidal cavities, all of the dimensions were comparable

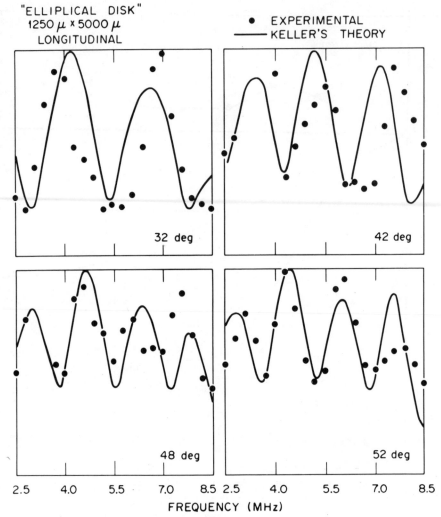

Figure 8.   Comparison between experiment and Keller's theory for
the intensity of a longitudinal scattered wave as a
function of frequency for a 1200 x 5000 thin elliptical
disk-shaped cavity embedded in titanium. The incident
wave was normal to the disk. The scattered wave was at
45° to the normal with the azimuthal angles as shown.

to the wavelength, and the Born Approximation proved accurate. For
the other defects, such was not the case, and further models are
needed. For all the cases, the experimental data clearly demonstrate
the ability of the scattering measurements to differentiate between
defects of different shapes.

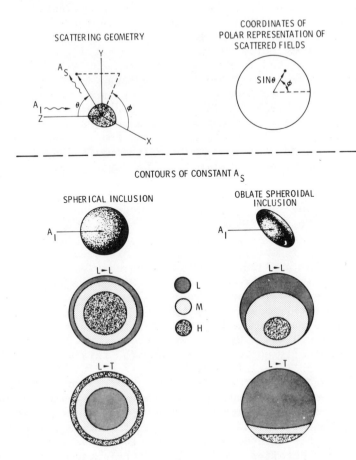

Figure 9. Polar plots of scattering from ellipsoidal inclusions.

Another approximate technique has been studied which is complementary to the Born Approximation in region of applicability.
Figure 8 shows a comparison of the geometrical diffraction theory
of Keller [12] to experimental measurements by Adler [13] for the
scattering from an ellipsoidal, crack-like cylindrical cavity in
the regime ka > 1. The agreement is for the most part quite good.

In order to compare the scattering patterns of different
objects, it has been found convenient to represent them in two
dimensional contour plots. Figure 9 shows results of a theoretical
calculation by Krumhansl, et al. [14]. A transmitting transducer

illuminates a defect in its far field and a second receiving transducer is used to detect the scattered fields as illustrated. Equal
scattering amplitude contours are shown in a polar representation
of the scattering angle. The specific calculation shown in for the
case of an aluminum inclusion in a titanium matrix, and the Born
Approximation, averaged over a range of frequencies characteristic
of a broadband ultrasonic pulse, has been used in the analysis.
For the spherical inclusion, the scattering is symmetric. Longitudinal scattering is greatest in the direct backscattering direction
whereas the backscattered shear wave is zero by symmetry. However,
for the oblate spheroidal inclusion at oblique incidence, the
situation is quite different. The longitudinal energy is primarily
scattered downwards, but not quite at the angle predicted by geometrical acoustics. Again, the shear wave scattering is somewhat
different. From plots such as these, it will be possible to choose
the relevant engineering parameter of a defect characterization
system such as the required aperture and bandwidth.

At this time it should be noted that excellent work in elastic
wave scattering as applied to NDE is being carried on elsewhere as
well. Included is the work of Pao [15], Mao [15], Sachse [16],
Datta [17], Beissner [18], and Kraut [19].

The next step in recovering defect information is the detection
of the ultrasonic signals by a transducer or transducer array.
Although conceptually quite straightforward, this step is often
the point at which much information is lost due to variabilities
in the performance of commercially available transducers. Lakin [20]
has developed improved techniques for characterizing piezoelectric
transducers which are proving useful both in the development of
improved transducer designs and in the quantitative evaluation of
presently available transducers. Figure 10 illustrates the use of
his system. The detailed displacement distribution across the face
of a transducer has been deduced from measurements of the phase and
amplitude of the radiation pattern in the far field. In this case,
a small wax spot that was placed on the face to simulate a defect
is clearly seen.

The ideal receiving transducer is an array which can provide
a real time readout of the fields over a wide aperture. Considerable technological advances in array design and construction have
been made by Kino [21] as part of his imaging program that will be
discussed below.

Once the ultrasonic signals have been detected with transducers
of known characteristics, some additional electronic processing
may be necessary to maximize the signal-to-noise and bandwidth of
the signals before final data interpretation. White [22] has
recently demonstrated that surface acoustic wave devices can be

used as filters which compensate for the narrow bandwidth of some transducers. Figure 11 is one example of the improved resolution that can be realized with the greater bandwidth. Many other applications of modern signal processing techniques to NDE appear likely [23].

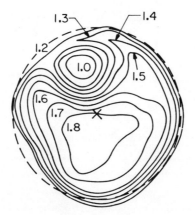

Figure 10.   Equal vibration amplitude contours on 0.7 in. diameter 5 MHz transducers with surface defect. Dashed circle represents assumed edge of the transducer.

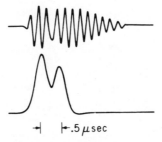

Figure 11.   Demonstration of SAW filter to improve resolution of reflections from aluminum-glass-aluminum sandwich. Top: NDT transducer output; input for SAW filter. Bottom: Demodulated SAW filter output showing resolved signals from the closely spaced impedance discontinuities.

COMPARISON OF IMAGING AND SCATTERING APPROACHES

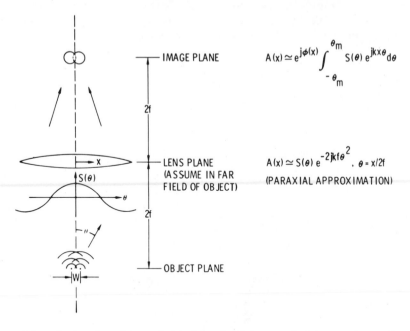

$$A(x) \simeq e^{j\phi(x)} \int_{-\theta_m}^{\theta_m} S(\theta)\, e^{jkx\theta}\, d\theta$$

$$A(x) \simeq S(\theta)\, e^{-2jkf\theta^2}\,, \quad \theta = x/2f$$

(PARAXIAL APPROXIMATION)

Figure 12.   Relationship of fields in lens and image plane in two dimensions.

The crucial step in the analysis of the ultrasonic information is the quantitative interpretation of the ultrasonic signals in terms of fracture related properties of the defect, e.g., its size, shape, and orientation. A familiar approach to this is the formation of an image, in which one tries to process the signals in such a way that an outline of the defect is produced. There are, however, alternate ways by which the same objective of defect characteriza-ţion can be obtained.

Figure 12 presents a generalized view of an imaging system which illustrates the relationship of an image to the scattered ultrasonic field. For simplicity, we consider a two-dimensional example of an object consisting of two parallel lines being pro-cessed by a cylindrical lens under monochromatic illumination. It is further assumed that the dimensions of the object, i.e., the distance between the two lines in this simple example, are suffi-ciently small that the lens is in the far field of the scattered elastic wave radiation. Then in the plane of the lens, the spatial distributions of the ultrasonic field will be proportional to the far field scattering pattern of the object, $S(\theta)$ modulated by a

TABLE II

Flaw Characterization Techniques

Type of Measurement	Goal	Conditions for Best Performance
Imaging	Directly define geometric outline	•Wavelength < Dimensions •Surface'ultrasonically diffuse •Defect has no resonances
Scattering	Deduce key geometric features from parti-cular details of scattered fields	•Wavelength $\sim$ Dimensions •Flat defects • Inclusions with internal resonances

parabolic phase shift. $S(\theta)$ is itelf related to the spatial Fourier transform of the fields that would be measured at the object. An imaging system, as modeled here by a simple lens, is designed to compensate to this phase shift and take the inverse spatial Fourier transform to obtain a reconstruction of the scattered fields that would be measured at the object.

Although the formation of an image represents the most familiar way to process scattered ultrasonic fields, it is not always the optimum approach. As shown in Table II, image formation is most successful when the ultrasonic wavelength is small with respect to the dimension of interest, and when the object is a diffuse reflec-tion. Under different conditions, other techniques for processing the scattered fields are needed. For example, Figure 13 compares the intensity distributions that would be measured in the configura-tion of Figure 12 in both the lens and the image planes. The specific calculation assumes an aperture of 27° and line sources separated by $h/\lambda = 0$, 1/2, and 2. It is clear that the differences in separation can be more accurately determined directly from the scattered fields when h/λ is small.

In the present program, both imaging and scattering approaches are being developed. Physical lenses have a number of practical limitations, which are overcome by electronically focussed imaging systems as developed by Kino [24]. Figure 14 shows a novel appli-cation of this system to produce a Rayleigh wave image of three holes drilled in a plate. Similar results have been obtained using longitudinal waves, angle shear waves, and Lamb waves. The important practical advantage of this electronically scanned approach is that the image is formed in real time and, hence, the system is compatible with the high inspection rates required in many applications.

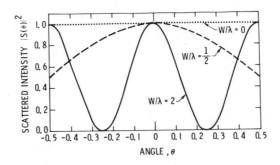

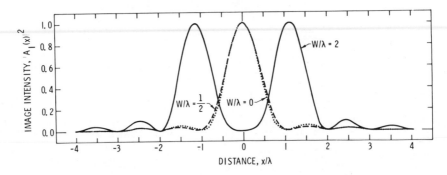

Figure 13.   Comparison of scattered fields and image for $W/\lambda = 2$,
              1/2, 0. The image intensity is normalized to a maximum
              value of unity and an aperture $\sin\theta_m = 1/2$ is assumed
              for both cases.

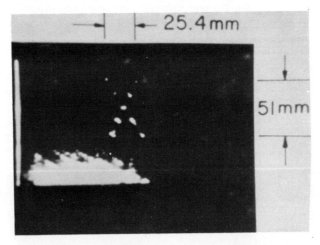

Figure 14.   2.5 MHz surface wave image of array of 2 mm diameter
              holes in aluminum block.

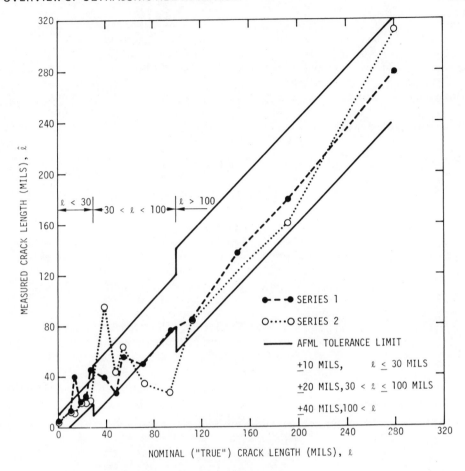

Figure 15.   Performance of adaptive learning network quantiative
surface/subsurface fatigue crack length measurement
system.

A variety of procedures for processing scattered fields, and
their variations with frequency and angle, are also being sought
as means to identify defect parameters. Most of these are based
on insights gained in the previously discussed fundamental studies
of the scattering process. Although not strongly emphasized in
this chapter, mode converted signals are believed to play an
important role in such procedures.

An alternate, empirical approach has recently been shown by
Mucciardi [25] to have promise. An adaptive learning process is
used to generate nonlinear operators which have had considerable
success in predicting sizes of flaws from scattered ultrasonic

data. Figure 15 shows what is probably the most impressive demonstration to date. Here, the size of a crack growing out of the
side of a fastener hole was successfully measured despite the
obvious difficulty with interference from reflection from the shank
of the hole. A future objective of the ARPA/AFML program is to
combine the power of this empirical approach with the generality
of more analytical approaches to obtain optimized techniques for
interpreting scattered fields.

APPLICATIONS

Although the research program is just finishing its second
year, two important applications have already been demonstrated.
The first is in the area of ultrasonic standards. Tittmann [25]
has proposed that the spherical defect, either a ball in water or
a cavity or inclusion embedded in a solid, be used as a standard
for calibrating instruments (and operators) for quantiative ultrasonics. One advantage of this approach is the fact that the
expected scattering is known theoretically from first principles.
Hence, it is not necessary to rely on empirical calibration curves
provided by a manufacturer. Furthermore, the angular dependence
of the scattering can be used to provide responses at a variety of

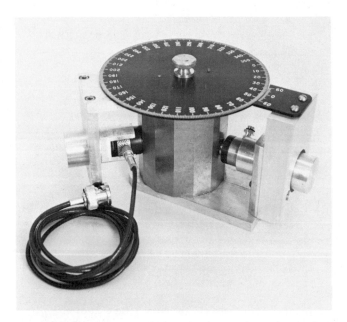

Figure 16. Photograph of measurement fixture for calibration of
            NDE system. Included are a polygonal sample containing
            spherical cavity and pair of commercial transducers on
            goniometer mount.

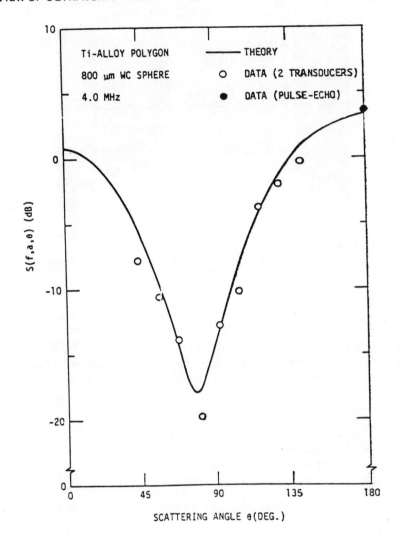

Figure 17.   Absolute comparison of experimental and theoretical
scattering from tungsten carbide sphere embedded in
Ti-6%Al-4%V alloy.

levels, hence replacing an entire set of flat bottom holes standard
by one sample with a single spherical scatterer. Figure 16 shows a
prototype titanium calibration sample and measurement goniometer.
The sample contains a spherical tungsten carbide inclusion and its
surface has been machined into a polygonal shape so that the
scattering at a variety of angles can be measured. Figure 17 shows
results of a calibration run in which <u>absolute</u> agreement between

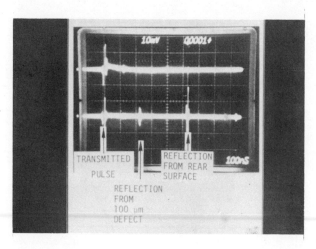

Figure 18.   Ultrasonic reflection from 100µ diameter defect in MgO
             at 200 MHz.

theory and experiment were obtained. In this work it was possible
to completely account for the prepagation attenuation of the titanium
sample and the conversion losses in the transducers.

     Application of the scattering principles to the NDE of ceramics
is being studied by Evans [27]. As was shown in Table I, the critical flaw sizes are quite small in these materials and high frequencies are needed for defect detection and characterization.
Fortunately, the materials are fine-grained and, when they are
prepared with close to their theoretical density, their ultrasonic
attenuation is quite low. Figure 18 demonstrates the detection of
a 100 µm flaw using a 200 MHz broadband inspection system developed
by Kino [28].

                              SUMMARY

     These research and development results demonstrate the excellent
progress that has been made toward the realization of a system for
quantitative defect characterization. The relationship of this work
to the development of a total accept/reject criteria is illustrated
in Figure 19 [29]. The present program is providing design criteria
whereby defect characterization systems can be built in either
scattering or imaging modes. The result will thus be procedures
whereby fracture related defect parameters such as size, shape,
and orientation, can be determined experimentally. To complete the
inspection system, this new capability must then be combined with

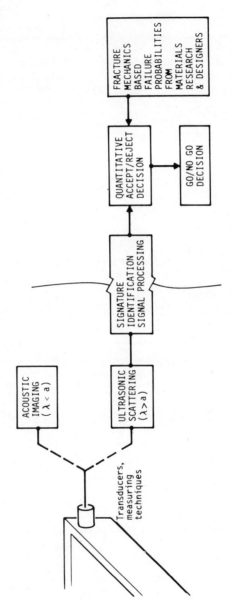

Figure 19. Schematic decision making ultrasonic inspection system.

failure analysis from other programs to develop a quantitative accept/reject system.

ACKNOWLEDGEMENT

This research was sponsored by the Center for Advanced NDE operated by the Science Center, Rockwell Internationa, for the Advanced Research Projects Agency and the Air Force Materials Laboratory under Contract No. F33615-74-C-5180.

REFERENCES

1. Firestone, F.A., "The Supersonic Reflectoscope for Interior Inspection", Metal Progr., 48 (1945) 505.

2. Thompson, D.O. and M.J. Buckley, "New Opportunities and Challenges for NDE", in ASM Handbook, Vol. 11 (in press).

3. Coffin, M.D. and C.F. Tiffiny, "How the Air Force Assures Safe, Durable Air Franes", Metal Progr. (March 1976) 26-32.

4. The most recent results of this program are fully described in the Proceedings of the ARPA/AFML Review of Progress in Quantitative NDE, August 31-September 3, 1976, Asilomar, Calif., to be published as a technical report by the Air Force Materials Laboratory.

5. Buck, O. and R.B. Thompson., "Acoustic Interactions with Internal Stresses in Matals", in Nondestructive Characterization of Materials, Proceedings of the Sagamore Army Research Conference 23, this volume.

6. Buckley, M.J., "The Future Economic Role of NDE", IEEE Trans. on Sonics and Ultrasonics, SU-23, No. 5 (1976) 287-92.

7. Thompson, R.B. and A.G. Evans, "Goals and Objectives of Quantitative Ultrasonics", IEEE Trans. on Sonics and Ultrasonics, SU-23 (1976) 292-99.

8. Paton, N.E., "Ultrasonic Samples Using Diffusion Bonding Techniques", in Proceedings of the ARPA/AFML Review of Quantitative NDE, AFML-TR-75-212 (1976) 89-106.

9. Tittmann, B.R., E.R. Cohen and J.M. Richardson, "Scattering of Longitudinal Waves Incident on a Spherical Cavity in a Solid", J. Acoust. Soc. Amer. (in press).

10. Gubernatis, J.E., E. Domany, M. Huberman and J.A. Krumhansl, "Theory of the Scattering of Ultrasound by Flaws", 1975 Ultrasonics Symposium Proceedings (IEEE N.Y., 1975) 107-10.

11. Tittmann, B.R., "Measurement of Scattering of Ultrasound by Ellipsoidal Cavities", Ref. 4 (in press).

12. Keller, J.B.,"Diffraction by an Aperture", J. Appl. Phys., 28 (1957) 426-43.

13. Adler, L. and D.K. Lewis, "Models for the Frequency Dependence of Ultrasonic Scattering by Real Flaws", Ref. 4 (in press).

14. Krumhansl, J.A., E. Domany, P. Muzikar, S. Teitel, D. Wood and J.E. Gubernatis, "Interpretation of Ultrasonic Scattering Measurements by Various Flaws from Theoretical Studies", Ref. 4 (in press).

15. Pao, Y.H. and C.C. Mao, "Theory of Normal Modes and Ultrasonic Spectral Analysis in the Analysis of the Scattering of Waves in Solids", J. Acoust. Soc. Amer., 59, No. 5 (1976) 1046-56.

16. Sancar, S. and W. Sachse, "Determination of the Geometry and Mechanical Properties of a Fluid Filled Bi-inclusion in an Elastic Solid", 1976 IEEE Ultrasonics Symposium Proceedings (IEEE, N.Y., 1976) (in press).

17. Datta, S.K., "Diffraction of Elastic Waves by Cavities and Cracks: Scattered Far Field", in Proceedings of Workshop on Application of Elastic Waves in Electrical Devices, Non-Destructive Testing, and Seismology, Evanston, Ill., May 24-26, 1976 (to be published by National Science Foundation).

18. Beissner, R.E., private communication.

19. Kraut, E.A., IEEE Trans. on Sonics and Ultrasonics, SU-23, No. 3 (1976) 162-67.

20. Lakin, K.M. and A. Fedotowsky, "Characterization of NDE Trans-Ducers and Scattering Surfaces Using Phase and Amplitude Measurements of Ultrasonic Field Patterns", IEEE Trans. on Sonics and Ultrasonics, SU-23, No. 5 (1976) 317-22.

21. De Silets, C.S., J. Fraser and G.S. Kino, "Transducer Arrays Suitable for Acoustic Imaging", 1975 Ultrasonic Symposium Proceedings (IEEE, N.Y., 1975) 148-52.

22. Kerber, G.L. and R.M. White, "Surface-Wave Inverse Filter for Non-Destructive Testing", 1976 Ultrasonics Symposium Proceedings (IEEE, N.Y., 1976) (in press).

23. White, R.M., "Some Device Technologies Applicable to Nondestructive Evaluation", IEEE Trans. on Sonics and Ultrasonics, SU-23, No. 5 (1976) 306-12.

24. Waugh, T.M., G.S. Kino, C.S. De Silets and J.D. Fraser, "Acoustic Imaging Techniques for Nondestructive Testing", IEEE Trans. on Sonics and Ultrasonics, SU-23, No. 5 (1976) 313-17.

25. Mucciardi, A.M., "Measurement of Subsurface Fatigue Crack Size Using Nonlinear Learning", Ref. 4 (in press).

26. Tittmann, B.R., D.O. Thompson and R.B. Thompson, "Standards for Quantitative NDE", Proceedings of the Symposium on Non-destructive Testing Standards, May 19-21, 1976, Gaithersburg, Md. (to be published by ASTM).

27. Evans, A.G., B.R. Tittmann, G.S. Kino and P.T. Khuri-Yakub, "Ultrasonic Flaw Detection in Ceramics", Ref. 4 (in press).

28. P.T. Khuri-Yakub and G.S. Kino, "High Frequency Pulse Echo Measurements of Ceramics and Thin Layers", 1976 IEEE Ultrasonics Symposium Proceedings (in press).

29. Thompson, D.O., private communication.

Chapter 12

ELECTROMAGNETIC ULTRASONIC TRANSDUCERS

T. J. Moran

Air Force Materials Laboratory

Wright-Patterson AFB, Ohio

ABSTRACT

Direct electromagnetic generation is a relatively recent addition to the list of techniques available for the generation of ultrasonic waves for nondestructive evaluation purposes. The history and physics of the generation technique will be discussed. In addition, transducer designs for the generation of various types of sound waves (longitudinal, shear and surface) will be presented. In particular, surface wave transducer designs will be shown to be very similar to those used in piezoelectric SAW devices. This similarity has been exploited to obtain EM transducers with signal processing capabilities similar to the piezoelectric devices. Other applications of these devices as well as possible directions of future research efforts will be discussed.

INTRODUCTION

One of the major problem areas in ultrasonic nondestructive testing is the transducer. At the present time the piezoelectric transducers commonly used have characteristics which vary tremendously between supposedly identical units and there is the additional problem that the characteristics vary with age. In addition to the reproducibility problems, the requirement that the transducer be physically coupled to the part being inspected leads to difficulties. Since most parts must be scanned, the coupling medium is required to be a liquid such as water or oil which will only transmit longitudinal sound waves. For shear or surface wave inspections, mode conversion techniques must be used. These and other problems with piezoelectric transducers have produced the impetus behind the research into NDE applications of electromagnetic generation of ultrasound since the technique has the potential to provide a reproducible, contactless transducer.

The history of electromagnetic generation is rather short relative to other methods of ultrasound generation. The earliest mention in the literature of direct ultrasonic generation occurred in 1955 when Aksenov, Vikin and Vladimirskii [1] reported an interfering resonance effect in their NMR coil in the presence of a magnetic field. They correctly identified the effect as being use to the direct generation of an ultrasonic standing wave. After this short note was published, the effect aroused little interest and it was not until 1967 when Gantmakher and Dolgopolov [2] and a second group Larsen and Saermark [3] observed the effect at low temperatures that work began in earnest in this field. Most of the early research efforts were devoted to explaining the low temperature effects. However, it was noted again [4-6] that the effect also occurred at high temperatures and that the high temperature effect had a classical explanation.

In addition to the ultrasonic case, it is also possible to generate sonic elastic vibrations using the same mechanism. This was first noted in 1939 by Randall, Rose and Zener [7].

In the next section a brief discussion of the classical theory of electromagnetic generation of ultrasound will be presented. Since NDE is only concerned with the noncryogenic temperature region, this theory is adequate to describe the process. Typical NDE applications will then be described with emphasis on transducer design. Finally, the remaining problem areas for future research and application efforts will be discussed.

THEORY OF OPERATION

The typical experimental set-up for electromagnetic generation of ultrasound consists of an rf coil placed in close proximity to the surface of a conductor. rf current in the coil induces an eddy current distribution within the skin depth layer. A steady magnetic field is required to act on the eddy currents and produce the mechanical force on the lattice. It has been shown [8] that at high temperatures (>50K) in nonmagnetic metals, the driving force on the lattice is, to a good approximation, equal to the Lorentz force acting on the eddy currents.

If the z axis is taken to be normal to the surface, the Lorentz force is given by:

$$\vec{F}(z) = \vec{J}(z) \times \vec{B} \tag{1}$$

where $\vec{J}(z)$ is the eddy current density and $\vec{B}$ is the static magnetic field. It follows from Maxwell's equations that the current density is given by [9]

$$J_x(z) = \frac{-\hat{x}(1+i)}{\delta}h_y \exp[-(1+i)\frac{z}{\delta} - i\omega t] \tag{2}$$

where x is a unit vector, h_y is the rf magnetic field produced by the coil, $\omega = 2\pi f$, where f is the frequency and δ is the classical skin depth given by

$$\delta = (\mu_o K_m \pi f J_o)^{-1/2} \tag{3}$$

where $\mu_o = 4\pi \times 10^{-7}$ Hm^{-1}, is the magnetic permeability of the metal and J_o is the dc conductivity. In a good conductor such as 7075 Al, $\delta \simeq 85$ μm at 1 MHz.

Given the force, Equation (1), one can then write the lattice equation of motion for the displacement, $\vec{U}$.

$$\rho_m \frac{\partial^2 \vec{U}(z)}{\partial t^2} = \frac{1}{v^2} \frac{\partial^2 \vec{U}(z)}{\partial z^2} + \vec{F}(z) \tag{4}$$

where ρ_m is the mass density and v is the velocity of sound. Using this equation of motion, Maxfield and Hulbert [9] have shown that for a free surface, the solutions for the shear wave and longitudinal wave cases are

SHEAR WAVES, $\vec{B} = B_z$

$$U_s(z=o) \simeq \frac{B_z h_y}{\rho_m v_s \omega}[\frac{i}{1-i\beta}] \tag{5}$$

LONGITUDINAL WAVES, $B = B_y$

$$U_l(z=o) \simeq \frac{B_y h_x}{\rho_m v_l \omega}[\frac{i}{1-i\beta}] \tag{6}$$

where

$$\beta = \frac{2\pi^2 \delta^2}{\lambda^2}$$

Equations (5) and (6) refer to shear and longitudinal waves propagating normal to the surface and describe the generation of ultrasound.

In a typical experimental set-up, either the same or a second transducer is used to detect the ultrasonic waves. In the case of detection, the motion of the metal surface in the presence of a static magnetic field induces eddy currents in the surface which are detected by the receiver coil. Maxfield and Hulbert [9] have shown that the voltage in the pick-up coil can be calculated either from first principles or by invoking the principle of reciprocity, which was first used by Thompson [10] to calculate the total response for a surface acoustic wave electromagnetic transducer (EMT). In either case, the result is expressed as a transfer impedance which is defined to be the ratio of voltage induced in the detector to the current in the drive coil. For the case of bulk wave generation, the result is [9]

$$Z = \frac{V_R}{I_T} = \frac{B^2}{\rho_m v} \frac{G_T G_R A}{(1+\beta^2)} \tag{7}$$

where v is the appropriate sound velocity, G_R and G_T are approximately the wire winding densities in the transmitter and receiver coils, respectively, and A is the effective area over which generation and detection takes place (assumed equal for both coils). The acoustic attenuation was not included in the derivation to obtain Equation (7). It may be included by multiplying the result by exp $(-\alpha z)$, where α is the attenuation coefficient and z is the distance traveled.

The previous discussion for bulk waves is applicable to the electromagnetic generation of surface waves as well, with one additional feature which must be taken into account in the surface wave case. This feature is the unique design of the surface wave transucer coil. A typical coil which is referred to as a meander line is shown in Figure 1. Most efficient Rayleigh wave generation occurs when the spacing between the lines, L, is equal to 1/2 the wavelength of the Rayleigh wave. Thompson [10] has shown that the transfer impedance for a pair of these devices operating in transmit and receive modes is

$$Z = 2\omega^2 B^2 n^2 W \exp(-2\pi h/L) \left| \frac{\sin\theta}{Ym_1^{1/2}} - \exp(-i\phi) \frac{\cos\theta}{Ym_3^{1/2}} \right|^2 \tag{8}$$

where h is the lift-off distance of the coil to the surface; θ is the angle between the surface plane of the sample and the static magnetic field, B; Ym_1 and Ym_3 are the mode admittances relating

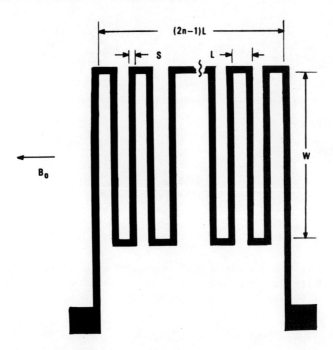

Figure 1.   Meander line coil geometry for surface wave generation.
            Number of periods is n.

power carried by the mth mode to surface displacements Um_1 and
Um_3, the shear and longitudinal displacements respectively; and
ϕ is the phase angle of the complex ratio, Um_3/Um_1.

TRANSDUCER DESIGN AND APPLICATIONS

     The first bulk acoustic wave EMT coils were either flat
spiral or rectangular shapes. These are shown in Figure 2(a).
The flat spiral geometry is the more efficient of the two since
all of the flux lines generated by this design intersect the surface and induce eddy currents. In contrast only a fraction of the
rectangular design's flux lines intersect the surface. While the
spiral coil is the more efficient, its eddy current pattern [shown
in Figure 2(b)] is probably the least desirable. This is especially
true for shear wave generation where the polarization is radial,
while the rectangular design produces linearly polarized waves.

     A second deficiency which is shared by both the spiral and
rectangular patterns is an axial null in the radiation pattern.
This occurs in the piral case due to the circular distribution of
eddy currents which requires an 180° phase shift across the coil

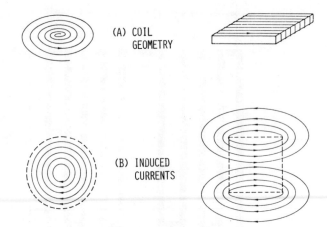

Figure 2.   (a) Basic spiral and rectangular coil geometries for
            bulk wave generation. (b) Induced eddy current
            patterns for these geometries.

for both the longitudinal and shear cases. In the case of the
rectangular coil, the origin of the axial null is not as obvious.
It arises from the fact that the return eddy current pattern is
out of phase with that directly under the coil. In a uniform
magnetic field, these return currents generate ultrasonic waves
which will cancel with the main pulse on the axis, creating the
null. Maxfield and Hulbert [9] have measured and calculated the
acoustic radiation pattern of the spiral geometry coil. Their
results are shown in Figure 3. Note that there is excellent
agreement between theory and experiment. The figure also shows
very explicitly the axial null mentioned above.

     Various attempts have been made to improve both the efficiency
and beam pattern. One rather successful approach is the use of a
metal shim with a small rectangular window cut in it to mask off
all but one side of a spiral coil. If the coil is elongated to
expose only a straight section through the mask, the eddy current
pattern of the rectangular design is then obtained with the larger
flux density characteristic of the spiral design.

     The acoustic radiation pattern of the device can be improved
by the use of permanent magnets to produce the static magnetic
field instead of a large electromagnet. Recent developments in
SmCo technology have made available permanent magnets which produce
fields greater than 5kG, a reasonable level at which to operate

EMT's. Configurations for the generation of either longitudinal
or shear waves are shown in Figure 4. The coils used in these
cases are elongated spirals which are made sufficiently long that
the segments under or between the magnet pole faces in the high
field region are straight and parallel. The remainder of the
coil is far enough from the magnet that the eddy currents are not
acted on by the static field and the spurious signals are significantly less than the main acoustic signal. Kawashima and McClung

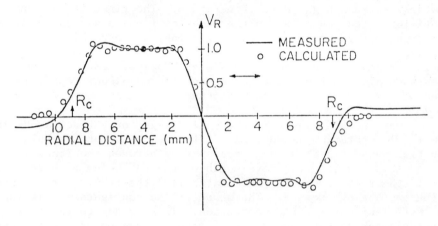

Figure 3.   Displacement pattern of a 1.8 cm diameter spiral coil
            0.49 cm from the source. f = 5.0 MHz, B = 13.8 kG.

(A) LONGITUDINAL WAVE EMT

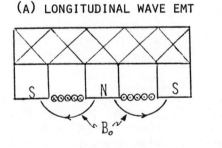

(B) SHEAR WAVE EMT

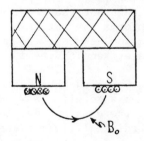

Figure 4.   Configurations of longitudinal and shear wave EMT's using
            SmCo bar magnets coupled with iron keeper plates to pro-
            duce suitable static magnetic fields.

[11] showed that such a design would produce acceptable signals
for NDE purposes.

Maxfield and Hulbert [9] have shown that with typical permanent
magnet fields (5kG) and coil dimensions suitable for NDE (area
$\sim$1cm^2) one could expect a transfer impedance on the order of
25μV/amp. Kawashima and McClung [11] showed that by using a spark
gap to produce drive currents on the order of 100 amps, one could
obtain signal levels from flaws which were comparable to those
obtained using piezoelectic transducers. There are two serious
drawbacks to such high current operation. The first is the low
repetition rates possible, on the order of 1 pulse/sec and secondly,
the large voltages generated cause ringing in the receiver for
relatively long times which would make the inspection of thin
samples or near the surface impossible. However, for many applications such high currents are not essential and higher repetition
rates and less ringing can be achieved.

As was seen in Figure 1, the coil design for surface acoustic
wave (SAW) generation is entirely different than that for bulk
acoustic waves. In this case, the spacing between the lines is
set equal to one half the wavelength of the SAW being generated to
fulfill the requirement to phase match [12] the transducer geometry
to the mode being excited. The meander line configuration is the
most efficient one for SAW generation. This is due to the fact
that the alternating phase of the forces exerted by adjacent lines
results in a minimum coupling to bulk modes. It should be noted
that a meander line transducer will produce a pair of surface waves
traveling in opposite directions.

Early devices for low frequency operation used wire meander
lines for the coils [10]. However, for operation at frequencies
above 1 MHz it is difficult if not impossible to accurately
position and align the wires in the coil. More efficient coils
are made using printed circuit techniques [13] or by cutting
commercial flat cables and connecting the lines in a meander
fashion [14]. The printed circuit devices are the most versatile
although not the least expensive since arbitrary frequencies and
waveforms are obtainable. One example of this flexibility is the
focussed SAW coil shown in Figure 5. By focussing the beam,
greater sensitivity is obtained [15].

Detailed theoretical and experimental analyses of the response
of a meander line SAW EMT were carried out by Szabo [16] who showed
that there is a great deal of similarity between the EMT and an
interdigital (IDT) transducer. Szabo [16] also demonstrated that
the frequency response may be tailored somewhat by apolization
(varying the length of each element in the coil as a function of
position). In addition, he suggested that most of the signal
processing functions performed by IDT's may also be performed by

Figure 5.   Focussed meander line.

EMT's. At the same time it was shown that an EMT designed to
produce a chirp output (a tone burst in which the frequency varies
with time) exhibits improved range resolution to the point where
it is comparable to that obtainable with piezoelectric transducers
[17]. To understand the importance of this latter development,
the output of a typical single frequency EMT pair must be considered.
Figure 6 shows the output of a pair of 32 line, 5 MHz devices when
the transmitter is excited by a 50 nsec wide pulse. In this case
the output is 6 μsec in duration which corresponds to a spatial
extent of 1.8 cm. For NDE purposes, this 1.8 cm is the minimum
distance a defect would have to be from a strong reflector such
as a hole or edge for it to be detectable. Such a limitation
makes EMTs virtually useless for many important NDE applications.

    There are two possible means to improve the spatial resolution
of EMTs. The first is the brute force approach which involves a
reduction in the number of periods in the meander line. From
Equation (8), we note that the response is proportional to the
square of the number of periods; thus, this approach would increase

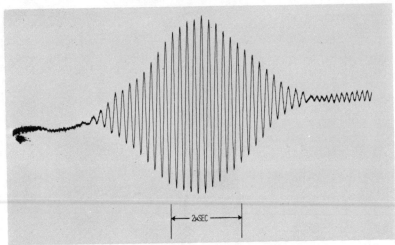

Figure 6. Output from a pair of 5 MHz, 16 period EMT's when the transmitter is excited with a 50 nsec pulse. Markers are 2 μsec apart.

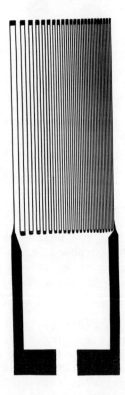

Figure 7. Dispersive EMT coil design.

the insertion loss considerably. For most applications, the
dynamic range achievable with a 15 period EMT is on the borderline
of being acceptable. If the number of periods were reduced to the
1 or 2 required for good range resolution, the device would not be
sufficiently sensitive to detect most flaws of interest. The
second approach is to utilize techniques originally developed for
radar to compress a frequency modulated rf burst and produce an
output pulse sufficiently short to give the desired resolution [17].

Figure 7 shows the coil pattern for a dispersive SAW EMT designed
to produce an rf burst in which the frequency varies linearly with
time from 2 to 6 MHz when operated in aluminum. When this transducer
is pulse excited, the waveform shown in Figure 8 results. The
amplitude modulation of this waveform is a result of the nonuniform
static magnetic field which is produced by a horseshoe shaped permanent magnet of the type shown in Figure 3(b). When this waveform
is received by a second identical transducer, the receiver has a
small output due to frequency mismatch except when there is an exact
correlation. This compressed output is shown in Figure 9. In this
case, the input is 3.5 μsec and the output is less than 0.25 μsec,
giving an effective compression ratio of approximately 14 in agreement with the theoretical value which is equal to the product of
the pulse length and the bandwidth.

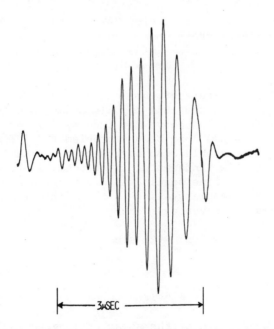

Figure 8.   Output waveform of the coil shown in Figure 7 when
            excited with a 50 nsec pulse. Receiver is a broadband
            piezoelectric transducer. Markers are 3 μsec apart.

Figure 9.   Response of a dispersive SAW EMT to the acoustic waveform
             shown in Figure 8.

        One additional feature of the meander line EMT remains to be
discussed. This is the coupling to bulk modes when the transducer
is operated at frequencies above the fundamental for SAW generation.
Bulk wave generation has long been a problem for IDT's and it was
noted [12] that this should also be the case for EMT's. At frequencies above the SAW fundamental it is possible to phase match
the projection of the line spacing in a direction away from the
surface to a bulk mode propagating in that direction. The phase
matching condition is given by

$$n\lambda = 2L \sin \theta$$

or

$$f = nv/2L \sin \theta \qquad (9)$$

where λ is the wavelength of the shear or longitudinal wave being
generated, v is the velocity of that wave; f, the frequency; L,
the line spacing and θ, the angle between the surface normal and

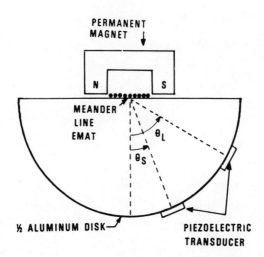

Figure 10.   Test set-up to determine the frequency dependence of
             the propagation direction of bulk waves generated by
             a SAW defect.

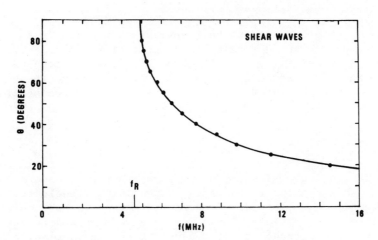

Figure 11.   Frequency dependence of the propagation direction of
             shear waves generated by a 4.5 MHz SAW EMT.

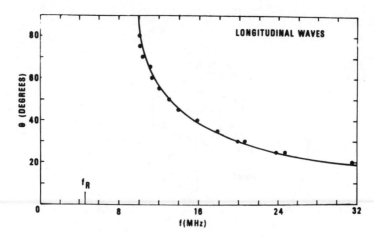

Figure 12. Frequency dependence of the propagation direction of
            longitudinal waves generated by a 4.5 MHz SAW EMT.

the propagation direction. The relation given in Equation (9) was
experimentally demonstrated using the set-up shown in Figure 10
[18]. The results of that investigation are given in Figures 11
and 12. It should be noted that for bulk waves propagating near
the surface, there is a great deal of beam spreading due to the
fact that the transducer is extremely narrow, almost a line
source, when viewed from such an angle.

    While the generation of bulk waves is a problem in the IDT
case, it adds another dimension to the use of EMT's in NDT. It
provides the possibility of inspecting both the surface and interior
using a single transducer. By varying the frequency, the beam may
be steered into the interior and in many cases may reach hard to
inspect points.

CONCLUSIONS

    In conclusion, EMT's are currently rather well understood
devices. Work remains in the area of providing the electronics
needed to give the large drive currents these devices require, as
well as the low impedance, low noise receivers needed to make the
dynamic range comparable to piezoelectric devices. Also, other
signal processing operations which are presently feasible with
IDT's might be carried over to EMT's to enable more specific
defect identifications.

REFERENCES

1. Aksenov, S.I., B.P. Vikin and K.V. Vladimirskii, "The
    Excitation of Ultrasonic Vibrations by Ponderomotive Forces",
    J. Exper. Theoret. Phys. USSR, 28 (1955) 762. (Eng. Trans. -
    Sov. Phys. JETP, 1, 1955 p. 609)

2. Gantmakher, V.F. and V.T. Dolgopolov, "Excitation of Standing
    Sound Waves in Bi by an Electromagnetic Method", Sov. Phys.
    JETP Lett., 5 (1967) 12. (Eng. Trans.)

3. Larsen, P.K. and K. Saermark, "Helicon Excitation of Acoustic
    Waves in Aluminum", Phys. Letters, 24A (1967) 374.

4. Betjemann, A.G., H.V. Bohm, D.J. Meredith and E.R. Dobbs,
    "R.F. Ultrasonic Wave Generation in Metals", Phys. Letters,
    25A (1967) 753.

5. Gaerttner, M.R., W.D. Wallace, and B.W. Maxfield, "Experiments
    Relating to the Theory of Magnetic Direct Generation of Ultra-
    sound in Metals", Phys. Rev., 184 (1969) 702.

6. Grubin, H.L., "Direct Electromagnetic Generation of Compres-
    sional Waves in Metals in Static Magnetic Fields", IEEE Trans.
    on Sonics and Ultrasonics, SU-17 (1970) 227.

7. Randall, R.H., F.C. Rose and C. Zener, "Intercrystalline
    Thermal Currents as a Source of Internal Friction", Phys. Rev.,
    56 (1939) 343.

8. See: E.R. Dobbs, "Electromagnetic Generation of Ultrasonics",
    Physical Acoustics, Vol. X, W.P. Mason and R.N. Thurston, eds.,
    Academic Press, New York, 1973. Chapter 3 and the references
    contained therein.

9. Maxfield, B.W. and J.K. Hulbert, "Electromagnetic Acoustic Wave
    Transducers (EMATS): Their Operation and Mode Patterns", Proc.
    10th Symp. on NDE, San Antonio, Texas, April 1975, p. 44.

10. Thompson, R.B., "A Model for the Electromagnetic Generation
    and Detection of Rayleigh and Lamb Waves", IEEE Trans on Sonics
    and Ultrasonics, SU-20 (1973) 340.

11. Kawashima, K. and R.W. McClung, "Electromagnetic Ultrasonic
    Transducer for Generating and Detecting Longitudinal Waves
    (with a Small Amount of Radially Polarized Transverse Wave)",
    Mat. Eval., 34 (1976) 81.

12. Burstein, E., H. Talaat, J. Schoenwald, J.J. Quinn and
    E.G.H. Lean, "Direct EM Generation and Detection of Surface
    Elastic Waves on Conducting Solids I. Theoretical Considera-
    tions", Proc. 1973 IEEE Ultrasonics Symp., Cat. No. 73,
    CHO 807-8 SU, p. 564, 1973.

13. Moran, T.J., M.J. Lin, F. Bucholtz and R.L. Thomas, "Electro-
     magnetic Generation of Rayleight Waves at MHz Frequencies",
     Rev. Sci. Instru., 46 (1975) 931.

14. Frost, H.M., J.C. Sethares and T.L. Szabo, "Applications for
     New Electromagnetic SAW Transducers", 1975 Ultrasonics
     Symposium Proc., IEEE Cat. No. 75 CHO 994-SU, 604, 1975.

15. Moran, T.J., R.L. Thomas, G.F. Hawkings and M.J. Lin,
     "Electromagnetic Generation of Bulk and Surface Sound Waves
     at MHz Frequencies for NDE", Proc. 10th Symposium on NDE,
     San Antonio, Texas, April 23-25, 1975, p. 105.

16. Szabo, T.L., "Advanced SAW Electromagnetic Transducer Design",
     1976 Ultrasonics Symposium Proc., IEEE Cat. No. 76 CH1120-
     5SU, 29, 1976.

17. Moran, T.J., "Use of Pulse Compression Techniques to Improve
     the Range Resolution of Electromagnetic Surface Wave Trans-
     ducers", 1976 Ultrasonics Symposium Prog.,   IEEE Cat. No.
     76 CH1120-5SU, 26, 1976.

18. Moran, T.J. and R.M. Panos, "Electromagnetic Generation of
     Electronically Steered Ultrasonic Bulk Waves", J. Appl. Phys.,
     47 (1976) 2225.

Chapter 13

ULTRASONIC SPECTROSCOPY

O. R. Gericke

Army Materials & Mechanics Research Center

Watertown, Massachusetts

ABSTRACT

The fundamental concept of ultrasonic spectroscopy is discussed which is based on a time-domain to frequency-domain conversion of ultrasonic signals by means of Fourier transform. Technical detail pertaining to transducer characteristics and design considerations of the electronic equipment utilizing this method is presented. Furthermore, practical applications in the area of nondestructive testing are discussed, such as the determination of microstructure, the assessment of defect geometry, and the detection of delaminations in composite structures. An extensive bibliography illustrating the current state-of-the-art of ultrasonic spectroscopy is included.

INTRODUCTION

Ultrasonic spectroscopy offers an advanced approach of ultrasonic testing by introducing the frequency-domain or "color" of an ultrasonic signal as in inspection criterion. The spectral signature of an ultrasonic signal represents characteristic data augmenting the information available from conventional test techniques which consist only of the ultrasonic pulse amplitude and pulse transit time. By applying the method of ultrasonic spectroscopy one can determine the frequency-dependence of ultrasonic propagation losses in a material, the frequency-selective reflectivity of internal discontinuities, and thickness resonances introduced by reflections from back surfaces.

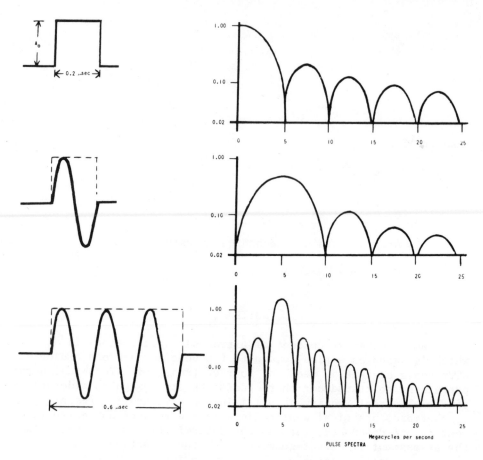

Figure 1. Various types of idealized pulses and their associated
amplitude spectra (absolute values are plotted on an
arbitrary logarithmic vertical scale).

FUNDAMENTAL CONCEPTS

The theoretical basis for ultrasonic spectroscopy is furnished
by the Fourier principle which states that any time-dependent
function such as an ultrasonic pulse is associated with a frequency
spectrum.

As an illustrative example, Figure 1 shows idealized pulses
and portions of their actually infinite spectra. The first time
function is a rectangular, single polarity pulse which means that
its carrier frequency is zero. The other two pulses are 5 MHz
carrier-frequency signals with rectangular envelopes that differ

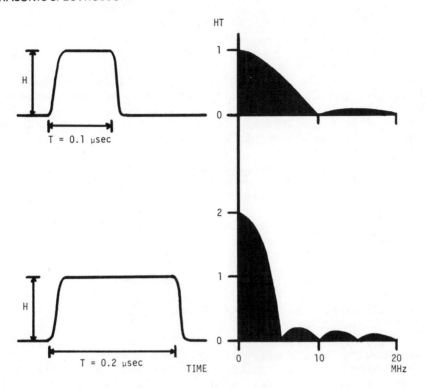

Figure 2.   Pulses with finite rise and decay time that differ in
            duration and their associated spectra (absolute values
            of amplitude are plotted on a linear vertical scale).

in duration. The spectra of these pulses obtained by mathematical
procedure exhibit amplitude maxima at their respective carrier
frequencies. The spectral amplitude has nulls occurring at equi-
distant frequency points. The spacing of these nulls as well as
the width of the main spectral lobe whose center frequency is the
carrier frequency are found to be inversely proportional to the
pulse duration.

To illustrate a more practical case, Figure 2 portrays two
rectangular pulses with finite rise and decay times. In contrast
to the previous case, the spectra of these pulses are finite, their
upper frequency limit being governed by the pulse rise or decay
time. In this figure, spectral amplitudes are plotted on a linear
scale rather than the logarithmic scale used in Figure 1 to
emphasize the fact that most of the pulse energy is concentrated in
the main spectral lobe. Since, as in Figure 1, the width of this
lobe is inversely proportional to the pulse length, this means

that the shorter the pulse, the wider its usable frequency spectrum.
But as Figure 2 also shows, shortening of the pulse lowers its
energy, and therefore, results in reduced spectral amplitudes.

It can therefore be concluded, that the minimum permissiable
pulse length, and thus, the maximum obtainable spectral range will
depend on the signal strength required to penetrate the test object.
This means that the ultrasonic losses in the test specimen are an
important factor in ultrasonic spectroscopy.

If piezoelectric transducers are utilized, short pulse dura-
tions necessitate high damping constants which reduce the conversion efficiency of the transducer and impose further limitations.
In practice, a trade-off between frequency coverage, which governs
the amount of test information obtainable, and the signal amplitude
requirements will be necessary.

TRANSDUCERS FOR SPECTROSCOPY

The ultrasonic transducer represents an essential link
between the test specimen and the electronic equipment used for
generating and analyzing signals. At this stage of development,
the shortcomings of piezoelectric transducers which are the ones
used exclusively for spectroscopy are considered more restrictive
than the limitations imposed by the electronic instrumentation.
To illustrate the importance of transducer characteristics,
Figure 3 shows the time-domain and frequency-domain traces of two
types of transducers that differ with respect to the resonance
frequencies of their unmounted piezoelectric plates and, more
importantly, the degree of damping introduced by the mounting
procedure. The two upper traces of the figure depict the results
obtained for a transducer yielding poor resolution in the time-
domain and having a narrow frequency range. The two lower traces,
on the other hand, illustrate the type of transducer that will
produce a relatively short pulse and a much broader spectrum. The
latter type of transducer would be regarded as superior for time-
domain as well as frequency-domain work if its lower efficiency
(traces shown on Figure 3 were amplitude-normalized) does not pose
a problem.

Figure 4, which shows additional frequency-response curves of
commercially produced transducers, indicates that considerable
differences exist in overall spectral coverage as well as response
amplitude. This emphasizes the need for an accurate analysis of
transducer characteristics as a first step in setting up a spectroscopic test procedure.

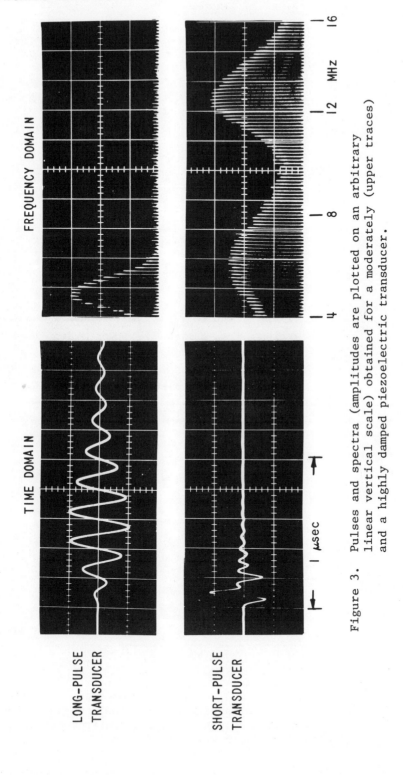

Figure 3.   Pulses and spectra (amplitudes are plotted on an arbitrary linear vertical scale) obtained for a moderately (upper traces) and a highly damped piezoelectric transducer.

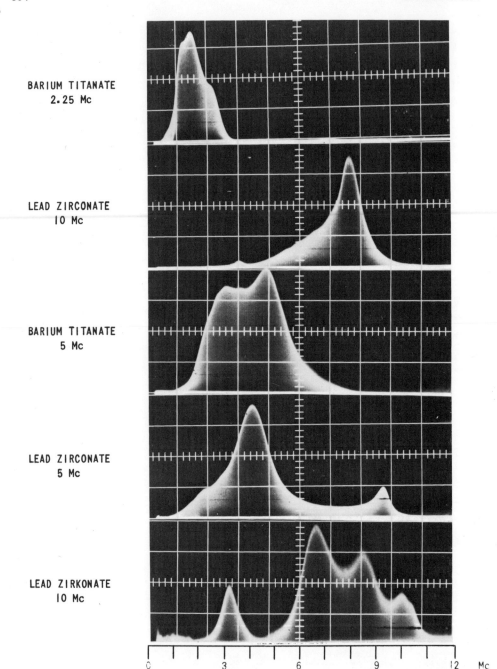

Figure 4.   Loop responses obtained for various types of commercially made ultrasonic transducers. Nominal resonance frequencies according to specifications are from top to bottom: 2.25 MHz, 10 MHz, 5 MHz, 10 MHz.

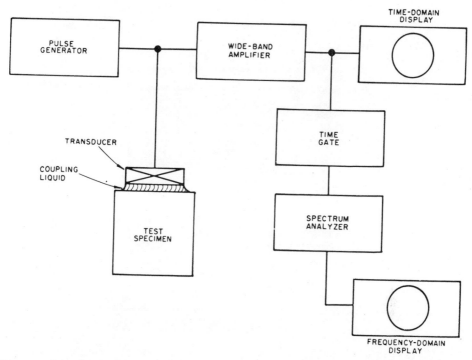

Figure 5.   Simplified block diagram of an ultrasonic pulse-echo
             spectroscope.

EQUIPMENT

Today adequate electronic instrumentation for ultrasonic
spectroscopy is readily available. Figure 5 illustrates the basic
block diagram of a typical system. The circuitry contains the
same elements found in many conventional ultrasonic pulse-echo
instruments designed for the attainment of high resolution in the
time-domain. The additional components needed are an electronic
time gate and a spectrum analyzer. The latter can be of the type
which is essentially an automatically tuned radio receiver with
a synchronously swept signal output display. Another possibility
is the use of a real-time analyzer which samples the spectral
amplitude at various discrete frequencies by employing a number
of tuned filters. A further approach is to use a computer with
a fast Fourier transform program.

Figure 6 shows an ultrasonic spectroscope of commerical manufacture which utilizes a swept-frequency electronic spectrum analyzer.
In addition to the circuitry outlined in Figure 5, the unit contains

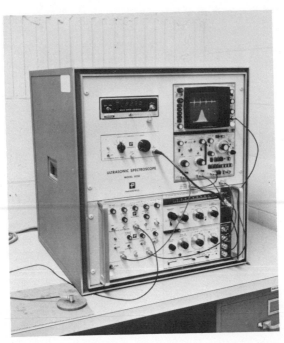

Figure 6.   Commercially produced ultrasonic spectroscope provides
             simultaneous display of time-domain and frequency-
             domain.

a tunable oscillator and frequency counter which enables the superposition of precise frequency markers on the spectrum. Pulse-echo
train and spectrum are displayed on the same storage cathode-ray-
tube.

APPLICATIONS IN NONDESTRUCTIVE TESTING

The practical utilization of ultrasonic spectroscopy for the
purposes of nondestructive testing is still somewhat limited due
to the relatively high equipment cost and the resultant lack of
familiarity of field inspectors with this technology. The following
five examples illustrate the diverse applicability of ultrasonic
spectroscopy:

1. <u>Thickness Gaging</u>. The ultrasonic spectroscope can read
out thickness resonances over a wide range of frequencies and under
conditions that may not be amenable to other ultrasonic test
methods commonly used for this purpose. Particularly in the case
of highly corroded specimen surfaces, a spectroscopic evaluation

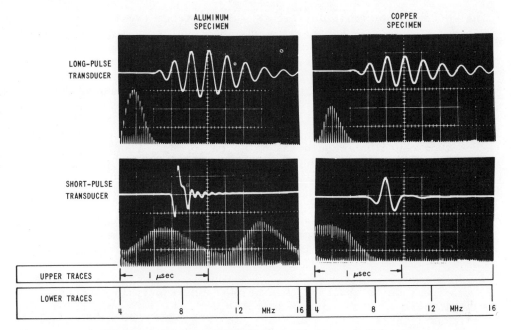

Figure 7. Ultrasonic back-echoes obtained for an aluminum and a
           copper block and associated spectra (amplitudes are
           plotted on identical, arbitrary vertical scales) show
           the effect of transducer performance on the detectability
           of frequency-dependent ultrasonic attenuation exhibited
           by the microstructure of a polycrystalline copper
           specimen.

will sometimes be the only successful method because it permits an
evaluation of broadened resonance indications as to their center
frequencies from which the average wall thickness can then be
computed.

     2. <u>Determination of Microstructure</u>. The attenuation of
ultrasound in a material depends on its microstructure but is also
a function of the ultrasonic frequency. In the case of polycrystalline substances such as metals, losses are due to scattering at
grain boundaries. The average grain size in terms of ultrasonic
wavelength is one determining factor for the rate of attenuation
being encountered. Another factor is the elastic anisotrophy in a
single crystal of the material because the degree of this anisotropy governs the acoustical mismatch at grain boundaries.

     Figure 7 provides an example in which a copper specimen
exhibiting high anisotropy is contrasted with aluminum having low

anisotropy. In addition, Figure 7 illustrates the important role of
the ultrasonic transducer by comparing results obtained with the two
transducers discussed earlier in connection with Figure 3. Using
the long-pulse transducer, the two types of microstructures can not
be differentiated at all because the narrow spectrum of this transducer limits the test to a frequency range where spectral signatures
exhibit no pronounced diversity. With the short pulse transducer,
on the other nad, the situation is quite different. The copper
spectrum is seen to extend to only about 8 MHz wile the aluminum
spectrum contains frequencies of up to 16 MHz. Thus, even a qualitative inspection of the two spectral signatures will reveal the
difference in microstructure.

Similar results involving steel samples are shown in Figure 8
which shows micrographs and back-echo spectra obtained for three
specimens having different average grain sizes. In this figure, a
logarithmic vertical scale was chosen to accommodate a larger range
of spectral amplitudes.

The results indicate that below 5 MHz the grain structure has
little effect on the spectral amplitude. The influence becomes
more noticeable in the range of 5 to 8 MHz and very obvious at
frequencies beyond 8 MHz.

The two examples show that the method of ultrasonic spectroscopy can be used as an effective technique for fast determinations
of microstructure. In addition, the spectroscopic method can be
employed to examine the transmissivity of materials at various
ultrasonic frequencies with the purpose of selecting suitable test
frequencies for conventional ultrasonic inspection methods.

3. Determination of Defect Geometry. Perhaps the most vexa-
tious problem encountered with ordinary ultrasonic pulse echo testing
techniques is the correlation of defect echo amplitudes with actual
flaw sizes. Good correspondence can only be obtained if the geometry of flaws is flat and parallel to the test surface and can
thus be represented by flat-bottom holes. Commercial efforts to
promulgate test blocks containing flat-bottom holes and associated
calibration curves have been considerable, but have never had the
same impact as, for instance, the introduction of radiographic
penetrameters. The reason for the limited usefulness of these test
blocks is that they are based on the assumption of simple and constant defect geometry.

This serious problem can be alleviated by the use of ultrasonic
spectroscopy which provides a method for differentiating various
defect geometries that is independent of the relative magnitude of
reflected echoes. An example is given by Figure 9, which shows
spectra obtained for flaws with different geometries. The specimens

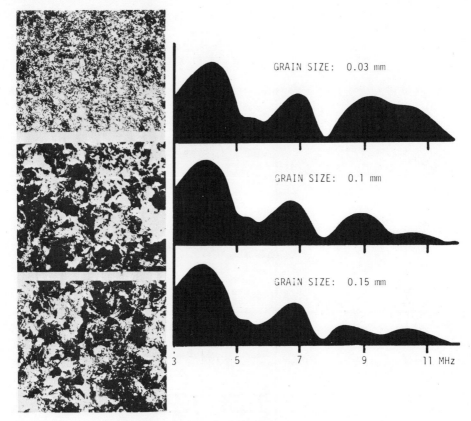

Figure 8.   The effect of average grain size on the pulse-echo
            amplitude-spectra (plotted on a logarithmic vertical
            scale with a range of about 40 db).

used in this test were made of aluminum whose microstructure
exhibits little frequency dependence in the analyzed range of 0
to 20 MHz. In addition, the specimens and the transducer used for
the test were made large enough to avoid geometrical attenuation
due to beam spreading which would otherwise introduce a frequency-
dependence. Four of the analyzed defects were accurately machined
slots whose bottoms had different angles with respect to the test
surface which was the top of the block. These cuts which diametri-
cally penetrated the test blocks represent cracks of 0°, 10°, 20°
and 40° orientation with roughly equal magnitude. The fifth defect
is a cylindrical hole with a diameter of about 1/4 of the width
of the above.

The spectra depicted in Figure 9 were obtained from amplitude-

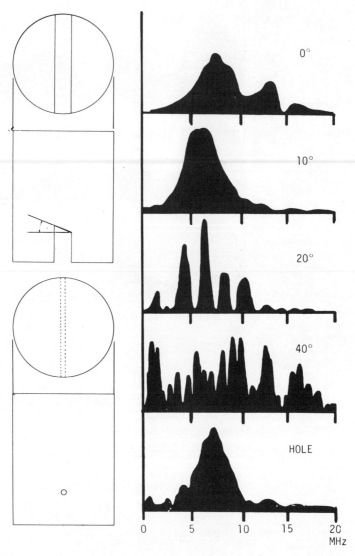

Figure 9.   Pulse-echo spectra (amplitudes are plotted on an
            arbitrary linear vertical scale) obtained for discontin-
            uities representing cracks of different orientations and
            a cylindrical hole. The transducer is coupled to the
            top of each cylindrical test specimen.

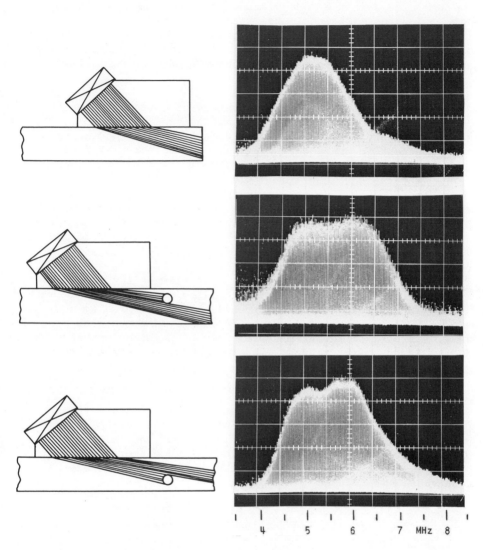

Figure 10. Pulse-echo spectra normalized amplitudes are (plotted on a linear vertical scale) obtained with an angle-beam transducer producing transverse ultrasonic waves. The first trace shows a reflection from the edge of the plate used as a test specimen. The second and the third trace show reflections from cylindrical holes of 1 mm diameter in different locations.

normalized defect echoes because although all defects were of about
the same size defect-echo amplitudes varied as much as 34 db.   The
very obvious disparities in the spectral signatures of Figure 9
indicate that differences in defect configuration rather than defect
size must be the cause for the variations in defect echo amplitude.
On the basis of the spectroscopic data, gross errors in flaw size
assessment can thus be avoided.

    4.   Angle-Beam Testing.   The method of ultrasonic spectroscopy
is not limited to the use of the compressional (longitudinal) wave
mode used in the previous examples.   Intermediate mode conversions
which are not frequency-dependent will not affect test data.   An
illustrative example is given by Figure 10, which shows the spectra
of echo returns obtained from a plate inspected with transverse
waves generated by beam angulation.   Even though the frequency
range of the transducer used covers only about 3.5 MHz noticeable
differences in spectral signature can be observed for the reflection
from the end of the plate and those from cylindrical holes drilled
parallel to the plate surfaces at different depth levels.

    5.   Detection of Delaminations.   In view of the growing interest
in composite structures, the applicability of spectroscopic methods
for NDT in this area has been investigated.   The problem attacked
was the determination of disbonds in laminated structures made of
three layers of different materials bonded together by very thin,
uniform layers of glycerine.   The tests were conducted with the
equipment shown in Figure 6 using a transducer whose frequency
response is given by Figure 11.   The physical arrangement of the
test is illustrated on the left side of Figure 12.   The transducer
element which is marked with a cross hatch is fitted with a wear-
plate which is coupled to the specimen; the bond being indicated
by the first heavy vertical line.   The second and third heavy
vertical lines indicate the bond lines between the three layers of
the laminate.   An interruption of the heavy line represents a
disbond.

    As is illustrated in Figure 12, three different cases were
examined: a completely bonded laminate, a laminate with a second
interface disbond, and a laminate with a first interface disbond.
The spectra shown on Figure 12 were obtained for a laminate made
of a .76 mm thick layer of Plexiglas (Pl), a .62 mm thick layer
of stainless steel (S), and a 3.0 mm thick layer of polycarbonate.
Above each spectral trace, the corresponding time-domain pulse
train is depicted with the down step indicating the end of the
time gate.

    In evaluating the three spectral traces of Figure 12, one
finds that the first and the second spectrum are not materially
different except in the immediate vicinity of 13.5 MHz where a

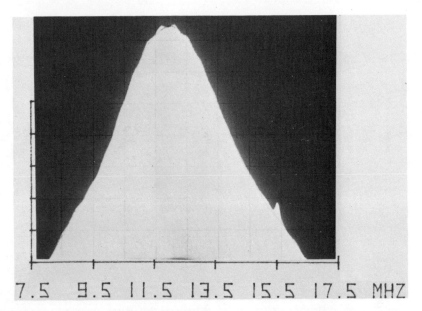

7.5   9.5   11.5   13.5   15.5   17.5   MHZ

Figure 11.   Loop response of transducer used for ultrasonic spectro-
              scopy of laminated structures.

significant variation in spectral amplitude is observed. The first
interface disbond, on the other hand, which is shown at the bottom
of Figure 12 is associated with marked frequency shifts for the
spectral maxima.   One can therefore conclude, that in the case of
this laminate, a first interface disbond is relatively easy to
determine, while the detection of second interface disbonds requires
careful examination of the spectral range between 13 and 14 MHz.

Figure 13 shows results obtained for two additional laminates
with spectra arranged in the same manner as in Figure 12.   The
laminates tested consisted of 1.55 mm thick Plexiglas (P2, 1.0 mm
thick aluminum (A), and 3 mm thick polycarbonate (P4) or 3.5 mm
thick Plexiglas (P3), respectively.

For both these laminates, spectral signatures exhibit obvious
differences permitting a detection of first as well as second interface bond defects.

The superiority of the spectroscopic method over narrow-band
techniques becomes more evident if one makes the futile attempt to
discern bonding conditions of the P2/A/P4 laminate of Figure 13 on
the basis of amplitude measurements in the narrow range of 13 to
14 MHz.

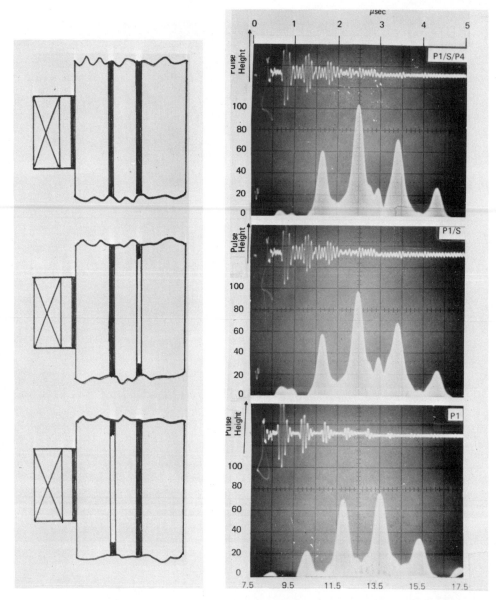

Figure 12.   Transducers with wear-plate is coupled as indicated by a
              heavy vertical line to a three-layer laminate with the
              heavy vertical lines indicating bond lines. Characteris-
              tic amplitude spectra (plotted on identical linear verti-
              cal scales) are produced for an intact specimen shown at
              the top and for first and second interface disbonds in-
              dicated by the absence of heavy vertical lines.   The
              laminate consists of .76 mm thick Plexiglas (P1), .62 mm
              thick stainless steel (S), and 3.0 mm thick polycarbonate
              (P4).

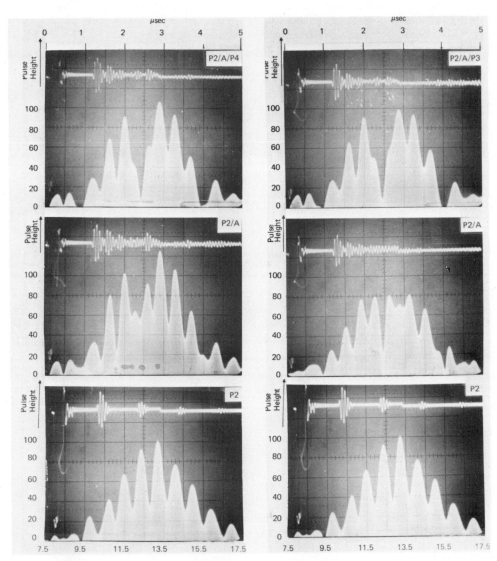

Figure 13.   Similar to Figure 12 spectra illustrates the various
             bonding conditions for laminates consisting of 1.55 mm
             thick Plexigla (P2), 1.0 mm thick aluminum (A), and
             3 mm thick polycarbonate (P4) or 3.5 mm thick plexiglas
             (P3), respectively.

However, Figure 13 also shows that in this particular case narrow-band techniques could be used successfully if a 12 to 13 MHz range were chosen.   Thus a spectroscopic examination may be employed as an initial step in determining suitable test parameters for methods requiring less sophisticated equipment.

CONCLUSION

The described applications of ultrasonic spectroscopy illustrate the potential of this novel method for nondestructive testing.   The bibliography following is a compilation of publications relevant to the subject and covers theoretical as well as applied work.

BIBLIOGRAPHY

1. Gericke, O.R., "Determination of Defect Geometry and Material
    Microstructure by Ultrasonic Pulse Analysis Testing", Proc.
    3rd Nat'l. Symposium on NDT of Aircraft and Missile Components,
    San Antonio, Texas (1962) 199-211.

2. Gericke, O.R., "Determination of the Geometry of Hidden Defects
    by Ultrasonic Pulse Analysis Testing", J. of Acoust. Soc. of
    Am., 35, No. 3 (March 1963) 364-68.

3. Redwood, M., "A Study of Waveforms in the Generation and
    Detection of Short Ultrasonic Pulses", Appl. Mat. Res. (April
    1963) 76-84.

4. Gericke, O.R., "Dual-Frequency Ultrasonic Pulse-Echo Testing",
    J. of Acoust. Soc. of Am., 36, No. 2 (February 1964) 313-22.

5. Gericke, O.R., "Ultrasonic Spectroscopy of Steel", Mat. Res.
    and Stds., 5, No. 1 (January 1965) 23-30.

6. Gericke, O.R., "Ultrasonic Spectroscopy - A New Inspection
    Tool", IEEE, 1965, International Convention Records, Part II,
    46-51.

7. Gericke, O.R., "Experimental Determination of Ultrasonic
    Transducer Frequency Response", Mat. Eval., XXIV, No. 8 (August
    1966) 409-11.

8. Gericke, O.R., "Defect Determination by Ultrasonic Spectro-
    scopy", J. of Metals, XVIII, No. 8 (August 1966) 932-37.

9. Gericke, O.R., "Differential Spectroscopy for Defect and
    Microstructure Identification", Proc. of Sixth Symposium on
    NDT of Aerospace and Weapons Systems Components and Materials,
    And Antonio, Texas, April 1967.

10. Tokarev, V.A. and I.I. Averbukh, "Inspection of the Structure
    of Thin-Walled Parts by the Ultrasonic Spectrometry Method",
    Defektoskopiya (May-June 1967) 68-77.

11. Gericke, O.R., "Ultrasonic Pulse-Echo Spectroscopy", Proc. of
    Symposium on the NDT of Welds and Materials Joining, ASNT,
    March 1968, 267-87.

12. Baryshev, S.E., "The Spectral Composition of Reflected Signals",
    Defektoskopiya, No. 4 (July-August 1968) 47-55.

13. Gericke, O.R., "Theory and NDT Applications of Ultrasonic
    Pulse-Echo Spectroscopy", Proc. of Symposium on the Future of
    Ultrasonic Spectroscopy, British Non-Ferrous Metals Research
    Association, 1970.

14. Gericke, O.R., "Ultrasonic Spectroscopy", in Research Techniques
    In Nondestructive Testing, ed. by R.S. Sharpe. London:
    Academic Press (1970) 31-61.

15. Gericke, O.R., "Ultrasonic Spectroscopy", Encyclopedic Dictionary of Physics, Oxford: Pergamon Press (1970) 502-06.

16. Aldridge, E.E. and G.S. Mitchell, "Ultrasonic Spectroscopy at the N.D.T. Centre, Harwell", Proc. of Symposium on the Future of Ultrasonic Spectroscopy, British Non-Ferrour Metals Research Association, October 1970.

17. Clipson, W.R., "Ultrasonic Spectroscopy Development for Inclusion Cloud Assessment", Proc. of Symposium on the Future of Ultrasonic Spectroscopy, British Non-Ferrous Metals Research Association, 1970.

18. Mitchell, R.T., "Wide-Band Acoustic Bulk Wave Transducers", Proc. of Symposium on the Future of Ultrasonic Spectroscopy, British Non-Ferrous Metals Research Association, 1970.

19. Lloyd, E.A., "Wide-Band Ultrasonic techniques", Proc. of symposium on the Future of Ultrasonic Spectroscopy, British Non-Ferrous Metals Research Association, 1970.

20. Whaley, H.L. and K.V. Cook, "Ultrasonic Frequency Analysis", Mat. Eval. (March 1970) 61-66.

21. Merkulov, L.G. and V.A. Tokarev, "Physical Basis of Special Method of Measuring the Damping of Ultrasonic Waves in Materials", Defektoskopiya, No. 4 (July-August 1970) 3-11.

22. U.S. Patent No. 3,538,751. November 10, 1970. Ultrasonic Thickness Gage. O.R. Gericke to the U.S.A. by the Secretary of the Army.

23. U.S. Patent No. 3,538,753. November 10, 1970. Ultrasonic Spectroscope. O.R. Gericke to the U.S.A. by the Secretary of the Army.

24. Papadakis, E.P. and K.A. Fowler, "Broad-Band Transducers: Radiation Field and Selected Applications", J. of Acoust. Soc. of Am., 50 (1971) 729-45.

25. Whaley, H.L. and L. Adler, "Flaw Characterization by Ultrasonic Frequency Analysis", Mat. Eval. (August 1971) 182-88.

26. Chaskelis, H.H., "Transducers - Fact and Fiction", Nondestructive Testing (December 1971) 375-79.

27. Adler, L. and H.L. Whaley, "Interference Effect in a Multifrequency Ultrasonic Pulse-Echo and Its Application to Flaw Characterization", J. of Acoust. Soc. of Am., 51 (1972) 881-87.

28. U.S. Patent No. 3,662,589. May 16, 1972. Ultrasonic Flaw Determination by Spectral Analysis. L. Adler and H.L. Whaley to the U.S.A. by the Atomic Energy Commission.

29. Papadakis, E.P., K.A. Fowler and L.C. Lynnworth, "Ultrasonic Attenuation by Spectrum Analysis of Pulses in Buffer Method and Diffraction Corrections", J. of Acoust. Soc. of Am., 53 (1973) 1336-43.

30. Brown, A.F., "Materials Testing by Ultrasonic Spectroscopy", Ultrasonics (September 1973) 202-10.

31. U.S. Patent No. 3,776,026. December 4, 1973. Ultrasonic Flaw Determination by Spectral Analysis. L. Adler and H.L. Whaley to the U.S.A. by the Atomic Energy Commission.

32. Sachse, W., "Ultrasonic Spectroscopy of a Fluid-Filled Cavity in an Elastic Solid", J. of Acoust. Soc. of Am., 56 (1974) 891-96.

33. Simpson, W.A., "Time-Frequency-Domain Formulation of Ultrasonic Frequency Analysis", J. of Acoust. Soc. of Am., 56 (1974) 1776-81.

34. Lele, P. and J. Namery, "A Computer-Based Ultrasonic System for the Detection and Mapping of Myocardial Infarcts", Proc. of the San Diego Biomedial Symposim, 1974, Vol. 13.

35. Rose, J.L. and P.A. Meyer, "Signal Processing Concepts for Flaw Characterization", British J. of Nondestructive Testing (July 1974) 97-106.

36. Brown, A.F. and J.P. Weight, "Generation and Reception of Wideband Ultrasound", Ultrasonics (July 1974) 161-67.

37. Hudgell, R.J., L.L. Morgan and R.F. Lumb, "Nondestructive Measurement of the Depth of Surface-Breaking Cracks Using Ultrasonic Raleigh Waves", British J. of Nondestructive Testing (September 1974) 144-49.

38. Bell, J.C. and N.F. Haines, "Measuring Corrosion Layers on Inaccessible Surfaces", Ultrasonics (November 1974) 237.

39. Lloyd, E.A., "Nondestructive Testing of Bonded Joints", Nondestructive Testing (December 1974) 331-34.

40. Highmore, P.J., "Nondestructive Testing of Bonded Joints", Nondestructive Testing (December 1974) 327-30.

41. Whaley, H.L., K.V. Cook, L. Adler and R.W. McClung, "Applications of Frequency Analysis in Ultrasonic Testing", Mat. Eval. (January 1975) 19-24.

42. Sachse, W. and C.T. Chian, "Determination of the Size and Mechanical Properties of a Cylindrical Fluid Inclusion in an Elastic Solid", Mat. Eval. (April 1975) 81-88.

43. Yee, B.G.W., F.H. Chang and J.C. Couchman, "Applications of
     Ultrasonic Interference Spectroscopy to Materials and Flaw
     Characterization", <u>Mat. Eval.</u> (August 1975) 193-202.

44. Rose, J.L. and H. Schlemm, "Equivalent Flaw Size Measurements
     and Characterization Analysis", <u>Mat. Eval.</u> (January 1976) 1-8.

Chapter 14

ACOUSTIC INTERACTIONS WITH INTERNAL STRESSES IN METALS

O. Buck and R. B. Thompson

Rockwell International

Thousand Oaks, California

ABSTRACT

This chapter discusses two techniques that show promise of measuring different aspects of internal stress distributions in metals. The present knowledge of states of stress and their relation to failure mechanisms is first briefly reviewed, noting the distinction between long and short range stresses. Electromagnetic transducer efficiency tests for long range stresses are then discussed. It is demonstrated that the efficiency of the electromagnetic generation of ultrasonic waves is very sensitive to the deformations of ferromagnetic materials. Experiments defining the general features of the effect in three iron-nickel alloys, and its sensitivity to minor variations in material properties such as texture, composition, and heat treatment in low carbon steels are presented. The results are interpreted in terms of a magnetostrictive model of the generation process and the previously known strain dependence of the magnetostriction of these materials. Ultrasonic harmonic generation is then shown to be a technique sensitive to dislocation parameters. This measurement can thus be used to define the microscopic internal stresses which control such material properties as flow stress and state of fatigue. Some experimental data on harmonic generation in deformed and fatigued aluminum specimen, which demonstrates this effect are presented. Limitations of this technique, as observed in Al alloys, as well as results obtained on various other structural materials, are also discussed.

TABLE I

Classification of Internal Stresses

Kind	Range	X-Ray	Examples
1st	Macroscopic (Order μm to cm)	Line shift	Elastic deformation of a cut and rewelded toroid Thermal stresses
2nd	(Order μm) Microscopic	Line shift and line broadening	Particles of different phases in a matrix
3rd	(Order 100 – 1000Å)	Line broadening	Edge and screw dislocations (plastic deformation)

INTRODUCTION

A stress maintained in the bulk of a material without application of an external load is called an internal stress. Internal stress can be generated in a number of ways and some simple examples are given in Table I. It is most convenient to classify these internal stresses with respect to the range over which they are active. Such a simplified scheme [1] is shown in Table I, in which the internal stresses have been divided into three kinds. Also shown are the effects that the given examples exhibit using x-ray diffraction, which is the most popular measurement technique [2]. The following discussions will concentrate on internal stresses of the first and third kind only and examples of attempts to measure those stresses ultrasonically will be given.

In a recent meeting on nondestructive evaluation of internal stresses [3] the need for work in the area of stress detection became quite apparent. It is very important to know the presence of either compressional or tensile macroscopic internal stress in a structural component since such stresses determine crack initiation and crack propagation under external loads. Strong efforts are being made at present to develop techniques to measure this kind of internal stress. Attention in the past has been focussed on x-ray techniques and, recently, acoustic birefringence and magnetic techniques [3]. In the second section of this chapter a new magnetic technique is discussed that appears to have considerable advantage over the existing magnetic techniques.

The role of microscopic internal stresses on mechanical properties (in particular fatigue properties), on the other hand, seems to be not quite clear to the engineering world. The third part of this chapter, therefore, is addressed specifically to this subject. In particular, the internal stresses due to dislocations have been treated [1]. These dislocations are the defects responsible for plasticity effects in metals and alloys. Experiments have been performed using acoustic harmonic generation and it is this technique that is being examined as a possible nondestructive tool to study the microscopic internal stress due to dislocation or, in other words, plasticity effects. It is hoped that the results obtained will lead to a way to determine nondestructively the remaining life of a material in use.

ELECTROMAGNETIC TRANSDUCER EFFICIENCY TEST FOR MACROSCOPIC STRESSES

Background

As noted above, x-ray techniques are perhaps the most commonly accepted procedure for determining macroscopic material stresses. However, these suffer from a number of operational disadvantages which include bulky apparatus and required surface finish of high quality. Furthermore, x-rays in general sense the deformation of a thin surface layer a few tenths of a mil (2.5×10^{-3} cm) in thickness, and are influenced by texture and other material variations. In ferromagnetic materials, a number of magnetic properties have been considered as candidates for stress detection which overcome some of these problems. Included are the Barkhausen effect [4], permeability measurements [5], and multiple magnetic parameter measurements [6]. In each case, the physical origin of the stress intensity is the magnetoelastic effect, whereby deformations of the material produce changes in the relative energies of various magnetic domain configurations and consequentially influence the macroscopic magnetic response.

A fourth magnetic effect that is known to be very strongly influenced by the level of stress in a material is magnetostriction [7], as shown in Figure 1. Until recently, there has been no technique available by which this important parameter could be non-destructively measured on structural components. However, it has now been established that the efficiency of the electromagnetic generation of ultrasonic waves in ferromagnetic materials is proportional to the differential magnetostrictive coefficients [8]. Efficiency values can thus be directly related to the material stress [9-11].

Physical Origin of the Stress Effect

An accurate theoretical description of the complete magnetic response of a ferromagnetic polycrystal has not yet been reported

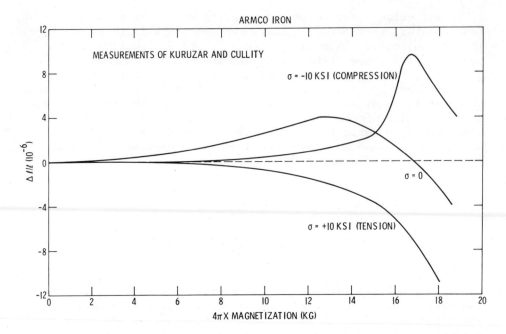

Figure 1.   Magnetostriction of iron as a function of stress.

to the authors' knowledge. However, qualitative arguements may be presented that provide insight into the phenomena of interest here.

The magnetic state is determined by the minimization of the appropriate energy functions subject to constraints imposed by impurities, dislocations, etc., which restrict the free motion of magnetic domain walls. Equation (1) includes selected terms in the energy expansion of a uniformly magnetized single crystal of a cubic material [12].

$$E = K_1(\alpha_1^2\alpha_2^2 + \alpha_2^2\alpha_3^2 + \alpha_3^2\alpha_1^2)$$

$$- \frac{3}{2}\sigma[\lambda_{100}(\alpha_1^2\gamma_1^2 + \alpha_2^2\gamma_2^2 + \alpha_3^2\gamma_3^3)$$

$$+ 2\lambda_{111}(\alpha_1\alpha_2\gamma_1\gamma_2 + \alpha_2\alpha_3\gamma_2\gamma_3 + \alpha_3\alpha_1\gamma_3\gamma_1)]$$

$$- M_sH(\alpha_1\beta_1 + \alpha_2\beta_2 + \alpha_3\beta_3) \qquad (1)$$

where α_i are the direction cosines of the magnetization M_s with respect to crystal axes, β_i are the direction cosines of an applied magnetic field H, γ_i are the direction cosines of an applied stress σ, K_1 is the first magnetic anisotropy constant, and λ_{100} and λ_{111} are the saturation magnetostrictions along the indicated axes.

The first term shows the crystalline influences on magnetic energy. This is minimized when α lies along a cube axis and $K_1>0$, or along a body diagonal and $K_1<0$, the so-called "easy axes" of magnetization for iron and nickel, respectively. The second term shows how this crystalline energy is changed by deformation of the lattice in an applied stress, while the final term gives the energy of the magnetization in an applied field. With no applied field or stress, minimization of the energy predicts that the magnetization will lie along one of the easy axes of magnetization, as determined by the crystalline anisotropy. As the magnetic field increases, the direction of magnetization will rotate from this easy asix until it is ultimately parallel to the field. Stress modifies this procedure through the second term.

Equation (1) is incomplete since we have neglected the magnetic exchange energy, which determines the size of the magnetic domain, and the magnetostatic energy, which demands the existence of domain closure to remove free poles. Hence we are considering a uniformly magnetized material. The magnetic behavior of a polycrystal can be estimated by considering it to be composed of a large number of randomly oriented grains in each of which Equation (1) applies. The macroscopic response is then an appropriate average over crystal orientation. This approach works best near saturation, and has been shown to lead to quantitative predictions of transducer efficiency when $\sigma = 0$ [8]. The approach also leads to qualitative insights at lower fields.

Figure 2 illustrates this by plotting the first two terms of Equation (1) for tensile stresses of 0 and 200 MN/m^2 (29 ksi) applied along the [100] axis or iron. Two important effects are illustrated. With no stress, the six axes have equal energy, and the domains can be expected to be equally distributed among these cube axes in the demagnetized state. Under stress, this degeneracy is removed. For the case shown, the demagnetized state will contain a greater number of domains aligned along the [100] and [100] axes, and fewer along the others. This strongly influences the magnetic response at low fields.

A second important effect is the stress induced change in the difference between the minimum and maximum energy points; i.e., the anisotropy. Close to saturation, magnetization changes by rotation away from the easy axes toward the applied field direction. Changes in the effective anisotropy that oppose this clearly influence the high field magnetic response.

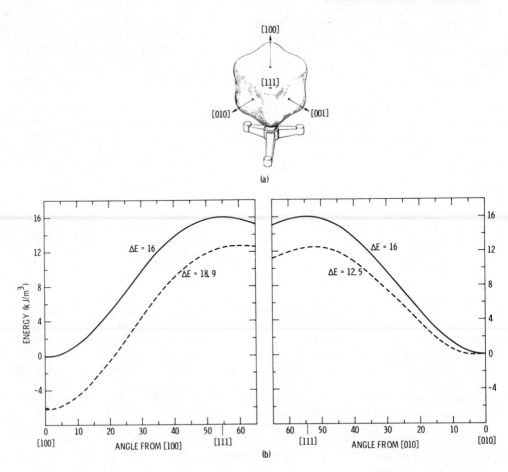

Figure 2.   Anisotropy energy of iron; (a) three dimensional plot of
             energy versus angle, (b) planar cuts of energy surface
             with zero stress (solid line) and a tensile stress of
             200 MN/m^2 (29 ksi) along the [100] axis (broken line).

All of the magnetic tests for stress are based on these physical
principles. The magnetostrictive response is believed to be one of
the most sensitive for reasons discussed elsewhere [9-11]. Among
these are the fact that magnetostriction differentiates between the
motion of 90° and 180° domain walls, the fact that the technique is
sensitive to the slope of the magnetostrictive response, and the
fact that measurement can be made in a regime of reversible magnetic
response where quantitative interpretations can be made.

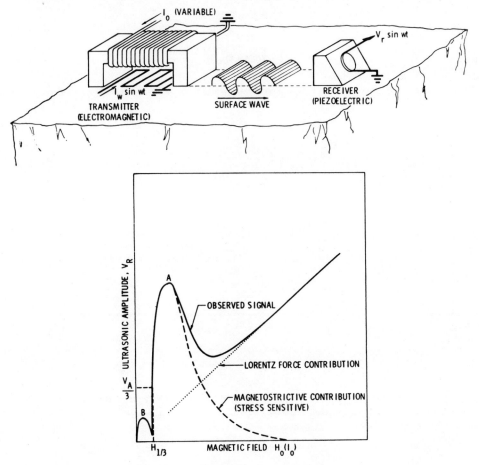

Figure 3.   Experimental apparatus for determining efficiency of
electromagnetic generation of surface ultrasonic waves
as a function of magnetic field. The major features
observed in ferrous materials are shown in the schematic
efficiency plot flow.

Measurement Technique

Figure 3 shows the experimental configuration. An electromagnetic transducer, consisting of an electromagnet and a meander
coil of wire carrying a dynamic current, is used to launch an ultrasonic surface wave. The amplitude of this wave, as detected by a
second transducer which can be a standard piezoelectric wedge, is
a measure of the efficiency of transduction. When the magnetic
bias produced by the electromagnet is changed, the efficiency varies

EXPERIMENT:

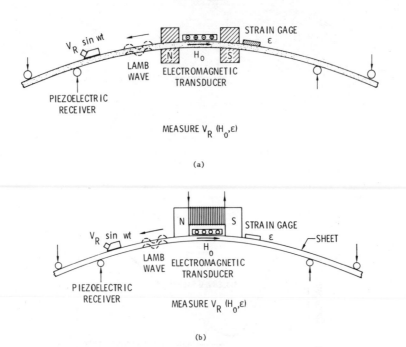

(a)

(b)

Figure 4.   Apparatus for measuring stress dependence of transducer
            efficiency: (a) bar configuration, (b) sheet configura-
            tion.

as shown. Such a graph will be referred to as an "efficiency plot"
in the remainder of this chapter. At high fields, for which the
material is magnetically saturated, the generated ultrasonic wave
amplitude is directly proportional to the static magnetic field.
The generation is caused by Lorentz forces on induced eddy currents,
and there are no significant magnetic contributions. At lower
fields, there are one or more magnetostrictive peaks in the
efficiency plot. It has been established that these efficiencies
are proportional to the dynamic, differential magnetostriction [8],
and hence can be expected to be sensitive to stress as discussed
above.

     Two experimental configurations for applying stresses have been
used as shown in Figure 4. Part (a) shows the apparatus used with
bar shaped samples, while Part (b) shows the apparatus used with
thin sheets. In each case, an electromagnetic transducer was
placed at the center of the sample and used to generate Lamb waves
which propagated a short distance and then were detected by a

wedge receiver. Stress was applied by four point bending and deformation was monitored by a resonance strain gage. The transducer coil was placed alternately on the convex and the concave sides, and the "efficiency plots" were obtained in each position using an X-Y recorder. Thus both tensile and compressive stresses could be studied for a single bending since the generation process was confined to the electromagnetic skin depth (a few mils) under the transducer. For the plate configuration, measurements both parallel and perpendicular to the stress axis were performed. Upon completion of the efficiency measurements, the load was increased and the process repeated. In each case, deformation was incrementally increased until strains of 3×10^{-3} were attained. In all except the invar sample, significant plastic deformation, as evidenced by a permanent bend in the sample, occurred.

Measurement on the 7/8 inch (2.22 cm) by 3/8 inch (0.95 cm) bar samples were made with 165 kHz flexural Lamb waves. Measurements on the 0.118 inch (0.30 cm) sheets were made with 0.95 MHz flexural Lamb waves. A minor difference in the two setups is the position of the electromagnet which was symmetrically positioned on the bars but was always on the tensile side of the sheets. To the extent that the sheet is thin, it should be uniformly magnetized, independent of the side on which the magnet is placed. This, in general, appeared to be true, but some effects were observed which could be attributed to demagnetization fields associated with the magnet position.

Responses of Different Iron-Nickel Alloys

Some of the basic mechanisms of the stress dependence of efficiency were studied by comparing the responses of iron, nickel, and invar, which represent three substantially different classes of magnetic behavior.

Figure 5 shows the efficiency plot of an Armco iron bar cut from a hot rolled plate at loads of 0 and $\pm$ 414 MN/m^2 ($\pm$ 60 ksi). A number of differences are immediately evident. Compression narrows peak A, and moves it to lower fields. On the other hand, tension broadens peak A, moves its sharp lower edge to lower fields, and suppresses peak B.

Since efficiency is proportional to differential magnetostriction, it can be shown that peak B is associated with the initial expansion of the material as shown in Figure 1, while peak A is associated with the contraction that occurs at higher fields. The suppression of peak B by tension is consistent with the effect of tension on the static data in Figure 1. It can be explained in terms of the previously discussed removal of the energy degeneracy

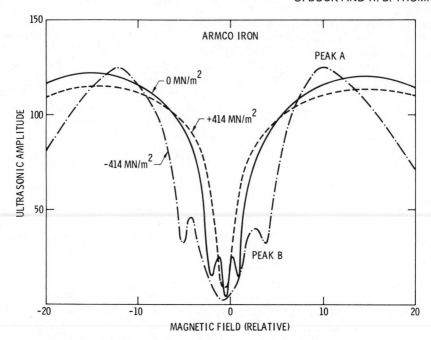

Figure 5.   Efficiency plot in Armco iron for zero applied stress
             and calculated stresses of ± 414 MN/m² (± 60 ksi).

of the easy axes [7]. The shift of peak A to higher fields is
consistent with an increase in the effective magnetic anisotropy
[13]. However, caution must be exercised in such an interpretation
since there is no single anisotropy constant that escribes the
directional dependence of the magnetic energy in a stressed crystal.

The responses of nickel and invar (Fe-36%Ni) under similar
conditions are shown in Figures 6 and 7. It will immediately be
noted that the effect of stress is considerably more substantial
in these materials. This follows since their magnetic anisotropy
constants are an order of magnitude smaller than that of iron.
Consequently, the stress induced energy changes play a more important
role in determining the total magnetic response.

                    Responses of Steel Alloys

One of the problems with magnetic techniques is the unwanted
sensitivity to minor material variations; e.g., texture, composition, heat treatment, and forming conditions. To determine the
extent to which these effects might degrade the transducer efficiency
test for stress, a series of survey experiments was performed. The

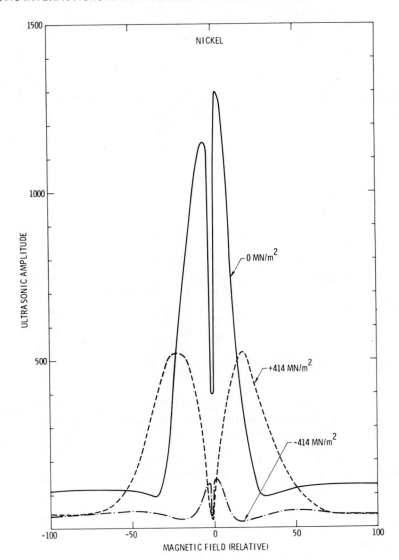

Figure 6. Efficiency plot in nickel for zero applied stress and calculated stresses of $\pm$ 414 MN/m^2 ($\pm$ 60 ksi).

samples studied were a 1018 cold drawn steel bar, an Armco iron bar cut from a hot rolled plate, a pari of A-366 cold rolled steel sheets cut parallel and perpendicular to the rolling direction, and a pair of A-569 hot rolled steel sheets cut parallel and per-pendicular to the.rolling direction. These latter two alloys have the same composition.

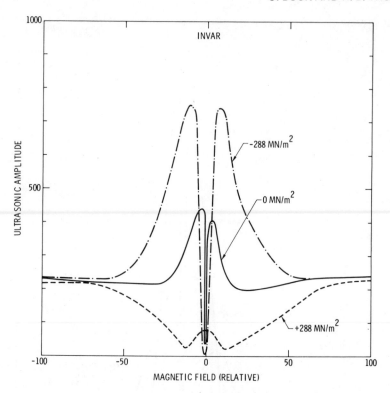

Figure 7.   Efficiency plot in Invar for zero applied stress and
            calculated stresses of $\pm$ 288 MN/m^2 ($\pm$ 42 ksi).

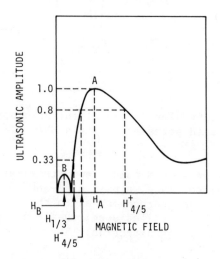

Figure 8.   Definition of stress sensitive parameters.

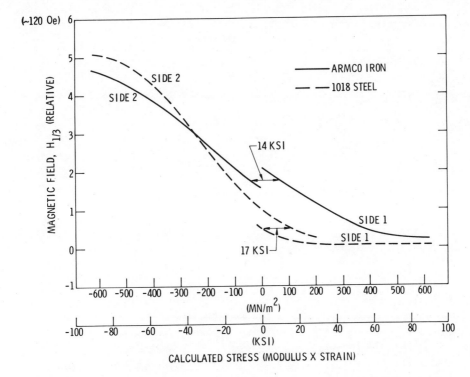

Figure 9. $H_{1/3}$ versus stress for Armco iron and 1018 steel.

Figure 8 schematically defines various parameters measured from the efficiency plots. In general, there is some uncertainty in absolute efficiency data because of possible variations in the efficiency of the piezoelectric wedge and in attenuation of the ultrasonic wave during propagation. Consequently, all efficiencies were normalized to unit value at peak A. The stress sensitive parameters were then chosen to be the magnetic field required to produce certain characteristic points on the efficiency curve. Thus, H_A is the field required to produce peak A, $H^+_{4/5}$ is the higher field at which the ultrasonic amplitude is 80% of that at peak A, etc. In all cases, the sample was first magnetically saturated and then measurements were performed while the field was decreasing.

Figure 9 shows a plot of the parameter $H_{1/3}$ versus stress for the Armco iron and 1018 steel bars. As noted previously [11], the offset in the data obtained on the two sides was corroborated by x-ray analysis and apparently is the result of stresses produced by the forming process.

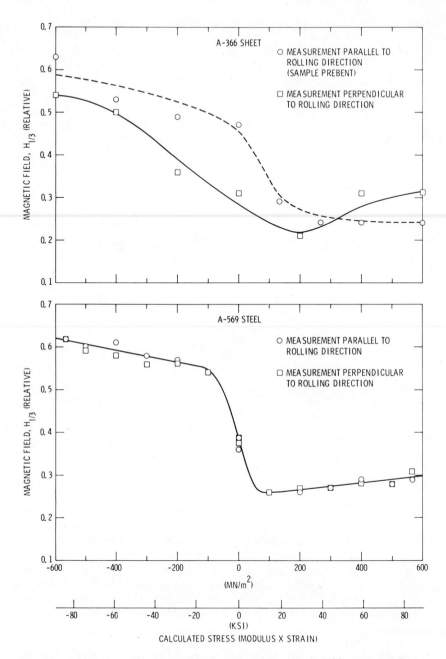

Figure 10. $H_{1/3}$ versus stress for A-366 and A-569 steel sheet cut both parallel and perpendicular to rolling direction. The parallel A-366 sheet was prebent as discussed in text.

Figure 10 shows similar $H_{1/3}$ data for the A-366 and A-569 sheets. For the case of the hot rolled A-569 sheet, virtually identical responses are obtained independent of the rolling texture. This is not the case for the A-366 sheet. However, one of the samples was accidentally bent well beyond the yield before the experiment began. Hence the elastic deformation produced during the experiment was superimposed on both a pre-existing elastic and plastic deformation. The shift and change in shape of the response curve is quite possibly a result of this rather complex history than the texture difference.

A comparison of Figure 9 th Figure 10 is made difficult by a change in the abscissa scale caused by the different magnetic con- figuartion used with the bars and beams. In each case, the measure of $H_{1/3}$ was the voltage applied to the particular magnet used. It appears clear, however, that the $H_{1/3}$ parameter has the same general stress dependence in all of these ferrous structural metals, but that the detailed shape is somewhat dependent upon the particular alloy under consideration. At least in A-569 steel, texture does not seem to have a major effect on the data.

Of the other parameters indicated in Figure 8, $H^{+}_{4/5}$ had the largest and most reproducible variation with stress. Figure 11 shows the results for the steel sheets. Note that $H^{+}_{4/5}$ increases with tensile stress up to the highest value applied, approximately twice the yield, with little saturation.

It is interesting to note that the differences in the $H^{+}_{4/5}$ responses for the two materials are considerably less that the differences in the $H_{1/3}$ responses. Even the present specimen has a response quite close to the others. A contributing factor appears to be the fact that $H^{+}_{4/5}$ is a magnetic bias that places the measurement in the reversible regime of magnetic response. Differences in dislocation structure and other material variabilities presumably have a much smaller effect than at lower fields such as $H_{1/3}$ for which the magnetic response is quite irreversible and large hysteresis effects occur. This data is also consistent with the results for Armco iron shown in Figure 5.

In all of the above measurements, the ultrasonic propagation direction coincided with the applied stress axis. Figure 12 com- pares that response to the response obtained when the ultrasonic wave propagates perpendicular to the stress. The particular sample shown is the A-569 plate cut perpendicular to the rolling direction, but similar results were obtained on the other samples. It is seen that sensitivity of the ultrasonic wave propagating perpendicular to the stress is quite small. The magnetic parameter is quite close to the stress-free value when the wave propagates parallel to the stress. This suggests that a stress whose axis is unknown can be

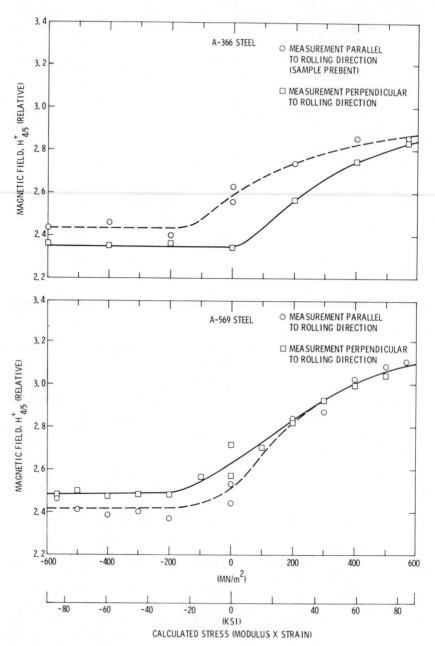

Figure 11. $H^+_{4/5}$ versus stress for A-366 and A-569 steel sheet cut both parallel and perpendicular to rolling direction.

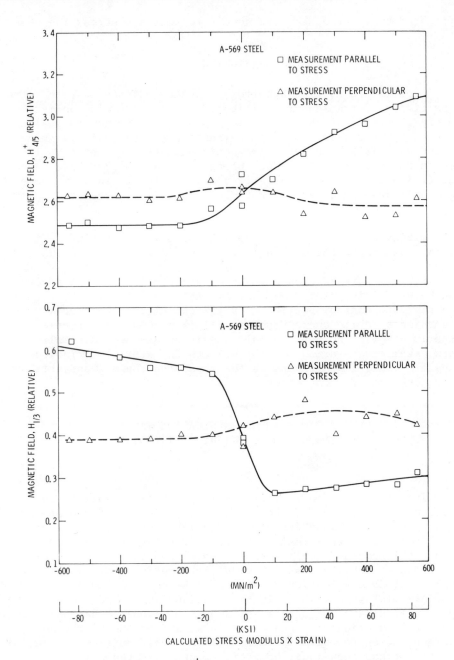

Figure 12. Comparison of $H^+_{4/5}$ and $H_{1/3}$ when measurements are made parallel and perpendicular to stress axis. Sample is A-569 steel sheet cut perpendicular to rolling direction.

evaluated by rotating the transducer for maximum stress response.

Conclusions

Experiments have been performed which indicate that the stress
sensitivity of transducer efficiency applies to a wide variety of
materials, and that the magnitude of the response can be qualitatively understood in terms of the fundamental magnetic parameters
of the materials such as the magnetic anisotropy constant. In iron,
two features of the efficiency plots, $H_{1/3}$ and $H^+_{4/5}$, have been
identified as being particularly sensitive to stress. Of these,
$H_{1/3}$ has the greater sensitivity to compressive stresses. However,
it appears to be changed by material variations as well as stress,
and shows saturation effects at high stresses. $H^+_{4/5}$ has greater
sensitivity to tensile stresses, and shows less material variability.
It is, however, rather insensitive to compression.

Further work is needed in three areas. Theoretical analysis
can be useful in predicting sensitivity to stresses, texture, and
plastic deformation, particularly for the $H^+_{4/5}$ parameter which
occurs in a regime where Equation (1) is a good approximation.
Experiments with better controlled samples are needed to allow a
more detailed assessment of the effects of material variations.
Finally, minor improvements in the apparatus could substantially
reduce the measurement error.

ULTRASONIC HARMONIC GENERATION TESTS FOR
MICROSCOPIC INTERNAL STRESS

Background

In addition to macroscopic stresses, it is important to
measure the microscopic, short-range stress due to dislocations.
Although the relationships are not fully quantified, it is known
that these stresses strongly influence the fatigue behavior of
structural metals. Earlier experimental and theoretical studies
[3] indicated that acoustic harmonic generation could be a useful
tool to study dislocation parameters. Recent results on ultra-
soncially fatigued samples in the Russian literature [14], indicated
that in Ti-alloys and steels the amount of acoustic harmonic generation is a function of fatigue strain amplitude and of the number of
fatigue cycles to which the material is subjected (see Figure 13).
No attempt has been made, however, to quantize the effect in terms
of a change of the materials' properties and, in particular, in
terms of changes of the dislocation parameters, or possibly other
mechanisms, such as a stress induced metallurgical phase change.

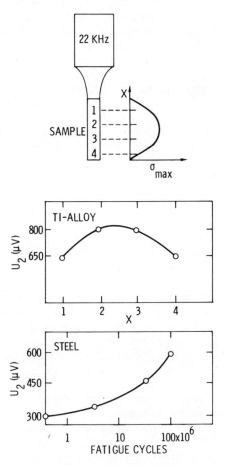

Figure 13.   Top: Experimental 20 kHz arrangement to generate
              fatigue. Middle: Amplitude of second harmonic (U_2)
              as a function of extent of fatigue (positions 1 and 4 –
              low stress amplitudes, positions 2 and 3 – high stress
              amplitudes). Bottom: U_2 as a function of fatigue at
              maximum stress amplitude. [14]

Theory

Mathematically, all internal stresses can be treated as being
due to an "Incompatibility tensor" [15] which is determined by the
deviations from the compatibility conditions governing the classical
theory of elasticity. In particular, the third kind of internal
stresses, due to dislocations, has been treated successfully using
the above theory (continuum theory of dislocations). However, since
this model has not yet enjoyed great success in describing the

complex material hardening phenomena during deformation or fatigue, theoretical efforts have concentrated on studying simple dislocation arrangements whose internal stress fields can be calculated using classical theory of elasticity [16]. The materials' response of plastic deformation and fatigue can then be described by dislocation effects [17]. Work hardening in this picture is mainly due to the interaction of dislocations via internal stresses of the third kind surrounding each dislocation (or dislocation pile-up) [18]. Since the subject has been discussed recently [18], the interested reader is referreh to that paper. The basic conclusion is that the second harmonic amplitude due to dislocations, U_{2d}, is related to the microscopic internal stresses, σ_G, and the flow stress, σ, as

$$U_{2d} \propto (\sigma_G)^{-2} \propto (\sigma - \sigma_o)^{-2} \tag{2}$$

where σ_o is the yield stress of the material. Both σ_G and σ are related to the dislocation density.

Experimental Results and Discussion

The apparatus used to measure harmonic generation has been described extensively in Reference 19. A quartz transducer (30 MHz) was mounted to one side of the specimen and the ultrasonic wave was detected on the opposite side using a capacity microphone. Measurements were made on an aluminum single crystal of [100] orientation. Deformation of this crystal was accomplished in compression in an MTS system between two parallel, flat plates. Harmonic generation was determined using the above apparatus after various stages of compression (given in multiples of the yield stress, σ_o). As shown in Figure 14, harmonic generation increases with increasing applied stress for stresses below σ_o with the situation being reversed above σ_o.

Harmonic generation is defined here as the ratio of the total received second harmonic amplitude U_2 over the fundamental amplitude U_1. U_2 is the sum of the harmonic amplitude due to a lattice contribution and the amplitude produced by dislocations (U_{2d}) [20].

The lattice contribution to U_2/U_1 should be independent of applied stress and has been drawn as the horizontal dashed line in Figure 14. The variations of the dislocation contribution, U_{2d}/U_1, can then be understood in the following way. The increase below σ_o is due to the dislocations bowing out under the applied stress [20]. The decrease of U_{2d}/U_1 above σ_o follows qualitatively the relation given by Equation (2). Using dislocation parameters, this decrease is due to a decrease of the dislocation loop length. Qualitative verification of Equation (2) is the major point made here. After the compression tests shown in Figure 14, the crystal

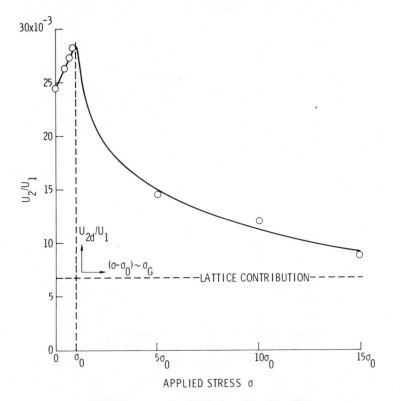

Figure 14. Normalized second harmonic displacement versus compressive stress (in multiples of the yield stress σ_0) for
30 MHz longitudinal waves along the [100] direction of
aluminum. A second coordinate system has been introduced
to show these data according to Equation (2).

was fatigued at a maximum stress level of 7.5 σ_0 and harmonic generation determined as a function of fatigue cycles applied. Since
it was impossible to determine the flow stress of the material after
different degrees of fatigue directly, the surface hardness was
measured using a Knoop hardness indenter. Basically, the harmonic
generation increased and Knoop hardness decreased as fatigue pro-
ceeded. These results are shown in Figure 15 in the form of U_2/U_1
versus Knoop hardness to demonstrate the qualitative features
expected from Equation (2). In terms of the dislocation parameters
the result can be interpreted in the following way. The dislocation
loop length in the present case is quite short prior to fatigue.
Long loops develop during fatigue and yield an increased contribution to the second harmonic, thus harmonic generation reflects the
internal stresses (at least within the "dislocation cells" as fatigue
proceeds.

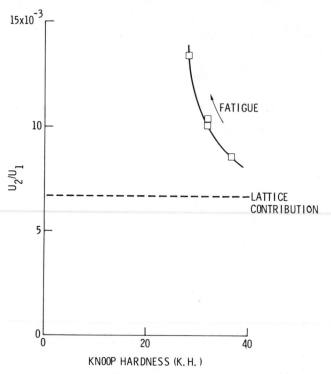

Figure 15. Normalized second harmonic displacement versus Knoop
            hardness during fatigue of a deformed Al single crystal.
            (σ_{max} = 7.5σ_o, σ_{min} = 0.8σ_o)

     Effects of fatigue on harmonic generation in Al 2219-T851 were
studied over a relatively wide range of maximum stress levels
(0.8$\sigma_y \leq \sigma_{max} \leq$ 1.1σ_y with σ_y being the yield stress). The results,
shown in Figure 16, indicate that within the accuracy of the
measurements fatigue produces no change in the second harmonic
amplitude. To be sure that the absence of an effect was not caused
by the (somewhat unusual) compression-compression type fatigue,
similar experiments were performed on Al 6061-T6 fatigued in tension-
compression at 20 kHz with an ultrasonic horn arrangement (similar
to the setup shown in Figure 13). Again, within the accuracy of our
measurements, no change in the second harmonic was observed.

     These results demonstrate the limitations of the method. It
is speculated that in this type of material the diffusion of inter-
stitials at room temperature cause repinning of dislocations, thus
effectively preventing changes in acoustic harmonic generation.
In materials like Ti-alloys and steels, such immediate repinning
is not expected. As can be clearly seen from the results on

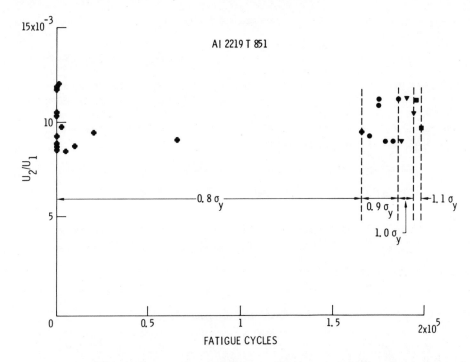

Figure 16. The effect of fatigue on the second harmonic generation at various maximum load levels.

Ti-alloy and steel in Figure 13, the generation of a second harmonic is a useful quantity to determine the remaining life of the specimen. Experiments using these materials are presently underway which are aimed at understanding the responsible mechanism(s).

Conclusions

The results on second harmonic generation in Al and Al alloys can be summarized as follows:

1. Theory and experiments demonstrate that acoustic harmonic generation in bulk waves could be a useful tool to nondestructively test the flow stress and the state of fatigue as long as processes are involved in which the free dislocation loop length is changed. Using an Al single crystal as a model system, theoretical semi-quantitative correlations have been verified. Limitations on the applicability of the method in the case of Al alloys have been studied. Alloy additions apparently repin the dislocations

immediately due to a low activation energy of motion and no effect can be observed.

2.   In materials where acoustic harmonic generation is sensitive to fatigue, the major part of the change occurs in the very early part of the fatigue life concomitant with a fatigue hardening or softening effect.   In the later part of the fatigue life where microcrack initiation and propagation takes place (saturation stage of fatigue), changes of the bulk properties are minor so that harmonic generation is affected very little.

Future work will continue as follows:

1.   Earlier studies on Ti-alloys and steels [14] using bulk wave experiments will be repeated.

2.   Surface wave measurements will be initiated to test its feasibility to monitor microcrack initiation, propagation, and coalescence (forming the macrocrack) in the saturation stage of fatigue where bulk wave harmonic generation becomes stationery.

ACKNOWLEDGEMENT

This research was sponsored by the Center for Advanced NDE operated by the Science Center, Rockwell International, for the Advanced Research Projects Agency and the Air Force Materials Laboratory under Contract F33615-74-C-5180.

REFERENCES

1. Kochendörder, A., Plastische Eigenschaften von Kristallen und metallischen Werkstoffen, Springer Verlag, Berlin, 1941.

2. Warren, B.E., X-ray Diffraction, Addison-Wesley Publishing Company, Reading, Mass., 1969.

3. Proceedings of a Workshop on Nondestructive Evaluation of Residual Stress, Report NTIAC-76-2, Nondestructive Testing Information Analysis Center, Southwest Research Institute, San Antonio, Texas (1975).

4. Gardner, C.G., Ref. 3, p. 211.

5. Cullity, B.D., Ref. 3, p. 227.

6. Williams, R.S. and J. Flora, Ref. 3, p. 197.

7. Kuruzar, M.E. and B.D. Cullity, Int. J. Magn., 1 (1971) 323.

8. Thompson, R.B., 1975 Ultrasonics Symposium Proceedings (IEEE, N.Y., 1975), p. 633, and Thompson, R.B., SC-PP-76-101, in preparation for publication.

9. Thompson, R.B., in Proceedings of the ARPA/AFML Review of Quantitative NDE, AFML-TR-75-212, p. 813.

10. Thompson, R.B., Ref. 3, p. 219.

11. Thompson, R.B., Appl. Phys. Letters, 28 (1975) 483.

12. Bozorth, R.M., Ferromagnetism, D. Van Nostrand, Inc., Princeton, 1951.

13. Thompson, R.B., Interdisciplinary Program for Quantitative Flaw Definition Special Report Second Year Effort, Science Center, Rockwell International, Thousand Oaks, Cal. (1976) 352.

14. Yermilin, J.V.,L.K. Zarembo, V.A. Krasil'nikov, Ye. D. Mezintsev, V.M. Prokhorov and K.V. Khilkov, Phys. Met. Metallogs., 36 No. 3 (1973) 174.

15. Kröner, E., Kontinuumstheorie der Versetzungen und Eigenspannungen, Springer Verlag, Berlin-Gottingen-Heidelberg (1958).

16. Nabarro, F.R.N., Theory of Crystal Dislocations, Clarendon Press, Oxford (1967).

17. Nabarro, F.R.N., Advances in Physics, 13 (1964) 193.

18. Buck, O., IEEE Transactions on Sonics and Ultrasonics, SU-23 (1976) 283.

19. Thompson, R.B., O. Buck and D.O. Thompson, J. Acoust. Soc. of Am., 59 (1976) 1087.

20. Hikata, A., B.B. Chick and C. Elbaum, Appl. Phys. Letters, 3 (1973) 195.

Chapter 15

OPTICAL PROBING OF ACOUSTIC EMISSION WAVES

C. H. Palmer and R. E. Green, Jr.

The Johns Hopkins University

Baltimore, Maryland

ABSTRACT

Piezoelectric transducers are commonly used to monitor
acoustic emission events. However, the information they yield is
modified by the response characteristics of the transducers themselves, as well as by those of the medium which acoustically
couples the transducer to the workpiece. Optical probing methods
offer many advantages over piezoelectric transducers for monitoring
acoustic emission events. Among these advantages are: (1) optical
probes require no contact with the specimen and, obviously, no
couplant; (2) they do not disturb the acoustic waves being
detected; (3) they have broad frequency response limited only by
the electronic amplifier and they are free from mechanical resonances; (4) because they can be absolutely calibrated, they measure
actual acoustic displacements; and (5) they can measure the instantaneous local displacements within a few hundredths of a millimeter
of a crack or other emission source. It is the purpose of the
present chapter to present a comparative evaluation of optical
methods for probing acoustic emission signals and to describe an
acoustic emission interferometer which is superior in many respects
to the other methods.

INTRODUCTION

Elastic waves play a major role in the non-destructive testing
of solid materials. The waves may be artifically generated by
transducers or naturally generated by the material itself under
stress. In the former case, pulses of ultrasonic waves are directed

through the material or across its surface, and their travel time
and attenuation measured [1-25]. Travel time measurements determine
the thickness and locate flaws in the workpiece. Attenuation measurements indicate alterations in microstructure, and may be used
to predict impending failure. Detection of these waves is greatly
simplified by the narrow range of frequencies involved, the sinusoidal nature of the waveform, the known pulse width, and the constant
pulse repetition rate.

Naturally generated acoustic emission waves, on the other hand,
exist over a broad range of frequencies, possess a variety of waveforms and pulse widths, and occur at apparently random time intervals
[26-31]. They propagate either as bulk waves or as surface waves,
both of which are detected by the transducer as surface disturbances.
To optimize the usefulness of acoustic emission monitoring for nondestructive testing applications, a detailed knowledge of the nature
of the emissions is required. This knowledge would permit correlation of the characteristics of the acoustic emissions with the
mechanism from which they originate.

The phenomenon of stress wave release from mechanically induced
microstructural alterations in solid materials, commonly known as
acoustic emission, has been the subject of an ever increasing number
of scientific investigations and technological applications for the
past 20 years. Nevertheless, despite innumerable measurements of
acoustic emission signals from a large variety of sources, the
amplitudes, frequency spectrum, and propagational characteristics
associated with specific sources remain substantially unknown. One
reason for the lack of proper understanding of the phenomena is
oversimplistic theoretical analyses based on unsubstantiated assumptions about thh nature of the acoustic emission generation, propagation, and detection processes. Figure 1 illustrates the elementary
manner in which acoustic emission is usually treated. The figure
shows acoustic emission signals propagating away from internal and
surface sources, which are subsequently detected by a transducer
located on the surface of the test piece. Note that, in this commonly used portrayal, the internal source emits a spherical wave
which propagates at constant speed in all directions. Its amplitude
decreases inversely with distance from the source so that the total
energy remains fixed. The surface source emits part of a spherical
wave which radiates into the interior with an inverse distance
amplitude decrease and, in addition, emits a Rayleigh wave which
propagates uniformly in all directions over the surface. The
surface wave amplitude decreases as the square toot of the distance
from the source.

Figure 2 illustrates schematically a somewhat more realistic
view of acoustic emission, although even this portrayal is oversimplified. In the case shown, the internal source emits a wave
which is non-spherical initially because of the shape of the

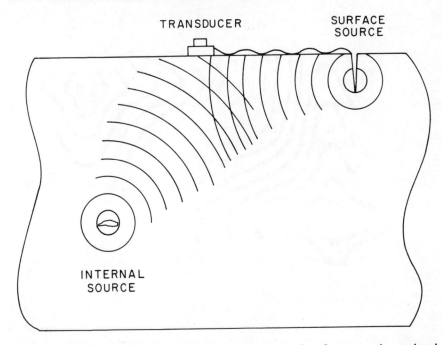

Figure 1.   Oversimplified schematic portrayal of acoustic emission
            bursts.

material defect which causes the emission. After leaving the
vicinity of the defect, the profile of the wavefront continues to
change because of the directional variation in wave speeds associated
with linear elastic wave propagation in anisotropic solids. Moreover,
the amplitude of the wave decreases with increasing distance from
the source, both as a result of the expanding wavefront and as a
result of attenuation caused by a number of mechanisms which are
active in real materials. Among these mechanisms are thermoelastic
effects, grain boundary scattering, acoustic diffraction, dislocation
damping, interaction with ferromagnetic domain walls, and scattering
by point defects. One possible result of this more realistic behavior is illustrated by comparison between Figures 1 and 2. In
Figure 1 it can be seen that the first acoustic emission signal
detected at the transducer from the internal source is due to the
spherical body wave arriving directly from the source. In Figure 2
it can be seen that the first signal to reach the transducer from
the same internal source has taken a longer path. Due to the anisotropy of the medium, the first portion of the body wave emitted by
the internal source to arrive at the specimen surface is that portion which propagates vertically from the internal source to the
surface. Upon arriving at the surface, this fast travelling wave
is partially converted to a Rayleigh wave which propagates along

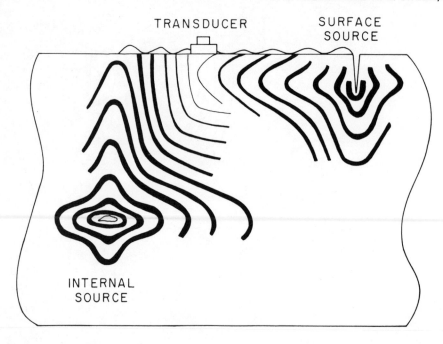

Figure 2.   Somewhat more realistic schematic portrayal of acoustic
           emission bursts.

the surface to the transducer. A source locating system based on
the first arriving signal being the direct body wave would give an
erroneous location for the source as would a source locating system
which assumed that the detected surface wave was due to a surface
source.

   In a similar manner, comparison between Figures 1 and 2 shows
that differences exist in location and identification of the surface
source depending on the physical model assumed. From the point of
view of Figure 1, the first acoustic emission signal detected at
the transducer from the surface source is due to the spherical body
wave arriving directly from the tip of the source, while from the
point of view of Figure 2, the first signal detected is close to
the fast travelling portion of the body wave which propagates hori-
zontally from the tip to the surface source. Since in the two
views the wave speeds and hence the travel times are different, a
source location system based on a given travel time will yield
different results.

   Other parameters which also influence the acoustic emission
signals detected on the surface of real materials include the fact
that actual sources generate signals which possess a distribution

of frequencies and amplitudes; these waves propagate with different
wave speeds, even in a single direction, and the higher frequency
components experience stronger attenuation than the lower frequency
components. Moreover, the elastic waves experience diffraction and
refraction (energy-flux deviation) mode conversion at boundaries,
and those with sufficient amplitude will be subject to non-linear
effects.

Since conventional piezoelectric transducers require a coupling
medium between transducer and test specimen, the signal detected
will be modified by the acoustical properties of this medium and by
the problems associated with non-uniformity in composition and thickness of the couplant. The necessity of using such a couplant also
prevents detection of acoustic emission signals at high temperatures.

Finally, the character of the piezoelectric transducer itself
exerts a major influence on the components of the acoustic emission
signal recorded, since conventional transducers do not respond as a
simple vibrating piston and have their own frequency, amplitude, and
directional response. In addition, the "ring" at their resonance
frequency and it is impossible to distinguish between the amplitude
excursions caused by this "ringing" and the amplitude variations
actually characteristic of the acoustic emission source.

Optical methods have proven to be ideally suited for probing
the detailed characteristics of ultrasonic waves [32-38], and can
be successfully adapted to probing the detailed characteristics of
acoustic emission waves as well. Optical methods permit the measurement not only of instantaneous local displacements of a surface as
a function of time, but also wave motions and disturbances within
transparent media. Furthermore, optical probes do not disturb the
waves being measured. There is also the possibility that acoustic
emission waves can be measured remotely in hostile high temperature
or radioactive environments.

There exists a variety of optical techniques developed for
specific problems, and we consider in the next two sections the
suitability of these techniques to the particular problems of
acoustic emission measurements.

COMPARISON OF OPTICAL TECHNIQUES

Assumptions and Problems

As a basis for the selection of the most promising optical
technique for the study of acoustic emission waves, we must make
reasonable assumptions about the nature of the phenomenon and about
the environment in which the measurements are to be made. We shall

assume the acoustic emission event is an elastic wave packet which
arises within a material as a stress relief mechanism and which,
upon reaching the surface, is at least partially converted to a
spreading Rayleigh wave group. Based on prior measurements with
commercially available acoustic emission transducers, we assume
that the important frequencies involved range from 10 KHz to 1 MHz.
The corresponding Rayleigh wavelengths are 280 mm to 2.8 mm on the
surface of aluminum. The final assumption is that typical acoustic
emission amplitudes lie in the range 1 to 100 Å (10^{-10} to 10^{-8} m).
It should be emphasized that these assumptions are made initially
as a basis of comparison of probing techniques. It is the purpose
of the research currently in progress to actually measure acoustic
emission events throughout their entire history and to determine
their actual frequencies, amplitudes, waveforms and wavelengths.

For a valid comparison of optical techniques we must also
consider the environment in which the measurements are likely to be
made. The chief problems are atmospheric turbulence, building
vibrations and instrumental vibrations. Atmospheric disturbances
generally involve minor changes in pressure and thus changes in
optical path which occur primarily at frequencies of perhaps 10 Hz
or less. Building vibrations range in frequency from perhaps 5 Hz
to, say, 50 Hz. Vibration isolation tables, for example, are
designed with a resonance frequency of 1 or 1.5 Hz to effectively
damp out such vibrations. Instrumental vibrations may include 60
Hz or 120 Hz from transformers (a very substantial effect), and
vibrations of a rigid instrument (shock excited in some way) mostly
at frequencies from roughly 1 to 5 KHz. All of these vibrations
may have amplitudes of the order of a micron or more. Thus the
measurement technique must sense 1Å movements at 0.1 to 1 MHz and
reject motions of perhaps 10,000 Å at frequencies below 5 KHz.

Optical Methods

We first consider several optical techniques which have
proved to be well suited to the study of ultrasonic waves. Some
of these methods will be shown to be unsatisfactory for the investigation of acoustic emission phenomena because of the basic differences between the two kinds of disturbances as mentioned at the
beginning of the chapter. In addition, it is apparent from the
outset that, because we need a bandpass of roughly a megahertz,
from 0.01 to 1 MHz, for acoustic emission sensing as contrasted
to the 1 Hertz bandwidth which is adequate for many ultrasonic
studies, the ultimate minimum detectable signal will be 1000 times
larger for acoustic emission.

Knife-edge Technique. The knife-edge technique of Adler,
Korpel, and Desmares [39] is probably the simplest useful optical
technique for sensing or visualizing Rayleigh waves. The technique

was devised for investigation of single frequency ultrasonic waves
and, as such, has excellent sensitivity. Figure 3 shows the basic
principle, adapted to the study of acoustic emission waves. The
scanning method used for visualization purposes does not apply here
and is omitted. Also, since we are here concerned with a wide range
of wavelengths, it is not possible to illuminate half an acoustic
wavelength which results in best sensitivity. In the figure, a
laser beam of diameter D is focused by lens L_1, having a focal
length F_1, on the acoustically perturbed surface. The non-uniform
(Gaussian) spot size is small compared to the shortest acoustic
wavelength. Light reflected from the unperturbed, polished specimen
surface is recollimated by lens L_2, having a focal length F_2. The
knife-edge, K, obstructs half the beam and allows the remainder to
be focused by lens L_3 on the photosensor S.

If the specimen surface is tilted by an acoustic disturbance
through an angle θ, as shown, the reflected beam is deviated through
an angle 2θ so that either more or less of the optical beam is ob-
structed by the knife-edge, and a signal results. Specifically, the
recollimated beam is displaced a distance $\Delta y = 2\theta F_2$, normal to the
knife-edge -- assuming the sound waves travel in a direction perpendicular to the knife-edge as implied in the insert in the figure.

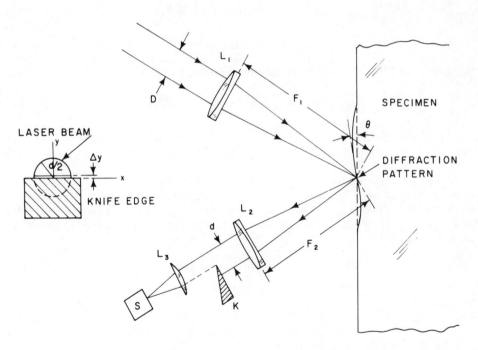

Figure 3. Simplified knife-edge technique.

Let the laser beam amplitude at the knife-edge have a Gaussian dependence

$$\tilde{u}(x,y) = (A_o/d) \sqrt{2/\pi} \exp [-(x^2+y^2)/d^2] \qquad (1)$$

where A_o is an amplitude, d the waist diameter as in the figure, and x and y are coordinates parallel and perpendicular, respectively, to the knife-edge. The total beam power is

$$P_{total} = \int_{-\infty}^{\infty}\int_{-\infty}^{\infty} |\tilde{u}(x,y)|^2 \, d x \, dy = A_o^2 = P_i \qquad (2)$$

where P_i is the incident beam power. The optical power represented by the strip Δy wide and d long (a strip across the center of the laser beam), i.e., the signal power, is

$$P_{sig} = \Delta y \int_{-\infty}^{\infty} |\tilde{u}(x,o)|^2 \, dx \qquad (3)$$

$$= \sqrt{2/\pi} \, P_i \, \Delta y/d$$

Thus the signal power, in terms of the surface tilt, is

$$P_{sig} = \sqrt{2/\pi} \, P_i \, 2\theta F_2/d = \sqrt{2/\pi} \, P_i \, 2\theta F_1/D \qquad (4)$$

since $d = (F_2/F_1)D$. Thus the signal power is independent of F_2, and any focal length will suffice. The photodiode signal current is proportional to the optical signal power,

$$i_{sig} = \alpha P_{sig} = \alpha \sqrt{2/\pi} \, P_i 2\theta F_1/D \qquad (5)$$

where α is the photodetector sensitivity in amperes current per watt optical power.

We can now express the signal current in terms of the acoustic wave properties. Let the sound wave propagate in the z direction with an amplitude δ_o, a radian frequency Ω_a, and a propagation constant $K = 2\pi/\Lambda$ with Λ the acoustic wavelength. The displacement is then

$$\delta = \delta_o \cos (\Omega_a t - Kz) \qquad (6)$$

The small surface tilt, θ, is

$$\theta = \partial\delta/\partial z = K \delta_o \sin (\Omega_a t - Kz) \qquad (7)$$

and the rms angle is:

$$\theta_{rms} = K \, \delta_o/\sqrt{2} = \sqrt{2} \, \pi \, \delta_o/\Lambda \qquad (8)$$

Thus the signal current, Equation (5), is given by

$$i_{sig} = \alpha \, \sqrt{2/\pi} \, P_i \, 2(F_1/D) \, \sqrt{2} \, \pi \, \delta_o/\Lambda$$

or

$$i_{sig} = \alpha \, P_i \, 4 \, \sqrt{\pi} \, (F_1/D) \, (\delta_o/\Lambda) \quad (rms) \qquad (9)$$

We can calculate the theoretical minimum detectable wave height with the knife-edge technique by noting that the ultimate limiting factor is predominatly the shot noise in the detector. This shot noise has an rms value $i_{noise} = \sqrt{2eBI}$ where e is the electronic charge, B the smplifier bandwidth, and I the steady dc photocurrent. Since the average light power on the detector is half P_i, $I = \alpha \, P_i/2$, and

$$i_{noise} = (2 \, e \, B \, \alpha \, P_i/2)^{1/2} \quad (rms) \qquad (10)$$

The smallest detectable signal is taken to be that for unity signal-to-noise ratio. If we set $i_{sig} = i_{noise}$, we find

$$\delta_{min} = [\frac{2eB}{\alpha \, P_i}]^{1/2} \frac{D}{F_1} \frac{\Lambda}{4 \, \sqrt{2\pi}} \qquad (11)$$

Reasonable values for the parameters are B = 1 MHz, α = 0.4 amp/watt, P_i = 1 mwatt, D = 1.6 mm, and F_1 = 100 mm. Calculation of the minimum detectable wave heights gives

$$\delta_m = 1.3 \times 10^{-6} \, mm = 13 \overset{o}{A} \, at \, \Lambda = 28 \, mm, \, f = 100 \, KHz$$

$$\delta_m = 13 \times 10^{-8} \, mm = 1.3 \overset{o}{A} \, at \Lambda = 2.8 \, mm, \, f = 1 \, MHz$$

The knife-edge technique, then does have the advantage of simplicity and it can be adapted to sensing bulk waves as well as surface waves, but it clearly lacks sensitivity at the lower frequencies. Since the method measures slope, the signal must be integrated to obtain the actual waveform. Furthermore, the technique will not sense waves moving in a direction at right angles to the knife-edge, and acoustic emission waves may come from any direction.

Diffraction Technique. Figure 4 illustrates the basic concept of the diffraction technique [40-44] which has proved so very effective in the study of ultrasonic waves - both for surface waves and

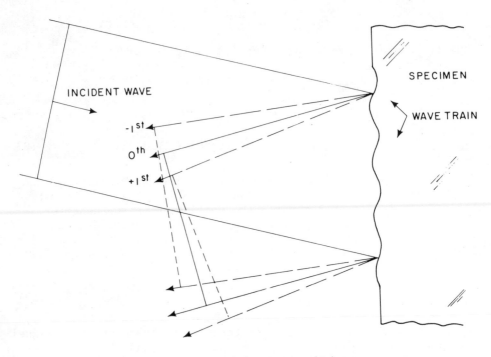

Figure 4. Diffraction technique.

for bulk waves. The principle is that the periodic distortion of
the surface in the case of Rayleigh waves or the periodic change in
refractive index for bulk waves creates an optical phase grating
which then disperses the incident collimated laser light into
spectral orders. Unfortunately, the technique is not applicable
to the study of waves having a wide range of frequencies and a
short wave train because the phase grating in such cases is not
periodic. The laser beam is still spread out, but in a rather
meaningless way. The angular order separation, $\theta = \lambda/\Lambda$ (that is,
the ratio of optical to acoustic wavelength), varies from about
22 microradians at $\Lambda = 28$ mm to 220 microradians at 2.8 mm. Since
the wave train is short, even for a single frequency the spectral
orders would be rather poorly defined. Clearly the method is
unsatisfactory for our purposes.

Optical Heterodyning. An interesting optical probing method
is based on the Doppler frequency shift produced in a laser beam
reflected from an acoustically perturbed surface. To avoid the
problem of "negative frequencies", it is convenient to frequency
shift the optical beams and mix them on the photosensitive surface
to obtain a heterodyne signal. Figure 5 illustrates the technique
of Whitman, Laub and Bates [45]. Frequency shifting is produced

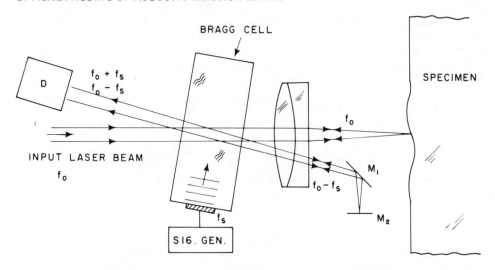

Figure 5.   Optical heterodyne technique.

by a Bragg cell in which bulk sound waves of frequncy f_s (23.5 MHz in Whitman's case) are generated at one end and absorbed at the other so as to make a moving optical phase grating.   The undeviated zero order optical beam emerges with no frequency change.   The positive first order, diffracted downward, has a frequency $f_o - f_s$, where f_o is the laser frequency; the negative first order, diffracted upward, has a frequency $f_o + f_s$.

In Whitman's system, the zero order laser beam, frequency f_o, is focused on the specimen surface and reflected.   The return beam has a frequency $f_o + f_{Doppler}$.   Upon passing through the Bragg cell again, the minus first order is selected, and the light incident on the photodetector has a frequency $f_o + f_s + f_m$.   The downshifted order, frequency $f_o - f_s$, for the input laser beam, is reflected by mirrors M_1 and M_2 and returns through the Bragg cell where only the zero order, still frequency $f_o - f_s$, is allowed to reach the photodetector.   The photodetector output contains a variety of frequency components including $2f_s \pm f_a$ which is electronically selected and gives information proportional to the square of the acoustic amplitude.

One difficulty with any Doppler sensing method is that the output signal contains information about the velocity of the surface, not the displacement.   Therefore electronic integration of the signal, containing a number of frequency components, must be performed. A further difficulty in the investigation of acoustic emission is that a very small amplitude, shortlived, randomly occurring disturbance must be measured while at the same time rejecting low frequency, random motions of very large amplitude (resulting from mechanical vibrations of the system). Both the acoustic emission and the mechanical vibrations produce Doppler shifts proportional to the velocity of the surface.

Although the heterodyning technique is sufficiently sensitive, there are clearly some serious difficulties to be overcome.

Optical Interferometry. Since optical interferometers measure optical path differences, they can be used to measure directly either instantaneous surface displacements as a function of time or position, or, alternatively, to measure the refractive index changes associated with dilatational or shear waves. Interferometers are not only very sensitive, but they can be absolutely calibrated. The primary difficulty in many applications is that their sensitivity for very small path changes depends on the operating point, being maximum when the optical output power is half the input power and zero at either the minimum or maximum power. There are various solutions to this operating point problem and each one has certain advantages. We discuss the application of interferometers to the study of acoustic emission in the next section.

COMPARISON OF INTERFEROMETERS

Two Beam Interference and Pathlength Sensitivity

The fundamental drawback of interferometric methods is the pathlength sensitivity problem. Figure 6 shows the essential optics of a Michelson two beam interferometer (minus the usual compensator plate which is needed to obtain white light fringes). Let a colli-mated laser beam be incident on the beam splitter BS as shown. Assume the splitter is lossless and that it divides the incident power P_i so that half this power is reflected toward mirror M_1 and half is transmitted to mirror M_2. The beams incident on these two mirrors thus have amplitudes $E_i/\sqrt{2}$ where E_i is the incident beam electric field amplitude. To describe the interference fringes we may ignore the common input and output paths which contribute only a common phase shift to the two beams and have no effect on the fringes. The beam reflected from mirror M_1 and returning to the beam splitter has an instantaneious electric field given by

$$E_1 = (E_i/\sqrt{2}) \exp \left[j(\omega t - 2kx_1 + \phi_1') \right] \tag{13}$$

in this equation x_1 is the distance between the center of the beam splitter surface and mirror M_1, k is the propagation constant in air, and ϕ_1' is a phase shift resulting from reflection at the beam splitter. For simplicity we assume that the beam splitter is

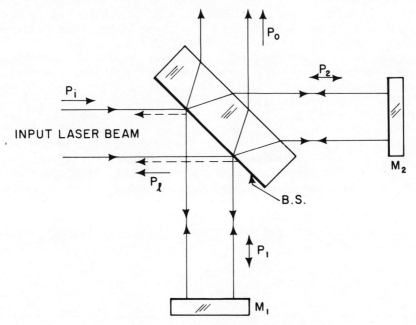

Figure 6.   Michelson interferometer optics.

exactly 45° from the incident ray direction and that mirror M_1 is
exactly normal to the incident light beam.   For the second beam
transmitted to mirror M_2, we have a distance x_2 from the rear surface of the beam splitter to the mirror (also assumed exactly
normal to the beam), and a traversal in glass through a distance s.
(We can calculate s for a given beam splitter, but we do not need
to know it.)   The instantaneous electric field describing this
second beam upon returning to the front surface of the beam splitter
is

$$E_2 = (E_i/\sqrt{2})\ \exp\ [j(\omega t - 2kx_2 - 2k's + \phi_2')] \qquad (14)$$

where ϕ_2' is the phase shift (= 0) on transmission through the beam
splitter, and k' is the propagation constant in glass.   The output
beam, after leaving the beam splitter, has an instantaneous electric
field given by the superposition of the two components:

$$E_{10} = (E_i/2)\ \exp\ [j(\omega t - 2kx_1 + \phi_1)] \qquad (15)$$

and

$$E_{20} = (E_i/2)\ \exp\ [j(\omega t - 2kx_2 - 2k's + \phi_2)] \qquad (16)$$

where ϕ_1 and ϕ_2 include additional phase shifts produced by the beam splitter at the second arrival of the light beams. The amplitudes are also reduced by another factor of $\sqrt{2}$ at the beam splitter. For the output beam we see that ϕ_1 is the phase shift produced by one external reflection (from a less dense to a more dense optical medium) and one transmission; ϕ_2 is the phase shift produced by one transmission and one internal reflection. According to Stokes treatment of reflection and transmission [46] internal and external reflection differ by π in phase shift, so that with otherwise equal optical path lengths to mirrors M_1 and M_2 there will be destructive interference. For the unwanted output beam directed toward the laser source, the equivalent phase shifts at the beam splitter will be ϕ_1'' involving two external reflections and ϕ_2'' involving two transmissions which together correspond to constructive interference. Thus the desired and undesired output beams produce complementary interference patterns (and thus conserve energy). The output beam then is described by

$$E_o = E_{10} + E_{20} = (E_1/2)e^{j\omega t}(e^{j\theta_1} + e^{j\theta_2}) \qquad (17)$$

where

$$\theta_1 \equiv -2kx_1 + \phi_1 \quad \text{and} \quad \theta_2 \equiv -2kx_2 - 2k's + \phi_2$$

The corresponding optical power, $P_o = A(E_o \cdot E_o^*)$ where A includes the reciprocal wave impedance. Then

$$P_o = (AE_i^2/2)[1 + \cos(\theta_2 - \theta_1)]$$

or

$$= (P_i/2)[1 + \cos(\theta_2 - \theta_1)] \qquad (18)$$

The unused power, P_ℓ, directed toward the laser is

$$P_\ell = (P_i/2)[1 + \cos(\theta_2 - \theta_1 + \pi)] \qquad (19)$$

so that $P_o + P_\ell = P_i$ as expected. We can express P_o more conveniently; let

$$\theta_2 - \theta_1 = -2k(x_2-x_1) - 2k's + \phi_2 - \phi_1 \equiv 2k\delta - \phi \qquad (20)$$

where δ is the physical displacement of mirror M_2 from its equilibrium position, and ϕ a phase difference we can set by adjusting x_1 or x_2. Thus

$$P_o = (P_i/2)[1 + \cos(2k\delta - \phi)] \qquad (21)$$

The optical power directed back into the laser is also given by
Equation (21) when we include the additional phase shift of $-\pi$ as
described above; this power varies with x_2 and thus may pull the
laser frequency or affect its power. Thus, in some cases it is
desirable to use an optical isolator (a combination of a polarizer
and a quarter wave plate) to block the disturbing return beam.
Since the objective is to measure small changes in δ, $\delta \ll \lambda$, we
see that the sensitivity is

$$dP_o/d\delta = + P_i \; k \; \sin \phi = (2\pi P_i/\lambda) \sin \phi \qquad (22)$$

That is, the sensitivity is directly determined by ϕ or the setting
of the reference beam mirror. Thus for optimum sensitivity we must
make ϕ and odd multiple of $\pm \pi/2$. There are several possible
solutions.

To get an idea of the magnitude of the path-length problem
resulting from atmospheric effects let us suppose that the distances
x_1 and x_2 are 150 mm ($\approx 6''$). The presence of air increases the
optical path for either beam by $2x_1(N-1)$ where N is the refractive
index of air. Under standard conditions $N \approx 1.00028$ for red lights
hence we find the optical path change is 0.084 mm or 133 wavelengths
or 266 fringes at 6328 Å. This difference is not very important
unless the path lengths x_1 and x_2 differ significantly. For example,
if one path used a plane mirror near the beam splitter rather than
a lens as in Figure 8, small atmospheric pressure changes would have
a very serious effect. However, even if $x_1 \approx x_2$, any atmospheric
changes which affect the paths differently will have serious effects.

Interferometer Systems

Differential Interferometer. One solution to the pathlength
sensitivity problem discussed in the previous section is differential interferometry [47]. Figure 7 illustrates the basic principles.
Two coherent point sources of light, s_1 and s_2, are derived from a
single laser using some sort of beam divider (several methods have
been used). Light from these sources is collimated by L_1 thus
forming two parallel beams making an angle of 2θ with each other.
In regions of overlap of the beams, interference fringes are
formed. Lens L_2 focuses the two sources at points A and B on the
specimen; this lens is made varifocal so that the separation of A
and B may be made just one-half acoustic wavelength. The specimen
surface is optically polished and reflects the light back through
lens L_2. The recollimated light again forms an interference fringe
pattern which is reflected onto the square wave (Ronchi) grid. If
the specimen as a whole moves toward or away from lens L_2, the
fringe pattern is unchanged; if the atmospheric pressure changes,
no shift is seen either. However, if a surface acoustic wave

passes under the probe, the light focused at A travels a shorter
distance (by twice the SAW amplitude) to the grid than does the
light focused at B. The optical fringes shift in the x direction
or across the grid accordingly, and the light flux transmitted by
the grid thus varies. Details are fully described in reference 47.
Here we merely state that the interferometer is highly sensitive –
it can detect 6×10^{-4} Å SAW on glass, it is unaffected by focal
errors up to $\pm$ 1 mm, it does not require an optically flat specimen
surface, it can measure dilatation or shear waves, and it yields
both amplitude and phase information. Nevertheless, the instrument
cannot measure acoustic emission waves because the probe points A
and B must be approximately half an acoustic wavelength apart and
acoustic emission bursts have a variety of wavelengths. The technique does, however, have application in the accurate calibration
of piezoelectric and other transducers at given frequencies.

Quadrature Dual Interferometer. A beam splitting interferometer whose output is essentially independent of the reference
optical path length was suggested as long ago as 1953 by Peck and
Obetz [48]. Recently it has been used to measure pulse picometer
displacement vibrations by Vilkomerson [49]. Suppose the input
beam of any standard two beam interferometer is plane polarized
in a direction $\pm$ 45° with respect to the horizontal. Let the
horizontal component give fringes described by Equation (21), that
is:

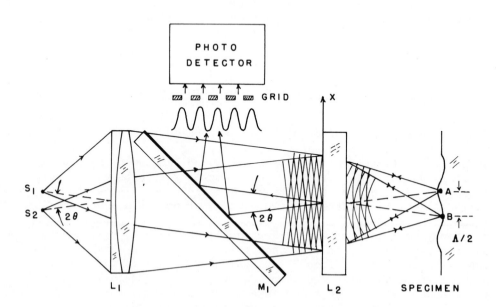

Figure 7. Differential interferometer.

$$P_H = (P_{iH}/2) \ [1 + \cos \ \{2k(x_2 - x_1) + \psi\}] \qquad (23)$$

where the subscript H means horizontal and ψ includes all phase factors except $2k(x_1 - x_2)$. Suppose x_2 varies by small amounts $\Delta x_2 \ll \lambda$ at high frequencies, then the optical power changes are given by

$$P_H = - P_{iH} \ k \ [\sin 2k \ \{(x_2 - x_1) + \psi\}] \ \Delta x_2 \qquad (24)$$

Suppose we have included in the reference path a $\lambda/8$ wave plate which retards the vertical component phase by $\pi/4$ with respect to the horizontal component. The returning vertical component is then delayed by $\pi/2$ in phase with respect to the horizontal component. Thus the vertical beam power is

$$P_v = (P_{iv}/2) \ [1 + \cos \ \{2k(x_2 - x_1) + \psi - \pi/2\}]$$

$$(P_{iv}/2) \ [1 + \sin \ \{2k(x_2 - x_1) + \psi\}] \qquad (25)$$

and the change in the vertical component is

$$\Delta P_v = P_{iv} k \ [\cos \ \{2k(x_2 - x_1) + \psi\}] \ \Delta x_2 \qquad (26)$$

where the subscript v means vertical. Also let $P_{iv} = P_{iH}$. If we use a polarizing beam splitter at the output, we can separate the two components and direct them to separate, identical photodetectors. The two detected optical signals are electronically squared and added to give the final output signal, thus

$$V_{out} \propto P_i^2 k^2 (\Delta x_2)^2 = P_i^2 \ (2\pi/\lambda)^2 \ (\Delta x_2)^2 \qquad (27)$$

independent of either x_1 or x_2. The system, therefore, maintains maximum sensitivity for any position of the reference beam mirror, and low frequency vibrations are of no consequence when detecting high frequency signals - a simple high pass filter removes the low frequency components before they have any effect. The disadvantage of the system is, clearly, that phase information is lost in this arrangement. Accordingly, we do not use this elegant technique for our preliminary investigations.

Swept Path Interferometer. Consider now an ordinary Michelson interferometer modified as shown in Figure 8 by the addition of lenses L_1 and L_2 which focus the collimated beams on the reference mirror M_1 and the reflecting surface of the specimen, M_2 as shown. The reference mirror is driven by a piezoelectric unit to which a sawtooth (ramp) voltage is applied [50]. Suppose the amplitude of the periodic motion is just $\lambda/4$, then at some time during every sweep the phase difference ϕ [Equation (22)] will be just $\pi/2$ and the interferometer sensitivity for small vibrations Δx_2 will be maximum. If the maximum value of the high frequency signal is observed, this system is also independent of the static path or low frequency difference $x_2 - x_1$ and the system can be used to measure ultrasonic waves. The objection for acoustic emission waves is rather evident – the acoustic emission wave is just as likely to pass under the optical probe at a time of low sensitivity as at one of high sensitivity. The chance of observing the desired wave is small.

Stabilized Optical Path Interferometer. Ideally, in the absence of any high frequency disturbance to be measured, the optical power incident on the photodetector should be constant, equal to $P_i/2$, and should thus produce a fixed DC voltage across the load resistor of the photodetector. The actual load voltage fluctuates at low frequency as a result of disturbances in the optical path. We can compare this fluctuating voltage with a fixed reference voltage equal to the ideal voltage to generate an error signal which, when suitably amplified, will adjust the piezo-electric driven mirror, M_1 in Figure 8, so as to maintain the correct

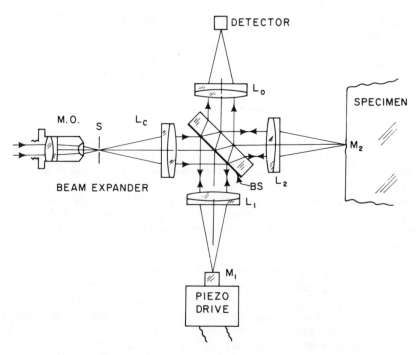

Figure 8. Modified two beam interferometer.

operating point. The correction circuit does not respond to high frequencies which constitute the signal to be measured, and these high frequencies are separately amplified and displayed or recorded. The construction and performance of such an interferometer is described in the next section.

STABILIZED OPTICAL PATH INTERFEROMETER

Optical Arrangement

The essential optical system is that shown in Figure 8. The input laser beam is expanded by a microscope objective, M.O., and collimating lens L_c. If desired, we can include a spatial filter S to reduce laser noise and improve the beam. Lenses L_1 and L_2 and mirrors M_1 and M_2 have already been described. Lens L_o focuses the output optical power on the silicon photodiode. This optical system is mounted on a substantial aluminum base plate, 14" x 16" x 3/4" resting on three sturdy supports. The 2 milliwatt laser is mounted below the platform and its light is reflected upward through a hole in the plate and again through the microscope objective. Not shown in the figure is an aperture in front of L_o which limits the beam incident on the photodetector; its purpose is described below. The lenses are clamped to rails attached to the platform so that they can be easily adjusted. Accurate focusing of lenses L_c and L_1 can be readily achieved using the beam splitter plate and observing shearing fringes as described by Lurie [51]. Lens L_2 is focused by observing the interference pattern near L_o.

Tilt and Focus Effects

Measurements on a slowly deforming specimen may cause two problems. First, the specimen may move so as to change the optical path x_2. Although the piezoelectric drive can correct for this change by maintaining the desired operating point, it does not control path x_2. As illustrated in the upper part of Figure 9, a displacement of the specimen (mirror M_2) from the focal point F causes the reflected beam to have a spherical wavefront which inter- feres with the plane wavefront from mirror M_1, giving a ring pattern equivalent to Newton's rings.

If a ray is incident on L_2 from the left (above the axis, upper part of Figure 9) it is refracted toward the foacl point F, but reflected through P by the mirror (distant Δ from F) before it reaches F. The ray emerges from L_2 as if it had come from point Q. The effective object distance S_1, measured from the first principal plane H_1 of the lens, is $S_1 = f - 2\Delta$. The radius of curvature of the returning wavefront, S_2, measured from the second principal plane, H_2, is:

$$S_2 = \frac{S_1 f}{S_1 - f} = \frac{(f - 2\Delta)f}{-2\Delta} \simeq \frac{-f^2}{2\Delta} \qquad \Delta \ll f \qquad (28)$$

We are interested in the distance ε between the spherical wavefront from mirror M_2 and the plane wavefront from mirror M_1. (Any uniform phase difference is corrected by the piezoelectric drive.) The maximum ε for a slightly curved wavefront, i.e., on small Δ, is $\varepsilon \simeq h^2/S_2$ where h is the beam radius. If we substitute for S_2 and solve for Δ, we find

$$\Delta \simeq - \varepsilon f^2/2h^2 = -2\varepsilon f^2/D^2 \qquad (29)$$

where D is the beam diameter.

The second problem caused by a deforming specimen is that the surface (i.e., mirror M_2) may tilt with respect to the normal as illustrated in the lower part of Figure 9. If the surface tilts through angle θ, the returning, recollimated beam is displaced transversely a distance $\gamma = 2\theta f$, so

$$\theta = \gamma/2f \qquad (30)$$

We are here concerned with small angles and large f-numbers so the approximation is good.

The expression for Δ, Equation (29), shows that for a given tolerance in ε we want a long focal length lens and small beam diameter. The expression for θ, Equation (30), dictates a small f and a large beam diameter so that $\gamma \ll D$. These conflicting requirements to give maximum tolerance in both position error Δ and tilt error θ can be resolved by using a stop, diameter d, at the output lens, L_0, Figure 8. We keep the beam diameter D large, but make diameter d small. Then,

$$\Delta \simeq - 2\varepsilon f^2/d^2 \qquad (31)$$

$$\theta = (D-d)/4f$$

and both Δ and θ can be reasonably large. As an example, let f = 120 mm, $\varepsilon = \lambda/10$, D = 20 mm and d = 5 mm. Then Δ = 73 micrometer and θ = 1.79°. If these values of Δ and θ are insufficient, we can increase them by changing the parameters, but at further expense of optical power.

Correction Circuit

The desired optical path difference for maximum sensitivity is $x_2 - x_1 = \lambda/8 \pm n\lambda/2$ or $3\lambda/8 \pm n\lambda/2$. With our correction circuit

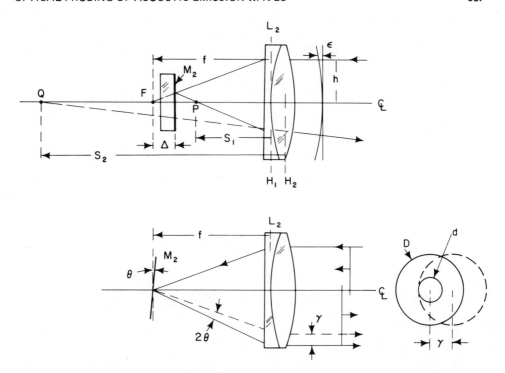

Figure 9.   Focus and tilt effects: effect of displacement of
             mirror M_2 from proper position (above); effect of tilt
             of mirror M_2 (below).

the second of these choices is unstable, and the piezoelectrically
driven mirror is fixed at $\lambda/8 \pm n\lambda/2$. Figure 10 shows a diagram
of the arrangement. The photodetector output E_i contains both a
low frequency (DC + Audio) component and an RF component. The low
frequency part (< 1 KHz) is compared with an adjustable reference
voltage, E_r, to generate an error signal E_e which charges or discharges the idling capacitor, C. The capacitor voltage, E_c,
determines the voltage output of the high voltage amplifier,
$20 < E_o < 280$ volts, and thus the position of mirror M_2 attached to
the piezoelectric device. If the voltage E_o is either too high or
too low for proper performance of the high voltage amplifier, this
over or under-range is sensed by a special switching circuit which
discharges capacitor C and resets the piezoelectric device at a
different position such that $x_2 - x_1$ changes by an integral number
of half wavelengths.

     The present circuit is designed to keep the output photo-
detector voltage E_i within a few percent of E_r, i.e., half the

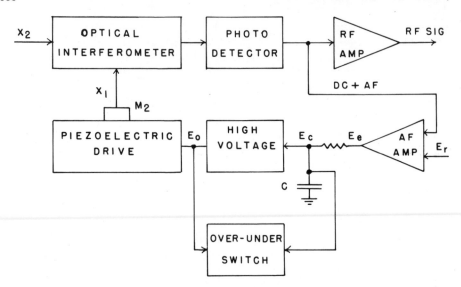

Figure 10.   Block diagram of path stabilization electronics.

maximum value of E_c. As shown by Equation (22), the interferometer
sensitivity to small signals, proportional to sin ϕ (with $\phi = \pi/2$
ideal), does not vary significantly for moderate changes in ϕ.
Thus an error of $\pm$ 10° from 90° gives less than 2% error; an error
of $\pm$ 15° less than 4%, and even $\pm$ 20° only 6%. Accordingly, it
is not necessary to have extremely tight position control in this
situation, and we allow errors of about 5% in voltage or $\pm$ 15° to
20° in phase. By allowing such errors the smaplifier gain can be
reduced and the stability improved. The only disadvantage is that
the calibration waveform varies somewhat according to the exact
operating point, but no error need result.

                              Calibration

     To obtain an absolute calibration for an interferometer such
as we have described, we substitute a special calibration unit for
the specimen (mirror M_2). The unit consists of a second piezoelectric disk and mirror which is driven by a high power sinusoidal
drive at its resonant frequency, 100 KHz, to achieve a mirror
motion $\delta = \delta_0 \cos \omega_a t$ where $\delta_0 > \lambda/8$. The resonant frequency is
chosen to be at the lower end of the electronic band pass both
because large amplitudes are easier to obtain at low frequencies
and because we can then see the desired waveforms. These waveforms, as will be shown, involve high harmonics of the drive

frequency, and with the 100 KHz fundamental, we include the ninth harmonic in the pass band (only odd harmonics exist).

Assuming the desired operating point, $\phi = \pi/2$, is maintained by the piezoelectric control system, the optical power incident on the photodetector according to Equation (22) will be

$$P_o = P_S [1 + \sin 2k\delta] \tag{33}$$

where P_S accounts for the reduction of optical power by the stop. With the drive voltage applied to the calibration unit, the RF optical power is

$$P_o = P_S \sin [(4\pi\delta_o/\lambda \cos (\omega_a t)] \tag{34}$$

where we have substituted for k and δ. The corresponding amplifier output voltage is

$$V = V_o \sin [(4\pi\delta_o/\lambda) \cos (\omega_a t)] \tag{35}$$

where the amplitude V_o depends on the optical power, the photo-detector sensitivity, the photodetector load resistance, and the amplifier voltage gain. Equation (35) shows that we need merely drive the calibration unit hard enough to make $4\pi\delta_o/\lambda > \pi/2$ for then the peak-to-peak output voltage seen on an oscilloscope will be 2 V_o. The waveform will of course be a function of δ_o, but not the peak-to-peak voltage. Some of the theoretical (and observed) waveforms are shown in Figure 11. Having determined V_o, the inter-ferometer is calibrated: Suppose $\delta_o \ll \lambda$, then we can approximate the sine function in Equation (35) by the angle itself, thus V = $V_o (4\pi\delta_o/\lambda) \cos (\omega_a t)$ or V = $V_o (4\pi/\lambda) \delta$. Hence

$$\delta = (\lambda/4\pi) V/V_o \tag{36}$$

which determines the instantaneous displacement from the rest position.

An alternative method of calibration is both more complicated and more accurate. It is fully described in the references. Essentially, voltage V is expressed as a series and we find that the fundamental frequency term is proportional to the J_1 Bessel function. Thus we can measure the fundamental frequency of V as a function of optical path change, determine the maximum value corresponding to $J_1(1.84) = 0.5819$ for which $\delta_o = 1.84\lambda/4 =$ 926.5 Å. From this information, accurate calibration for small signals is determined.

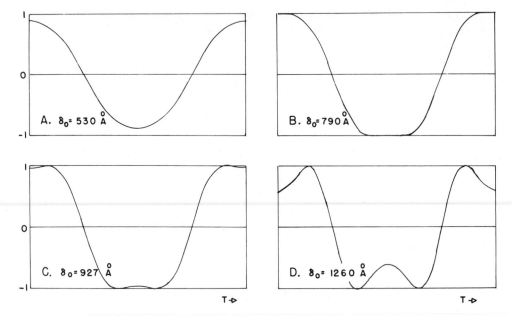

Figure 11.   Calibration waveforms using sinusoidal mirror drive.
A-amplitude below calibration level; B-calibration
level-read peak-to-peak voltage; C and D-large amplitudes still yield correct calibration voltage.

Minimum Detectable Disturbance

The theoretical minimum detectable disturbance amplitude is
calculated as in previous cases. From Equation (34) the RF optical
signal power is $P_o = P_S 4\pi\delta/\lambda$ where P_S is the power at the operating
point and we are assuming a small displacement. The noise gener-
ating optical power is P_S. Therefore the signal current is

$$i_{sig} = \alpha P_S 4\pi\delta/\lambda \tag{37}$$

and the shot noise current

$$i_n = (2eIB)^{1/2} = (2e\alpha P_S B)^{1/2} \tag{38}$$

and the minimum detectable displacement, S/N = 1, is

$$\delta_{min} = (2eB/\alpha P_S)^{1/2} \lambda/4\pi \tag{39}$$

Performance

The preliminary model of our interferometer operates
successfully. The low frequency optical path stabilization circuit
performs well and removes most of the mechanical vibration and
atmospheric turbulence effects. Calibration is easily achieved
with a piezoelectric driven calibration mirror unit. We obtain
$2V_O$ = 22 volt peak-to-peak signal at 100 KHz corresponding to the
calibration point displacement, $\lambda/8$ = 791 Å as in Equation (35).
The peak-to-peak noise voltage is approximately 0.05 volts corresponding to a minimum detectable disturbance amplitude of 1.8 Å
which is limited largely by the low light level (poor mirror
reflectance and low lens transmittance of the optical component)
and instabilities of the unpolarized laser. The optical signals
obtained in various tests indicate reliable amplitude and phase
measurements are possible.

PRELIMINARY RESULTS

Some preliminary tests were made using a cylindrical specimen
of 6061-T6 aluminum with one surface optically polished and the
rest of the surface smoothly machined. Figure 12 illustrates the
mounting of this disk. We mounted a 1/2" diameter, 1/4" thick
piezoelectric disk (PZT-5) at the center, A, of the machined surface

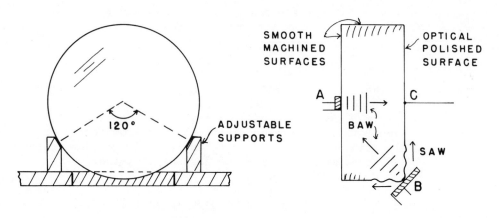

SPECIMEN: 6061-T6 ALUM.
        DIAM. 6.00"
        THICK. 2.500"

Figure 12. Aluminum test specimen arrangement.

PIEZOELECTRIC   OPTICAL

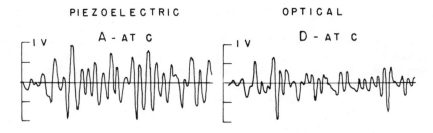

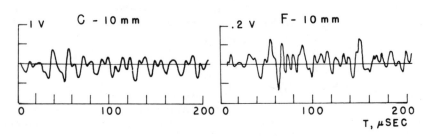

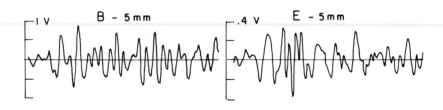

T, μSEC

Figure 13.   Tracings of optical (left) and piezoelectric (right)
             responses to dilatational waves at normal or near-
             normal incidence: (A), (D) sensor at center of speci-
             men; (B), (E) sensor displaced 5 mm from center of
             specimen; (C), (F) sensor displaced 10 mm from center
             of specimen.   Sweep speed 20 μsec/division.

and detected dilatational wave (BAW) signals at the center of the
polished surface, C.   Both optical and piezoelectric detection
methods were used.   First, the transducer at A was driven by
sinusoidal voltages and about forty resonant frequencies were
observed in the range 15 to 300 KHz.   A second PZT-5 transducer, 1"
diameter, 1/8" thick gave nearly all the same resonance frequencies.
Thus it is clear that to avoid signals characteristic of the specimen dimensions, we must limit the measurements to time intervals

which preclude the arrival of various reflected pulses at the
sensing transducer, either optical or piezoelectric - in our case,
approximately the first 25 microseconds.

The driving transducer (at A) was then driven by a high voltage
pulse of about 2 μsec duration and the signals observed at or near
C. Figure 13 shows both the optical and piezoelectric transducer
signals obtained. It is seen that the waveform changes, as might
be expected, as the point of observation moves from the center, C,
to 5 and 10 mm off center. Also it is seen, again as expected,
that the optical signal differs somewhat from the piezoelectric
signal because of the large sensitive area of the piezoelectric
device. It should be emphasized that the time scale is 20 μsec/div.
in this figure so that the waveforms include many reflected elastic
waves of various kinds. There exists an apparent 20 μsec period
in the waves (among others) corresponding to a round trip for a
wave along the specimen axis. The results are accurately reproducible for both optical and piezoelectric detection transducers.
When the driving transducer was waxed to the specimen edge, at B
in Figure 12, both bulk and surface acoustic waves (SAW) were
generated. With this arrangement, the piezoelectric sensor was
barely able to detect the disturbances whereas the optical probe
sensitivity was good.

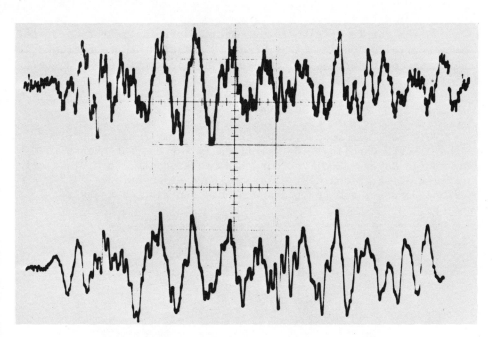

Figure 14.   Real acoustic emission in indium sensed optically
             (upper trace) and piezoelectrically (lower trace).
             Sweep speed 50 μsec/div.

We have also observed real acoustic emission due to twinning in indium. For this experiment we used an indium bar 25.4 x 6.25 x 150 mm optically polished on one side. We clamped this specimen about 25 mm from one end. A piezoelectric sensor was attached to one side of the specimen with the optical probe beam directed on the other side, opposite the piezoelectric sensor. Figure 14 shows a burst generated by bending the long end of the specimen (on the other side of the clamp from the sensors). The upper trace was recorded on video tape with the optical sensor, the lower trace was recorded on a second video tape with the piezo-electric sensor. The sweep speed was 50 µsec/div. Clearly the sensors both detect substantially the same waveform.

SUMMARY

In this chapter we have considered some of the characteristics of ultrasonic waves and acoustic emission bursts in non-destructive evaluation. Although more elaborate, optical probing techniques have several significant advantages over piezoelectric sensors: they are broad-band; they are local; they are sensitive; they can probe internally in transparent media; they do not disturb the system under test, and do not require any direct contact with the specimen.

In the future we plan to use optical probing methods both to characterize the phenomena of acoustic emission and to evaluate the response properties of commercial piezoelectric transducers.

ACKNOWLEDGEMENTS

This work was supported by the U. S. Army Research Office, Durham, North Carolina. We specifically express our thanks to Dr. George Mayer (AROD), to our graduate students, Boro Djordjevic, Steven E. Fick, Richard O. Claus, Richard Mignogna, to Corinne Harness, and to others who have contributed to this project.

REFERENCES

1. Hogarth, C.A. and J. Blitz, "Techniques of Non-Destructive Testing", London: Butterworths (1960).

2. Carlin, B., "Ultrasonics", New York: McGraw-Hill (1960).

3. Goldman, R.C., "Ultrasonic Technology", New York: Reinhold (1962).

4. Krasil'nikov, V.A., "Sound and Ultrasonic Waves", Israel Program for Scientific Translations, Jerusalem (1963).

5. McMaster, R.C., "Nondestructive Testing Handbook", Vol. II, New York: Ronald (1963).

6. Frederick, J.R., "Ultrasonic Engineering", New York: Wiley (1965).

7. Blitz, J., "Fundamentals of Ultrasonics", New York: Plenum (1967).

8. Schall, W.E., "Non-Destructive Testing", Brighton, England: Machinery Publ. (1968).

9. Gooberman, G.L., "Ultrasonics", New York: Hart (1968).

10. McGonnagle, W.J., "Nondestructive Testing", New York: Gordon and Breach (1969).

11. Krautkrämer, J and H. Krautkrämer, "Ultrasonic Testing of Materials", Berlin and New York: Springer-Verlag (1969).

12. Sharpe, R.S., "Research Techniques in Nondestructive Testing", New York: Academic Press (1970).

13. Ensminger, D., "Ultrasonics, The Low- and High-Intensity Applications", New York: Dekkar (1973).

14. Green, R.E., Jr., "Ultrasonic Investigation of Mechanical Properties", Vol. 3, Treatise on Materials Science and Technology, New York: Academic Press (1973).

15. Mason, W.P., "Physical Acoustics and the Properties of Solids", New York: Van Nostrand (1958).

16. Mason, W.P., ed., "Physical Acoustics", Vol. III A (1966), Vol. III B (1965), Vol. IV A (1966), Vol. IV B (1968), Vol. VIII (1971), New York: Academic Press.

17. Mason, W.P., Advanced Materials Research, 2 (1968) 287.

18. Truell, R., C. Elbaum and B.B. Chick, "Ultrasonic Methods in Solid State Physics", New York: Academic Press (1969).

19. Green, R.E., Jr., "Proc. Ultrasonic International 1963 Conference", Guildford, England: IPC Science and Technology Press, Ltd. (1973).

20. "Proceedings of Conference on Internal Friction Due to Crystal Lattice Imperfections", Cornell University, Ithaca, New York: (July 1961), _Acta Met._, 10 (1962).

21. Mason, W.P. and R.N. Thurston (eds), "Physical Acoustics", Vol. III, Part A, New York: Academic Press (1966).

22. "Proceedings of the International Conference on Ultrasonic Attenuation and Internal Friction in Crystalline Solids", Brown University, Providence, R.I. (September 1969), _J. Phys. Cham. Solids_, 31 (1970).

23. Nowick, A.S. and B.S. Berry, "An Elastic Relaxation in Crystalline Solids", New York: Academic Press (1972).

24. De Batist, R., "Internal Friction of Structural Defects in Crystalline Solids", North-Holland, Amsterdam, London, American Elsevier, New York (1972).

25. "Proceedings of the Fifth International Conference on Internal Friction and Ultrasonic Attenuation in Crystalline Solids", Aachan, Germany (August 1973) Springer, Berlin, New York (1975).

26. Hutton, P.H. and R.N. Ord, "Acoustic Emission" in R.S. Sharpe, ed., "Research Techniques in Nondestructive Testing", London and New York: Academic Press (1970) pp. 1-30.

27. "Acoustic Emission", R.G. Liptai, D.O. Harris and C.A. Tatro (eds.), ASTM STP 505, American Society for Testing and Materials, Philadelphia, Pa. (1972).

28. Pollock, A.A., "Acoustic Emission, A Review of Recent Progress and Technical Aspects", in "Acoustics and Vibration Progress", R.W.B. Stephens and H.G. Leventhall (eds.), Vol. 1, London: Chapman and Hall (1974) pp. 53-84.

29. Spanner, J.C., "Acoustic Emission, TEchniques and Applications", Evanston, Ill.: INTEX Publishing Co. (1974).

30. Lord, A.E., Jr., "Acoustic Emission" in "Physical Acoustics", W.P. Mason and R.N. Thurston (eds.), Vol. II, London and New York: Academic Press (1975) 289-353.

31. "Monitoring Structural Integrity by Acoustic Emission", J.C. Spanner and J.W. McElroy (eds.), ASTM STP 571, American Society for Testing and Materials, Philadelphia, Pa. (1975).

32. Whitman, R.L. and A. Korpel, "Probing of Acoustic Surface Perturbations by Coherent Light", _Appl. Opt._, 8 (1969) 1567-76.

33. White, R.M., "Surface Elastic Waves", _Proc. IEEE_, 58 (1970) 1238-76.

34. Stegeman, G.I., "III. Optical Probing of Surface Waves and Surface Wave Devices", _IEEE Trans. Sonics and Ultrasonics_, SU-23 (1976) 33-63.

35. Korpel, A., L.W. Kessler and M. Ahmed, "Bragg Diffraction Sampling of a Sound Field", J. Acoust. Soc. Am., 51 (1972) 1582-92.

36. Liu, J.M. and R.E. Green, Jr., "Optical Probing of Ultrasonic Diffraction in Single Crystal Sodium Chloride", J. Acoust. Soc. Am., 53 (1973) 468-78.

37. Cook, B.D. and E.A. Hiedemann, "Diffraction of Light by Ultrasonic Waves of Various Standing Wave Ratios", J. Acoust. Soc. Am., 33 (1961) 945-48.

38. Palmer, C.H., H.M. South and T.H. Mak, "Optical Interferometry for Measurement of Rayleigh and Dilatational Waves", Ultrasonics (1974) 106-08.

39. Adler, R., A. Korpel and P. Desmares, "Instrument for Making Surface Waves Visible", IEEE Trans. Sonics and Ultrasonics, SU-15 (1968) 157-61.

40. Korpel, A., L.J. Laub and H.C. Sievering, "Measurement of Acoustic Surface Wave Propagation Characteristics by Reflected Light", Appl. Phys. Lett., 10 (1967) 295-97.

41. Krockstad, J. and L.O. Svaasand, "Scattering of Light by Ultrasonic Surface Waves in Quartz", Appl. Phys. Lett., 11 (1967) 155-57.

42. Auth, D.C. and W.G. Mayer, "Scattering of Light Reflected from Acoustic Surface Waves in Isotropic Solids", J. Appl. Phys., 38 (1967) 5138-40.

43. Ippen, E.P., "Diffraction of Light by Surface Acoustic Waves", Proc. IEEE (Letters), 55 (1967) 248-49.

44. Alippi, A., A. Palma, L. Palmieri and G. Socino, "Surface Wave Amplitude Determination by Optical Technique", Nuovo Cimento, 1 (1971) 239-41.

45. Whitman, R., L.J. Laub and W. Bates, "Acoustic Surface Displacement Measurements on a Wedge-Shaped Transducer Using an Optical Probe Technique", IEEE Trans. Sonics and Ultrasonics, SU-15 (1968) 186-89.

46. Jenkins, F.A. and H.E. White, "Fundamentals of Optics", 4th ed., New York: McGraw-Hill (1976) p. 287.

47. Palmer, C.H., R.O. Claus and S.E. Fick, "Ultrasonic Wave Measurement by Differential Interferometry", Appl. Opt. (to be published).

48. Peck, E.R. and S.W. Obetz, "Wavelength or Length Measurement by Reversible Fringe Counting", J. Opt. Soc. Am., 43 (1953) 505-09.

49. Vilkomerson, D., "Measuring Pulsed Picometer - Displacement Vibrations by Optical Interferometry", Appl. Phys. Lett., 29 (1976) 183-85.

50. Mezrich, R., D. Vilkomerson and K. Etzold, "Ultrasonic Waves: Their Interferometric Measurement and Display", Appl. Opt., 15 (1976) 1499-1505.

51. Lurie, M., "Evaluation of Expanded Laser Beams Using an Optical Flat", Opt. Engr., 15 (1976) 68-69.

Chapter 16

RECOGNITION AND ANALYSIS OF DISTRIBUTED IMAGE INFORMATION

M. G. Dreyfus

Dreyfus-Pellman Corporation

Stamford, Connecticut

ABSTRACT

Visual inspection can be modeled as a data processing procedure involving analysis of distributed image information. This chapter will discuss the nature of the data processing involved, and will outline procedures for instrumenting visual recognition, orientation, and measurement for industrial process control. It will illustrate how modern data processing technology can be applied to automate visual inspection with speeds and accuracies far beyond human capabilities.

DISTRIBUTED IMAGE INFORMATION

The information in an optical image is generally distributed among a number of points scattered around the image. An inspector who wishes to extract meaningful information from the image must first recognize and orient on the array of information which it represents. The preliminary processes of recognition and orientation in themselves involve coarse image measurements and data processing. Thus we have an iterative information process involved in image recognition, orientation and measurement: initial coarse measurements are followed by successively finer measurements until the desired resolution has been achieved.

For example, the procedure involved in making measurements with an optical comparator will serve to illustrate the nature of the iterative visual process involved in analysis of distributed image information. When a comparator operator inspects a machined

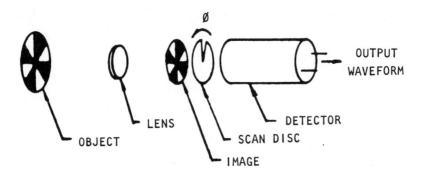

Figure 1

part, he performs the following sequence of operations:

1.   He first recognizes the part by comparing its appearance
with a master part or drawing, and decides whether it is similar
enough to be worth measuring.

2.   He physically reorients the part in the comparator so that
he can conveniently make valid measurements.

3.   He reads the measurements on the comparator micrometers,
and decides whether the part is within specified tolerances.

In Step 1, the inspector performs a mental recognition process
based on preliminary data concerning the size and shape of the part.
These data involve measurements such as linear dimensions and
angular subtenses. They are inexact, but nevertheless semi-
quantitative; the inspector knows whether the part is one inch long
or two inches long, and roughly where its projections and holes are
located. The inspector then compares these data with a memorized
set of data to recognize and orient the part.

Implicit in the inspector's recognition process is an algorithm
which he mentally applies to his visual data, matching up the various
characteristics of the part until he succeeds in orienting the part
to the drawing. The preliminary measurements which the inspector
makes by visual estimation are crude compared to the final comparator
measurements, and they are correspondingly tolerant to misorientation. Subsequently the inspector uses the crosshairs of the comparator as a visual aid to make the more precise dimensional measurements, and uses the micrometer spindles to furnish a more accurate measurement coordinate system.

Inspection of specific details in the image requires scanning

in those specific areas. If the image had not previously been
oriented, then detailed scanning would have to be performed over
a relatively large area in order to be assured of coverage of the
details desired; this enlarged coverage would lead to a massive
data processing procedure operating on a large assembly of information bits. Since the process control operator pays by the bit for
information processing, it is desirable to minimize the total
number of bits handled in the inspection procedure. Adaptive
control procedures [1] such as the iterative process described above
greatly improve the efficiency of inspection procedures. This
improved efficiency can be implemented by adding automatic logical
control circuits to the inspection equipment. High speed scanners
have existed for decades; however, only during the last few years
has microcircuit technology developed to a level which makes automated inspection with adaptive control practical in a wide range
of process control applications.

     In this context it should be noted that the human eye/brain
system is not, in general, an optimum instrument for processing
image information [2,3]; its ability to detect brightness changes,
image patterns, and image motions is limited by neurochemical
constraints which evolved to satisfy primitive animalistic survival
requirements. These constraints limit human visual performance in
a modern factory environment. Consequently, automatic equipment
can be designed for specific manufacturing tasks which far exceeds
the limits of human performance. It has been shown, for example,
that a comparator can be automated [4] with an electro-optical
scanner under minicomputer control to make dimensional measurements
on manufactured parts without human assistance, and that these
measurements can be made with ten times better precision than a
human can achieve, and at a rate of ten measurements per second.

     In scanning distributed image information, data symmetries can
often be used to reduce the complexity of the data processing task.
For example, scanners have been constructed [5] which scan arbitrary
imagery in a polar mode as shown in Figure 1 to simplify the process
of image orientation. The way in which this scan geometry operates
can be conveniently illustrated with the signal generated by scanning the four-bladed propeller shown in Figure 1. The scanner output
is monitored to determine the angular locations of the rising and
falling waveform edges due to the bright and dark shading in the
propeller image. The set of edge locations determined from one
scanner revolution operating on any scene is called the signature
of that scene. As long as the scanner is centered on the propeller
hub, the rising and falling waveform edges will be evenly spaced in
the scan rotation. However, motion of the propeller in its own
plane produces systematic shifts in the slope locations, advancing
the time of occurrence of some of the slopes, and retarding others.

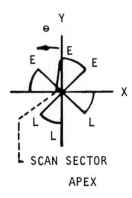

Figure 2

     To demonstrate how the polar scanner can recognize scene
movement, imagine that the scanner is centered on the propeller
when its signature is extracted and memorized, and then that the
propeller is moved to the right along the X axis as shown in
Figure 2. Assume that the scanning sector rotates counterclockwise
and that a reference pulse train starts counting when the sector
straddles the positive X axis. Propeller motion changes the scene
signature by advancing the timing of all edges in the first two
quadrants (i.e., between 0 degrees and 180 degrees) and by retarding the timing of all edges in the last two quadrants. Each edge
is compared in location with the corresponding memorized edge,
yielding positive edge shifts in the first two quadrants, and
negative edge shifts in the last two quadrants. The phase-
advanced edges in Figure 2 have been marked with an E (denoting
Early) and the phase-retarded edges with an L (denoting Late).

     X motion can be discerned automatically by standard phase
detection using the X axis as the phase reference. This yields
a positive number if the propeller moves to the right, or a negative number if the propeller moves to the left. The size of the
number depends on the amount of propeller shift. Similarly Y
motion can be determined by phase detection using the Y axis as
the phase reference. The only remaining scene motion in-plane is
θ motion (pure rotation); this can be discerned automatically
simply by adding all the edge phase shifts without any rectifica-
tion or sign reversal.

     The symmetries of polar scanning are particularly powerful in
general purpose orientation involving arbitrary imagery. One good
example of such imagery would be the patchwork quilt of crystal

patterns on the surface of metallic specimens observed through a
metallurgical microscope. Another example of such imagery is the
sequence of images photographed in mapping a strip of terrain with
an aerial camera; in fact, this polar scan technology was developed
for use in aerial reconnaissance to help stabilize terrain imagery
by means of automatic image memorization and tracking [6,7].

More recently we have shown that modern computer technology
can be applied to program and to analyze scans on various kinds of
imagery for recognition and measurement operations on specific
products [8,9]. Computer-controlled scanning is particularly
effective in inspection of manufactured parts such as cams, gears,
stampings, and printed circuit boards which have relatively simple
or regular images. We have recently designed an image data processor which controls a step in an industrial process by operating
directly on the output of a standard television monitor; the output
of this processor consists of automatic control decisions which
replace those previously made by a person watching the monitor.
We are currently exploring the application of these image processing techniques to the interpretation of moire patterns for measurement of surface contours. Similar design philosophies have also
been applied to the development of real-time color process control
equipment.

CONCLUSIONS

The far-reaching nature of the impact that automation of image
information processing is likely to have on industrial process
control will affect all of us. Human visual monitoring today is
directly involved in a wide variety of manufacturing operations.
An even wider variety of manufacturing operations can benefit from
the application of automatic image information processing. We are
currently applying new developments in this technology to proprie-
tary process control requirements in such diverse fields as production of roller bearings, microcircuits, razor blades, sewing
thread, and air pollution.

REFERENCES

1. Eveleigh, V.W., Adaptive Control and Optimization Techniques, McGraw Hill, New York, 1967.

2. Julesz, B. and R. Hesse, "Inability to Perceive the Direction of Rotation Movement of Line Segments", Nature, 225 (Jan. 17, 1970) 243-44.

3. Dreyfus, M., "A New Approach to Automation of Visual Perception", presented April 12, 1973, at a Symposium entitled "Measurement of the Workpiece: at Redstone Arsenal, Huntsville, Alabama, sponsored by the Standards and Metrology Division of the American Ordnance Association.

4. Dreyfus, M., "Automatic Image Alignment and Measurement", presented November 12, 1975, at the Electro-Optical Systems Design Conference-1975 in Anaheim, California, sponsored by the Laser Institute of America; see pages 791-793 in the Proceedings published by Industrial and Scientific Conference Management, Inc., Chicago, Illinois.

5. Dreyfus, M., "Visual Robots", The Industrial Robot, Volume I, No. 6, pp 260-64, December 1974.

6. Heen, H., W. Wilson, J. Widmer, D. Stone and E. Boase, "Lunar Orbiter Camera", J. Soc. Motion Picture and Television Engineers, 76, No. 8 (August 1967) 740-50.

7. Crabtree, J. and J. McLaurin, "The BAI Image Correlator", J. Amer. Society of Photogrammetry, 36, No. 1 (January 1970) 70-76.

8. Altman, N. and M. Dreyfus, "Electro-Optical Scanning System for Dimensional Gauging of Parts", U.S. Patent 3,854,822 dated December 17, 1974.

9. Dreyfus, M., "Non-Contact Dimensional Measurement Technique", U.S. Patent 3,941,484 dated March 2, 1976.

Chapter 17

MOIRE METHOD - ITS POTENTIAL APPLICATION TO NDT

A. S. Kuo and H. W. Liu

Syracuse University

Syracuse, New York

ABSTRACT

The moire method has been used to measure the in-plane and out-of-plane displacements. This method can be used under either static or dynamic conditions. The basic geometric principles of moire fringe formation and the essential experimental techniques are briefly reviewed. The potential application of the moire method to NDT is evaluated.

INTRODUCTION

The application of the moire technique to strain measurements was introduced by Kaczer and Kroupa [1]. Subsequently, Dantu [2] interpreted the moire patterns in terms of the components of displacements. Dantu's work laid the foundation of moire strain analysis. Durelli, Theocaris, Post, Sciammarella, and Chiang have made substantial contributions to the development of the fundamentals of the technique [3-16].

In addition to being used to obtain the two-dimensional in-plane strains, the moire method has been used to determine the following: the deflection of a plate under bending [16-19]; transverse bending moments in a cylindrical shell [20]; the large displacements of a uniformly loaded hyperboloid-parabolic shell [21]; the characteristics of shock-wave produced by explosives embedded in the central cavities of plexiglass spheres [22]; dynamic strain patterns near welds [23]; the complex strain distributions in solid propellants [24]; strains at elevated temperatures

[25]; surface contours [26-30]; and three dimensional strain distributions [13,31]. In recent years, the moire method was used
to measure the extensive crack tip deformation [32,33] and the
stress intensity factor [34]. It was also used to study the dynamic
crack propagation [35-37] and to analyze ductile fracture [38-40],
and fatigue crack growth [41-43].

For a given moire grille density, the resolution can be improved by the use of the fringe multiplication technique [7]. The
undesirable artifacts in the moire picture can be removed, and the
contrast of moire pictures can be improved by an optical filtering
method [3,14,44]. The introduction of a digital filtering method
not only reduces human error but also paves the way for automated
data processing [10].

Although the moire method has been used to measure displacements and strains, its applicability to NDT has not been explored.
In this chapter, a brief review of the basic geometric principles
of moire fringe formation and its essential experimental techniques
is made. Its potential application as a NDT method is evaluated.

BASIC GEOMETRIC PRINCIPLES

Moire fringes are formed by the geometric interference of two
sets of grille lines. A moire grille consists of alternating light
and dark parallel strips or lines. A "model grille" is printed on
a sample surface. This grille deforms with the sample surface as
a load is applied. The geometric interference between the model
grille and a rigid and undeformed "reference grille" forms moire
fringes.

Figure 1 illustrates the geometry of fringe formation caused
by axial deformation. In this figure, the slender rectangles
represent dark strips of the grilles. Assume that the model grille
and the reference grille have the same initial pitch, p. When
the model grille is uniformly deformed, either elongation or contraction in the direction perpendicular to the grille lines, the
pitch of the model grille is changed from p to p'. The interfer-
ences between the deformed model grille and the reference grille
forms moire fringes parallel to the grille lines. The dark fringe
forms in the region where the dark strip of the reference grille
covers the light strip of the model grille. The light fringe
forms in the region where the light strips of both the model and
the reference grilles coincide.

The magnitude of the relative displacement between two
neighboring fringes on the sample surface are constant equal to p.
The average strain over the gage length, i.e., the distance between

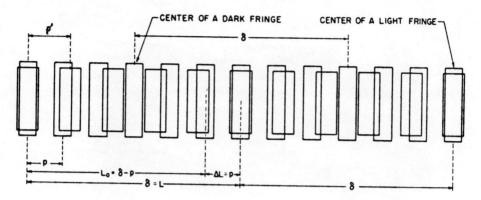

Figure 1. Fringe formation by axial deformation [9].

two neighboring fringes, is

$$\varepsilon = (L - L_o)/L_o = [\delta - (\delta \pm p)]/(\delta \pm p)$$

$$= \pm\, p/(\delta \pm p) \tag{1}$$

where L_o and L are the initial and final gage lengths, and δ is the fringe spacing. For small deformation, $\delta \gg p$, Equation (1) can be approximated by

$$\varepsilon = \pm\, p/\delta = \Delta v/\Delta y \tag{2}$$

where y is the direction perpendicular to the grille lines, and v is the displacement in the y-direction. The pitch and fringe spacing in Equation (2) can be interpreted as elongation and gage length. This equation clearly indicates that a high density grille is necessary to resolve a small strain over a short gage length.

Figure 2 shows the geometry of the moire fringes caused by the relative rotation between a model and a reference grille. Again, the magnitude of the relative displacement between two neighboring bringes on the sample surface is constant equal to pitch, p. The fringe spacing, δ, is related to the angle of rotation, θ, and the length of the side, d_1, of the rhombus as follows:

$$\delta = d_1 \cos\frac{\theta}{2} = p/(2\,\sin\frac{\theta}{2}) \tag{3}$$

For small angle rotation, the fringes are nearly perpendicular to the grille lines. In this case, the angle of rotation is:

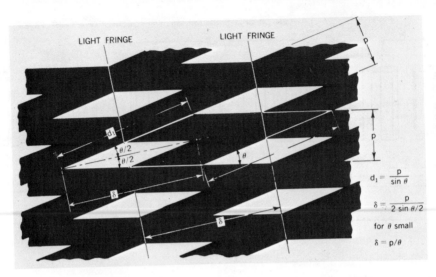

Figure 2. Formation of rotational fringe [3].

$$\theta = p/\delta = \Delta v/\Delta x \qquad\qquad (4)$$

where x is the direction of the grille lines, and v is the displacement perpendicular to the grille lines

Figure 3 shows the fringe lines of a double edge-cracked-sample with horizontal reference grille lines. In a general state of plane deformation, the fringes are neither parallel nor perpendicular to the grille lines. In this case, the displacment gradient components are given by the following equations

$$\Delta v/\Delta x = p/\Delta x \qquad \text{and} \qquad \Delta v/\Delta y = p/\Delta y \qquad (5)$$

where Δx and Δy are the fringe spacings measured in the x- and y-directions respectively (see Figure 4).

The four displacement gradient components of planar deformation, i.e., $\Delta u/\Delta x$, $\Delta u/\Delta y$, $\Delta v/\Delta x$, and $\Delta v/\Delta y$, can be measured by two sets of mutually perpendicular reference grilles and two sets of corresponding model grilles. In this way, Post [6] has used the grid-analyzer moire method to determine the complete two dimensional state of strains.

Moire fringes formed by a reference grille and its shadow can be used to measure out-of-plane displacements. The sample

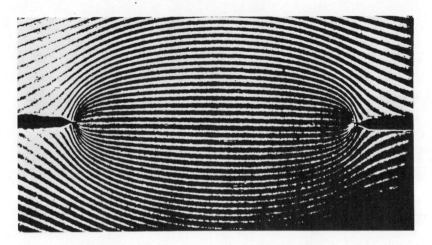

Figure 3.   Moire fringes of a double-edge-notched specimen under
           tension.

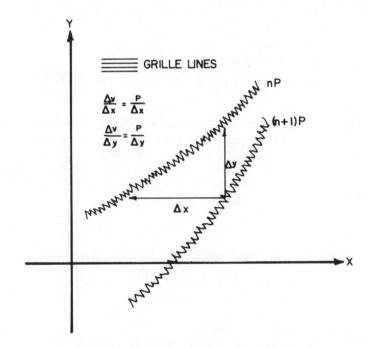

Figure 4.   Displacement gradients.

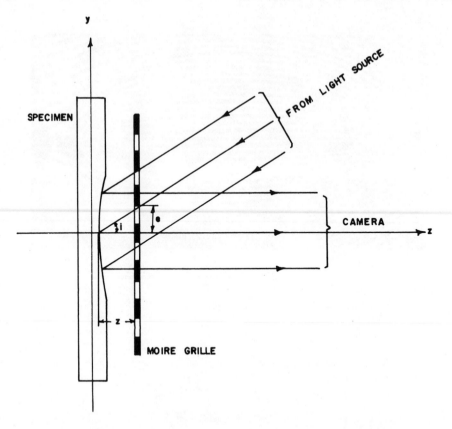

Figure 5.   Formation of shadow moire fringe for surface contouring.

surface is coated with a matte finish to obtain a distinctive
shadow.   Figure 5 illustrates the formation of the moire fringes.
The collimated incident light beam is inclined at the angle i,
the shadow is displaced by the distance e.   The separation, z,
between the reference grille and the sample surface is related to
e and i as follows:

$$z = np/\tan i \qquad\qquad (6)$$

where n is the fringe order given by the equation $n = e/p$. If i
is small, z can be approximated by the equation

$$z = np/i \qquad\qquad (7)$$

In their works on moire gauging of surface contours, Brooks
and Heflinger [27] and Shamir [30] used the interference lines of

two coherent, collimated laser beams to replace the reference
grille. Their method has the advantage of being able to alter
the pitch of the interference lines over a wide range. In addition,
with their method, there is no occurrence of diffraction effects
due to a small pitch as is encountered with the conventional moire
method.

Osgerby [20] has used shadow moire to determine transverse
bending moments in a cylindrical shell. Marasco [21] has used
shadow moire with a curved grille to investigate the displacements
of a uniformly loaded hyperboloid parabolic shell.

Moire fringes directly reveal the displacement field of a
deformed body. The strains are derived from the displacement
field by taking the gradients. For non-destructive testing, the
detection of the local displacements caused by cracks are more
important than the determination of the strain field. The crack
opening displacements of the sample in Figure 3 can be readily
measured. The opening displacement at the location of the n'th
order fringe is np with n = 1 for the first fringe behind the
crack tip.

EXPERIMENTAL TECHNIQUES

In order to make successful measurements, a suitable grille
density has to be chosen; a model grille must be affixed onto a
sample surface, and the moire interference fringes must be properly
formed and recorded. Grilles with a wide range of densities, up
to 10^6 lpm, are commercially available. Diffraction gratings can
suitably serve as high density grilles. Grilles with a pitch of
2.5 μm can be produced by the interference of two overlapping
coherent plane wave fronts. Stripping films with printed grille
lines can bondable grids are available commercially. The film
or the grid can be directly cemented onto a sample surface. The
desired accuracy and resolution determines the choice of grille
density.

A model grille can calso be photo-printed onto a sample surface. Several photo-sensitive emulsions are available commercially.
Photo-resist is especially convenient to use. The sample surface
is polished, cleaned, and dried. The properly prepared surface
is then coated with a photo-sensitive emulsion, which is exposed
to the negative of a model grille. After exposure, the model
grille is developed. A detailed description of the application of
photo-resist is given by Luxmoore and Herman [45] and Wadsworth
et al. [46]. The model grille thus produced is suitable for moire
measurements.

This process of model grille preparation can be carried one

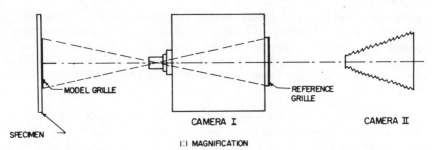

Figure 6. Experimental set-up to obtain moire fringe.

step further. After the development of the photo-resist, the model
grille consists of strips of bare metal surface and strips of metal
surface covered by the photo-resist. The bare metal strips can be
electro-etched. After etching, the residual photo-resist is
removed. The model grille is now made up of etched and unetched
metal strips. The grille thus prepared is more stable at elevated
temperatures. Similar results can be obtained if the electro-
etching process is replaced by electric deposition of a thin metal
film onto the sample surface [45,47].

Grille lines can be mechanically engraved onto a sample surface
with a modified biological microtome. After engraving, the sample
surface is mechanically polished to restore surface flatness and
reflectivity. Effective grille density of up to 0.23×10^6 1pm
has been successfully produced [32,33].

A grid of two mutually perpendicular lines or othogonal arrays
of uniformly spaced dots can be used as two orthogonal grilles.
Stencils can be used to produce such grids or arrays of dots [48].
A fine mesh of orthogonal arrays of holes is held in close contact
with a sample surface. A thin metal film is then vapor deposited
onto the sample surface in a vacuum. After deposition, the mesh
is removed, and orthogonal arrays of metallic dots remain on the
sample surface. The dots can also be printed onto the surface with
stencil and ink.

There are several methods used to form moire interference
patterns between the model and reference grilles. The simplest
method is to place the reference grille, which is often on a flat
glass plate, directly in contact with the model grille. The
geometric interference between these two grilles forms the fringes.
The spacing between the reference grille and the sample surface
must be small to avoid the diffraction effects. Small enough
spacing could be difficult to attain for a very fine pitch grille.

   Figure 6 shows the experimental set-up to project the image
of a model grille onto the reference grille of a ground glass plate.
A precision camera, which has a known and adjustable magnification
very close to one, was used. Any magnification other than one is
equivalent to an initial pitch mismatch of the model and reference
grilles. Mismatch is often used to increase the fringe density;
i.e., to shorten the gage length for strain measurements. The
mismatch method is discussed more fully later in this section.

   The double exposure method has also been used to obtain moire
fringes. The image of an undeformed model grille is recorded on a
photographic film in a camera as the reference grille, Before
development the same film is exposed a second time to the deformed
model grille. The interference of these two images forms fringes.
This double exposure method can also be used directly with the
photo sensitive emulsion on a sample surface. The image of a
grille is projected onto the photo sensitive emulsion on the
sample surface once before and once after deformation. After
double exposure, the emulstion is developed, and moire fringes are
directly formed onto the sample surface.

   Recently, Wadsworth et al. [46] have designed a system to
observe real time displacements with a coherent light source. The
schematic diagram of their system is shown in Figure 7. A set of
straight interference lines, i.e., grille lines, is formed by
superimposing two coherent collimated light beams produced by a
beam splitter and a laser. The sample surface coated with photo-
resist layer is exposed to the straight linterference lines.
After processing, a model grille is formed on the sample surface.
The sample surface is then re-illuminated by the two light beams
to form a moire pattern. The change of the moire pattern during
the course of loading can be observed directly. This method is
particularly adaptable to rought and curved surfaces. In addition,
the laser speckle effect has also been recently used to form
fringe patterns for displacement measurements [49].

   Short gage length is necessary to measure strains in a
region of high strain gradient such as in the vicinity of a crack
tip. As mentioned earlier, a high density grille is capable of
resolving a small strain over a short gage length. An alternative
method is to use either pitch mismatch or rotational mismatch to
increase the fringe density for pitch mismatch, the initial pitches
of the reference and the model grilles are different. The fringes
produced by the initial pitch mismatch represent a fictitious
strain. When the model grille is deformed in the same sense as
the initial mismatch, the fringe density is further increased.
The real strain in the sample is the difference between the strain
measured from the final moire fringe pattern and the fictitious
strain given by the initial mismatch pattern.

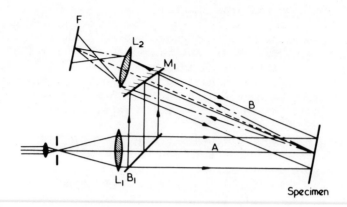

Figure 7.   Layout of apparatus for producing moire grille by the
            interference of laser beams [46].

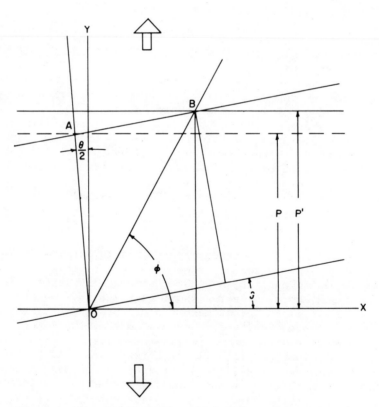

Figure 8.   Rotational mismatch moire fringe formation for axial
            strain measurement.

An initial rotational mismatch between the model grille and
the reference grille forms fringes which are nearly perpendicular
to the grille lines as illustrated in Figure 2. The orientation
of the fringes will be changed as the pitches of the model grille
are changed by deformation. Figure 8 illustrates the formation
of fringes due to pure tension with an initial rotational mismatch.
OA is the fringe due to rotation θ, while the pitch of the model
grille remains unchanged at p. OB is the fringe after a rotation
of θ and an elongation from p to p'. ϕ is the fringe inclination
angle. The following derivation of the strain, ε_y, is due to
Vafiadakis and Lamble [50].

$$OB = BC/Sin\ (\phi - \theta) = BD/Sin\ \phi$$

$$\varepsilon_y = (p/p_o) - 1 = (DB/BC) - 1 = [Sin\ \phi/Sin\ (\phi - \theta)] - 1 \qquad (8)$$

For this method, the necessary measurement in order to evaluate ε_y
is the angle ϕ. The exact location of the fringe center line is
not needed.

Luxmorre and Wyatt [40] have used the rotational mismatch
method to improve the resolution of their measurements of crack
opening displacement, COD, of notched-bend specimens. Figure 9
shows the moire pattern near a carck tip with rotational mismatch
for COD measurement. COD can be determined by plotting two
displacements versus distance curves for the upper and the lower

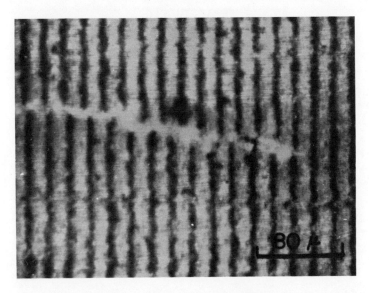

Figure 9. Rotational mismatch moire pattern near a crack tip.
Al 2024-T351, K_{min} = 4.3 ksi in.

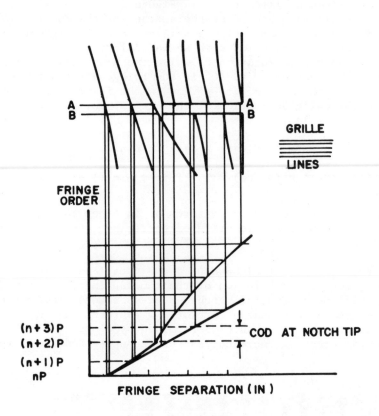

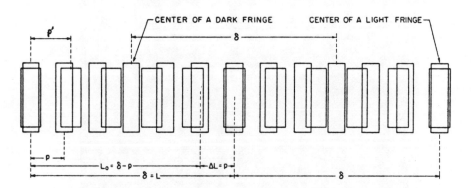

Figure 10.   Method to obtain COD by rotational mismatch moire
             technique [40].

notch surfaces as shown in Figure 10. The magnitudes of COD's
are the differences between these two curves. It should be noted
that the fringe density on one side of the notch is higher than
the other. Thus, an appropriate choice of moire grille pitch and
the rotation is important.

When a light beam passes through a transparent moire grille,
the grille acts like a diffraction grating. Based on the diffraction phenomenon, an optical filtering method has been developed
to sharpen the moire fringes [14]. It has also been used to
increase grille density [51].

Light travels as a wave. For the sake of simplicity, assume
that a monochromatic light of wavelength λ passes through a moire
grille, which acts as a diffraction grating. The light beams
leaving the grating propagate in all directions as if the grille
lines were very thin and elongated monochromatic light sources
emitting waves of the same wave length in phase with each other.
All the rays leaving the grille lines in a given direction of
angle α_1 are focused by a lens on a spot in the diffraction plane
(Figure 11). Because of the difference in path lengths, the rays
from various grille lines may arrive at the spot with different
phase angles. Therefore, the sum of all the rays is nearly zero
except at the specific locations where all the rays arrive in phase

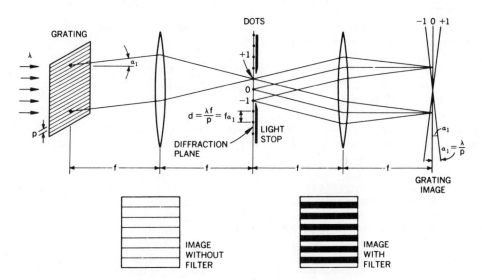

Figure 11.   Schematic diagram showing the principle of optical
             filtering [3].

forming successive dots, or diffraction spots, on the diffraction plane. The distance, d, between two neighboring dots can be readily derived [3],

$$d = \lambda f/p \tag{10}$$

where f is the focal length of the lens. All the rays in the direction of the optical axis are focused at the "0" dot. The successive dots, ± 1, ± 2, ... correspond to increasing angles, α.

The dots can be considered as new coherent light sources each emitting light of amplitude $A_n e^{i(2\pi/\lambda)nd/f} x$ with n = 0, ± 1, ± 2, ... If all the beams from all the dots are allowed to pass through a second lens, the resultant intergrating at the image plane of the second lens.

Higher harmonics can be "filtered" out by an aperture at the diffraction plane. Figure 11 illustrates the filter which allows the beams from + 1, 0, and - 1 dots to pass through. For this filter, the image of very thin lines of stepwise illumination is changed to one of broad bands of illumination varying sinusoidally but maintaining the same pitch.

In order to filter out the original grille lines from a fringe pattern, only the beam from the + 1 dot is allowed to pass through. The region containing grille lines will be illuminated uniformly at the image plane. The region without grille lines will be dark regardless of whether it is light or dark in the objective plane.

This diffraction phenomenon can also be used to multiply the grille density. For example, if only the beams from + 1 and - 1 dots pass through the aperture, the resulting amplitude is $2A_1$ cos $2\pi(x/p)$, which has a period of p. The observed light intensity at the image plane is proportional to the square of the amplitude. Therefore, the effective grille density is increased by a factor of two. Similarly, if the beams from + 2 and - 2 dots are allowed to pass through, the resulting amplitude is $2A_2$ cos 2 (2x/p), and the effective grille density is increased by a factor of four.

After a moire pattern picture is obtained, the fringe spacings can be measured manually, and the strains and displacements can be evaluated therefrom. The laborious process of fringe spacing measurements can be automated with a densitometer. In addition, a computer program can be used to eliminate the "noises" in a moire pattern picture, to decipher the true signal of the displacement field, and to perform the differentiation in obtaining the strains [10,52].

A photographic negative of a moire picture is scanned by a

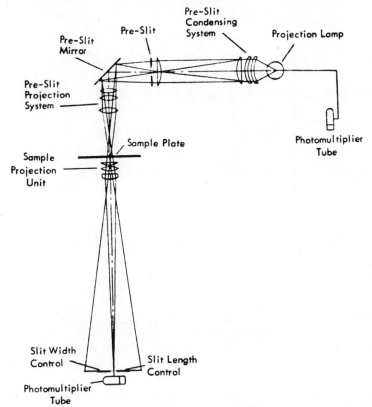

Figure 12.   Schematic diagram of microphotometer optical system [52].

densitometer as shown in Figure 12. The intensity of a small light
beam is measured by a photocell. A typical tracing of the recorded
intensity as the scanning beam traverses across the fringes is
shown in Figure 13. The tracing shows general sinusoidal characteristics of the true moire pattern together with high and low frequency noises. The noises are caused by the defects found in the
sample surface, in the photo-sensitive emulsion, in the optical
system or even in the densitometer. Noises are also caused by the
artifacts introduced by the printing and developing processes of
the grilles and the picture, and by the non-uniformity of the light
source. These noises can be eliminated by a digital filter.

An ideal sinusoidal variation of the light intensity can be
written as [11] follows:

$$I(x) = I_o + I_1 \cos (2\pi x/\delta) \tag{11}$$

where I_o and I_1 are constants. The fringe spacing δ can be written

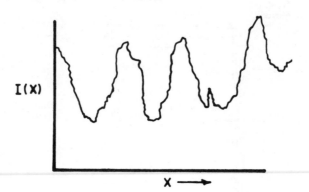

Figure 13.   Typical real intensity variation [52].

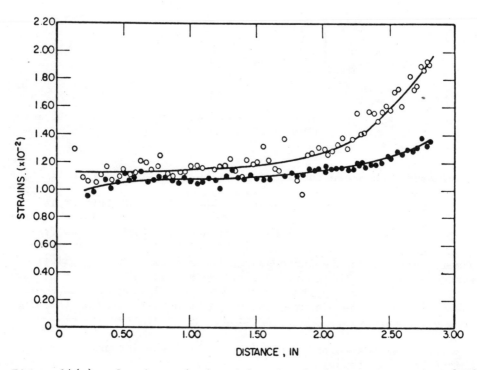

Figure 14(a).   Strains calculated by visual observation method [52].

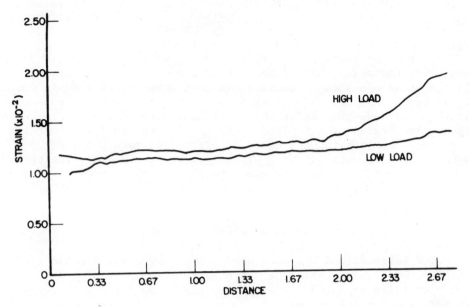

Figure 14(b). Strains obtained by the digital filtering method [52].

as follows:

$$\frac{1}{\delta} = \frac{1}{p} - \frac{1}{p'(x)} \tag{12}$$

where $p'(x)$ is the pitch of the deformed model grilles. Equation (11) can be written as follows:

$$I(x) = I_o + I_1 \cos \left[2\pi x \left(\frac{1}{p} - \frac{1}{p'(x)} \right) \right] \tag{13}$$

The instantaneous phase angle $\theta(x)$ is

$$\theta(x) = 2\pi x \left(\frac{1}{p} - \frac{1}{p'(x)} \right) \tag{14}$$

and

$$\theta(x) = \text{arc cos} \frac{I(x) - I_o}{I_1} \tag{15}$$

The strain ε_x is:

$$\epsilon_x = \frac{p}{2\pi} \frac{d\theta(x)}{dx} \tag{16}$$

If the constants I_o and I_1 can be evaluated and $I(x)$ can be measured, the phase angle θ and the strain ϵ_x can be calculated. In real tracings such as those shown in Figure 13, neither I_o nor I_1 appears constant. A digital filter can be designed to take care of the amplitude modulation and to eliminate the noises. Figure 14 shows the strains of a tapered plate tensile specimen in general yielding. The strains evaluated directly from manual fringe spacing measurements and by a digital filter are shown. The curves given by the filter are very close to the average curves derived from the manually measured points.

POTENTIAL APPLICATION TO NDT

The moire method has been used to measure displacements and strains, but its potential as a nondestructive testing method has not been fully explored. In order to apply the moire technique to NDT, two conditions have to be met. The moire method must be capable of resolving small displacement in order to detect small cracks and defects, and it must be applicable to rough and curved surfaces. It has been pointed out earlier that the method developed by Wadsworth et al. can be used with a rough and curved surface. Therefore, an elaborate surface preparation is no longer necessary. The method can be further developed for efficient and economical practical applications.

Figure 9 shows a picture of a moire pattern at a crack tip. The sample was made of 2024-T351 aluminum alloy. The K_{max} and K_{min} were 8.0 and 4.8 $Mnm^{-3/2}$ (7.1 and 4.3 $ksi\sqrt{in}$). The moire grille density was 528 lines/mm (13400 lpi). A rotational mismatch was used. Figure 15 shows the measured crack opening displacements at both K_{max} and K_{min}. The opening displacements in the order of 2.5×10^{-4} mm(10^{-5} in) can be readily measured.

For a crack of length 2a, in an infinite plate under a tensile stress, δ, the crack opening displacement at the mid-point of the crack length is [53]

$$COD = \frac{4\sigma a}{E} \tag{17}$$

The length of a detectable crack is directly proportional to the resolution of crack opening displacement measurement.

$$2a = \frac{E}{2\sigma} COD_R \tag{18}$$

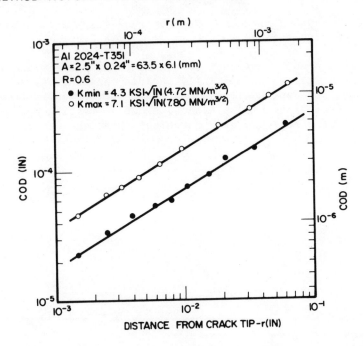

Figure 15.   COD versus distance from crack tip of fatigued
AL 2024-T351 specimen.

It is important that the stress intensity factor of a detectable crack at the working stress is below the threshold K for
crack growth, i.e.,

$$K_{\text{detectable crack}} < K_{\text{th}} \tag{19}$$

The stress intensity factor of a detectable crack at a working
stress, σ, with a displacement resolution COD_R is

$$K = \sigma\sqrt{\pi a} = \sqrt{\frac{\pi}{4} COD_R E\sigma} \tag{20}$$

with a working stress equal to $10^{-3} \pi E$, which is very close to the
endurance limits of many steels and aluminum alloys, the stress
intensity factor, $K = (\pi E/2) \times \sqrt{10^{-3} COD_R}$. For an aluminum alloy,
$E = 6.9 \times 10^4$ MN/m^2 (10^7 psi), and for a value of $COD_R = 6.25 \times$
10^{-4} mm (2.5×10^{-5} in), a crack length of 0.1 mm (0.004") can be
detected. The corresponding stress intensity factor, K, is
2.8 MNm$^{-3/2}$ (2.5 ksi$\sqrt{\text{in}}$), which is very close to ΔK_{th} for fatigue
crack growth. This result indicates that the current moire

technique is capable of detecting any crack which will propagate under a realistic fatigue load.

   In conclusion, the recent developments in the application of high density grilles, specifically the displacement resolution achieved and its applicability to a rough and curved surface, make the moire method feasible as a nondestructive testing method. It is likely that a reliable and convenient technique can be developed.

REFERENCES

1. Kaczer, J. and F. Kroupa, "The Determination of Strains by Mechanical Interference", Czechoslovak J. of Phys., 1 (1952) 80.

2. Dantu, M., "Recherches Diverses d'Extensometrie et de Determination des Contrainter", Conference faite au GAMAC, le 22 feurier, 1954.

3. Durelli, A.J. and V.J. Parks, Moire Analysis of Strain, Prentice-Hall, Inc., Englewood Cliffs, New Jersey, 1970.

4. Theocaris, P.S., "Moire Fringes: A Powerful Measuring Device", Appl. Mech. Rev., 15 (May 1962) 333–39.

5. Theocaris, P.S., Moire Fringe in Strain Analysis, Oxford, Pergammon Press, 1969.

6. Post, D., "The Moire Grid-Analyzer Method for Strain Analysis", Exp. Mech. (Nov. 1965) 368–77.

7. Post, D., "Sharpening and Multiplication of Moire Fringes", Exp. Mech. (1967) 154–59.

8. Morse, S., A.J. Durelli and C.A. Sciammarella, "Geometry of Moire Fringes in Strain Analysis", Proc. ASCE, Vol. 86, (August 1960).

9. Sciammarella, C.A. and A.J. Durelli, "Moire Fringes as a Means of Analyzing Strains", Proc. ASCE (Feb. 1961) 55–74.

10. Sciammarella, C.A. and D.L. Sturgeon, "Digital-filtering Techniques Applied to the Interpolation of Moire Data", Exp. Mech. (Nov. 1967) 468–75.

11. Sciammarella, C.A., "Basic Optical Law in the Interpretation of Moire Patterns Applied to the Analysis of Strains, Part I", Exp. Mech. (May 1965) 154–60.

12. Ross, B.E., C.A. Sciammarella and D. Sturgeon, "Basic Optical Law in the Interpretation of Moire Patterns Applied to the Analysis of Strains, Part II", Exp. Mech. (June 1965) 161–66.

13. Sciammarella, C.A. and F.P. Chiang, "The Moire Method Applied to Three Dimensional Elastic Problems", Exp. Mech. (Nov. 1964) 313–19.

14. Chiang, F.P., "Techniques of Optical Spatial Filtering Applied to the Processing of Moire Fringe Pattern", Exp. Mech. (Nov. 1969) 523–26.

15. Chiang, F.P. and G. Jaisingh, "A New Optical System for Moire Methods", Exp. Mech. (Nov. 1974) 459–62.

16. Chiang, F.P. and G. Jaisingh, "Dynamic Moire Methods for the Bending of Plates", Exp. Mech. (April 1973) 168–71.

17. Ligtenberg, F.K., "The Moire Method – A New Experimental Method for the Determination of Moments in Small Slab Models", Proc. SESA, XII (2) (Dec. 1955) 83–98.

18. Heise, U., "A Moire Method for Measuring Plate Curvature", Exp. Mech. (Jan. 1967) 47–48.

19. Bouwkamp, J.G., "The Moire Method and the Evaluation of Principal-Moment and Stress Direction", Exp. Mech. (May 1964) 121–28.

20. Osgerby, C., "Application of the Moire Method for Use with Cylindrical Surfaces", Exp. Mech. (July 1967) 313–20.

21. Marasco, J., "Use of a Curved Grating in Shadow Moire", Exp. Mech. (Dec. 1975) 464–70).

22. Theocaris, P.S., N. Davids, W. Gillich and H.H. Calvit, "The Moire Method for the Study of Explosions", Exp. Mech. (May 1967) 203–10.

23. Johnson, L., "Moire Techniques for Measuring Strains During Welding", Exp. Mech. (April 1974) 145–51.

24. Sampson, R.C. and D.M. Capbell, "The Grid-shift Technique for Moire Analysis of Strain in Solid Propellants", Exp. Mech. (Nov. 1967) 449–57.

25. Sciammarella, C.A. and B.E. Ross, "Thermal Stress in Cylinders by the Moire Method", Exp. Mech. (October 1964) 289–96.

26. Theocaris, P.S., "Moire Topography of Curved Surfaces", Exp. Mech. (July 1967) 289–96.

27. Brooks, R.E. and L.O. Heflinger, "Moire Gauging Using Optical Interference Patterns", Appl. Optics, 8, No. 5 (1969) 935–39.

28. Takasake, H., "Moire Topography", Appl. Optics, 9, No. 6 (1970) 1467–72.

29. Meadows, D.M., W.O. Johnson and J.B. Allen, "Generation of Surface Contours by Moire Patterns", Appl. Optics, 9, No. 4 (1970) 942–47.

30. Shamir, J., "Moire Gauging by Projected Interference Fringes", Optics and Laser Tech. (April 1973) 78–86.

31. Durelli, A.J. and I.M. Daniel, "A Nondestructive Three-Dimensional Strain-Analysis Method", J. Appl. Mech. (March 1961) 83–96.

32. Underwood, J.H. and D.P. Kendall, "Measurement of Microscopic Plastic-Strain Distributions in the Region of a Crack Tip", Exp. Mech. (July 1969) 296–304.

33. Underwood, J.H., J.L. Swedlow and D.P. Kendall, "Experimental and Analytical Strain in an Edge-Cracked Sheet", Eng. Fract. Mech., 2 (1971) 183–96.

34. Miyamoto, H., Y. Shirota and K. Kashima, "Determination of Stress Intensity Factors Using Moire Method", Proc. of 2nd Int. Conf. Materials Sci., 1976, pp. 1084-88.

35. Kobayashi, A.S., W.B. Bradley and R.A. Selby, "Transient Analysis in a Fracturing Epoxy Plate", Proc. 1st. Int. Conf. Fract., Vol. 3 (L965) 1809.

36. Kobayashi, A.A., D.O. Harris and W.L. Engstrom, "Transient Analysis in a Fracturing Magnesium Plate", Exp. Mech. (Oct. 1967) 434-40.

37. Kobayashi, A.S., W.L. Engstrom and B.R. Simon, "Crack Opening Displacements and Normal Strains in Centrally Notched Plates", Exp. Mech. (April 1969) 103-70.

38. Liu, H.W., W.J. Gavigan and J.S. Ke, "An Engineering Analysis of Ductile Fracture", Int. J. Fract. Mech., 6, No. 1 (1970).

39. Ke, J.S. and H.W. Liu, "The Measurements of Fracture Toughness of Ductile Materials", Eng. Fract. Mech., 5 (1973) 187-202.

40. Luxmoore, A. and P.J. Wyatt, "Application of the Moire Technique to Fracture Toughness Tests on Zirconium Alloys", J. of Strain Analysis, 5, No. 4 (1970).

41. Lehr, K. and H.W. Liu, "Fatigue Crack Propagation and Strain Cycling Properties", Int. J. Fract. Mech. (March 1969).

42. Liu, H.W. and N. Ino, "A Mechanical Model for Fatigue Crack Propagation", Proc. 2nd Int. Conf. Fract., April 1969.

43. Kang, T.S. and H.W. Liu, "Fatigue Crack Propagation and Cyclic Deformation at a Crack Tip", Int. J. Fract., 10, No. 2 (1974) 201-22.

44. Beranek, W.J., "Rapid Interpretation of Moire Photographs", Exp. Mech. (June 1968) 249-56.

45. Luxmoore, A. and R. Hermann, "An Investigation of Photo-resists for Use in Optical Strain Analysis", J. of Strain Analysis, 5 (1970) 162.

46. Wadsworth, N., M. Marchant and B. Billing, "Real-Time Observation of In-Plane Displacements of Opaque Surfaces", Optics and Laser Tech. (June 1973) 119-23.

47. McCall, J.L. and J.E. Boyd, "A Method of Applying Fine Grids to Metal Surfaces for Moire Microstrain Studies", J. of Sci. Inst., 43 (1966) 612-13.

48. Luxmoore, A.R. and R. Hermann, "The Rapid Deposition of Moire Grids", Exp. Mech. (August 1971) 375-77.

49. Leenderz, J.A., "Interferometric Displacement Measurement on Scattering Surfaces Utilizing Speckle Effect", J. of Physics E: Scientific Instruments, 3 (1970) 214-18.

50. Vafiadakis, A.P. and J.H. Lamble, "Mismatch Techniques to Strain Analysis", J. of Strain Analysis, 2, No. 2 (1967) 99-108.

51. Post, D., "Analysis of Moire Fringe Multiplication Phenomena", Appl. Optics, 6, No. 11 (1967) 1938-42.

52. Shah, D. and H.W. Liu, "Application of Digital Filtering to Moire Strain Measurement", Tech. Rep. Materials Sci. Dept., Syracuse U., August 1974.

53. Ang, D.D. and M.L. Williams, "Combined Stresses in an Orthotropic Plate Having a Finite Crack", J. Appl. Mech. (Sept. 1961) 372-82.

Chapter 18

IMAGE PROCESSING IN NONDESTRUCTIVE TESTING

D. H. Janney

University of California, Los Alamos Scientific Laboratory

Los Alamos, New Mexico

ABSTRACT

With increased emphasis on product reliability and with a
generally renewed emphasis on cost control, it is natural that the
users of nondestructive testing should turn to image processing
as a potential method of improving their radiographic techniques.
The questions being asked usually concern the applicability of image
processing for more certain detection of defects, sometimes in the
context of increasing the sampled population of an entire fabrica-
tion process at little increase in cost, sometimes in the context
of making it possible for less experienced personnel to make satisfactory examinations, sometimes simply in the context of obtaining
better radiographic resolution. In those applications where the
principal desire is for higher throughput, the problem often
becomes one of automatic feature extraction and mensuration.
Classically these problems can be approached by means of either an
optical image processor or an analysis in the digital computer.
Optical methods have the advantage of low cost and very high speed,
but are often inflexible and are sometimes very difficult to implement due to practical problems. Computerized methods can be very
flexible, they can use very powerful mathematical techniques, but
usually are difficult to implement for very high throughput. Recent
technological developments in microprocessors and in electronic
analog image analyzers may furnish the key to resolving the shortcomings of the two classical methods of image analysis.

Examples of applications of image processing to nondestructive
testing include weld inspection, dimensional verification of reactor
fuel assemblies, inspection of fuel pellets for laser fusion
research, and, somewhat peripherally, medical radiography.

INTRODUCTION

The manufacturer of any item is often caught between incompatible requirements. On the one hand he must build a product at a price which the customer will accept; on the other hand he not only desires, but is often legally bound, to produce a safe product which will perform its stated function. These requirements reach beyond the normally defined "consumer market". Note a recent court decision that a combat soldier is entitled to a "safe" artilliery shell [1]. Also, in many states there are efforts to restrict the installation of nuclear power reactors until safety is "proven" [2]. The manufacturer's attempts to meet these inconsistent demands are the basis of his quality assurance program, usually leading to a desire for more reliable or for less costly methods of nondestructive testing.

It is natural that the manufacturer in this situation should turn to image processing as a potential aid to his radiographic capabilities. His simply stated purpose is to make radiographically obscure defects more readily visible. In some instances he views this as a potential economy, for it may be possible to use less skilled labor for analysis of the radiographs; in other cases he may view it as a method for offsetting analyst fatigue and for increasing the number of films per hour which an analyst can view. In the ultimate application, image enhancement may be a precursor to automatice detection or mensuration of details in a radiograph.

Unfortunately, the potential of image processing has sometimes been oversold to the nondestructive testing community. This community has been led to believe that with image processing it is possible to make the invisible into the visible. It is this writer's belief, based on experience, that if the feature cannot be seen by a skilled radiographic analyst viewing the radiograph under good viewing conditions, then image processing will not make it visible.

CONCEPTS

Classically, modern image processing means one thing in the field of radiography: it means a boost of the high spatial frequencies in the space-film density surface. The reader is asked to visualize a three-dimensional surface in space. Two coordinates which are used to define this surface are the spatial coordinates of the radiographic image. The third coordinate, generally represented as a height above a base plane, is taken as the film density. Obscure details in a radiograph are usually represented in this surface as small but steep sided features on a broadly rolling contour. These features are represented by high frequencies in a

two-dimensional Fourier transform of this density surface [3,4].
Thus, conventional high-frequency boosting would make these
features show with increased clarity.

At this point it is desirable to distinguish between image
restoration and image enhancement. By "restoration" we usually
mean the removal of defects from an image so that the resultant
image is that which would have been obtained in the absence of all
defects. By "enhancement" we generally mean some process which
will make selected features in the image more readily visible for
detection or mensuration. Since degradations very often are
simply an attenuation of the high frequencies, restoration also
often becomes an effort at high-frequency boosting.

The simple approach of a high-frequency boost is often frustra-
ted by image noise. By noise we have in mind primarily film grain,
but it may also include potentially avoidable defects such as
developer streaks and scratches in the emulsion. Intuition and
experience both indicate that these defects will primarily affect
the high-frequency portion of the radiographic image. Thus, the
high-frequency boost which is desired for the revelation of fine
details will also reveal much noise, often to the extent that the
enhancement effort is made useless. The fact that the noise is a
statistically random aberration superimposed on a deterministic
image makes a substantial body of statistical theory immediately
applicable. Classically, the statistician and the signal processor
have sought to restore a signal rather than to enhance it. None-
theless, as will be indicated below, the available body of knowledge is useful for either task. It is similarly useful for
detecting a specified signal feature embedded in noise, even
though the feature may not be displayed as either enhanced or
restored.

Suppose, as an example, that a real radiograph is thought of
as a perfect radiograph deformed by x-ray scatter and corrupted
by random noise. The task is then one of restoring the real
radiograph to the perfect radiograph.

The Wiener filter is the classical approach with which to perform this restoration, for it will give a restored image which is
the best mean-square approximation to the perfect radiograph [5].
Design of a Wiener filter requires knowledge of both the degrading
function and of the power spectrum of the perfect image. This
type of restoration suffers two defects: the user may have
inadequate knowledge of the undegraded image and the human viewer
uses criteria other than mean-square-error in assessing image
quality [6-8].

These defects in application of the Wiener theory lead to

several other statistically based schemes for image restoration. Most of these schemes are dependent to varying degrees upon the image spectrum, the noise spectrum, and the statistics of both signal components [9-12]. The success of their application is dependent upon the nature of these data and to some extent upon the knowledge of these data.

All of these statistical restoration schemes make use of a point-spread-function (PSF) with which to describe the image degradation in passing from the ideal image to the actual image available to the radiographer. By using a model of PSF which exaggerates high frequencies it is possible to convert a restoration scheme into an enhancement scheme.

It is the author's personal opinion that these statistically based schemes offer the best possibilities for radiographic restoration and enhancement, for they furnish a theoretical basis for extracting signal from noise.

EQUIPMENT

Image processing equipment may consist of three basic types: optical, television-analog, and digital computer. Each will be discussed below.

Optical Processors

An early model of image formation in a microscope, proposed by Abbé, noted that at a selected plane in a projection optical system the Fourier transform is represented by light amplitude [13]. Thus, introduction of a filter at that plane would permit Fourier domain filtering. Though there are many possible variations on this principle, the most common implementation is shown in Figure 1. Systems of this type have potential for exceedingly high throughput, but have only been satisfactorily applied in relatively few instances. The limited acceptance results from a combination of experimental difficulties with extraneous interference effects, relative inflexibility due to a need for a new filter for each new filtering algorithm and due to difficulties in obtaining nonlinear image transformations. This latter difficulty is now being overcome [14-16], possibly the inflexibility will be overcome with programmable light modulators [17].

Television-Analog

It is natural to attempt using the image handling capabilities of a standard television system for image scanning and display. If

COHERENT PARALLEL LIGHT BEAM

INPUT TRANSPARENCY

LENS

FOURIER PLANE WITH FILTER

LENS

FILTERED IMAGE

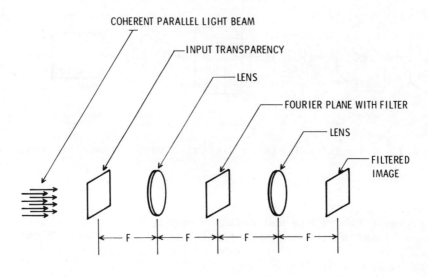

F = LENS FOCAL LENGTH

Figure 1.   Schematic of simple optical image processor.

it is desired only to manipulate the image along a direction parallel
to the scan line, the processing hardware then simply becomes a
suitable video analog signal shaper. Again, a high throughput is
possible. If the operator can rotate the film in its own plane, it
is possible to process readily in any number of directions. This
latter feature gives a variable "three-dimensional" shadow appearance which can be very useful for locating edges perpendicular to
the scan line. Several vendors now market equipment for this type
of processor. Accessories may include density level slicers,
pseudocolor generators and isometric displays of the density-space
curve described above in "Concepts". Systems of this type are
inexpensive. Their drawback, again relying on this author's observations, is that without more sophisticated processing functions it
is difficult to separate signal from noise. Often the sensitivity
to small low contrast features is inferior to the human eye (adverse
criticism relevant to this observation is expected; the author's
reply is simply to show otherwise on typical industrial radiographs).
Bold features can be made to stand out with great clarity for
reliable mensuration.

Digital Computer

Figure 2 shows in simplified schematic form how the density
values of an image may be digitized in a densitometer and the

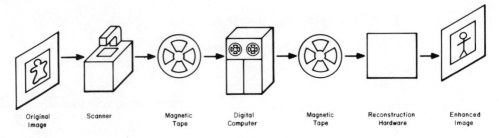

| Original
Image | Scanner | Magnetic
Tape | Digital
Computer | Magnetic
Tape | Reconstruction
Hardware | Enhanced
Image |

Figure 2.   Schematic of simple digital-computer image processor.

density-data matrix then manipulated in the computer for the purpose
of processing the image and redisplaying it.   That figure indicates
a simple image sharpening operation; however, the power of the
digital computer permits all types of manipulations: linear, nonlinear, space or frequency domain, isotropic or unidirectional.
The various statistically based schemes for estimating images in
the presence of noise often require iterative numerical solutions.
At present these can be performed only with the digital computer,
though optical processors with feedback offer future possibilities
[18].

Sometimes the TV-based approach and the digital computer
approach are combined into one processing system.

The digital computer system has relatively low throughput,
but is extremely flexible.   Very extensive processing routines can
be easily synthesized and controlled.   For this reason the systems
are well adapted to developmental activities, sometimes leading to
simpler implementations for high-throughput production work.

The recent dramatic drop in cost of programmable microprocessors may make it possible for laboratories which have in the past
considered themselves out of digital-computer image processing
how to enter the field.   Several vendors of specialized computer
hardware now have Fast-Fourier-Transform boxes which operate under
control of a minicomputer.   Other manufacturers have more general
purpose array processors.   One computer architect is proposing
complete, though small, general purpose computers operating in
parallel, each individually programmed from a control computer.
The advent of "computers on a card" makes many of these approaches
into feasible designs.   It is the author's belief that an installation satisfactory for many purposes can be implemented for no
more than $ 150,000.

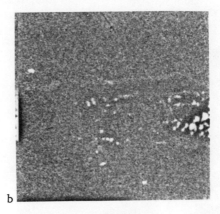

Figure 3.   (a) Radiograph of simulated weld defect.
            (b) Enhanced image of simulated weld defect of 3(a).

EXAMPLES

The purpose of this section is to show several applications
of image processing and analysis. No attempt will be made to
create a coherent unity from these examples.

Figures 3(a) and 3(b) shows the "before" and "after" images
of defective welds created by the U. S. Army Construction Engineering Research Laboratory. The metamorphosis results from frequency
domain processing for the purpose of flattening the background
field followed by extreme density stretching for improving the
contrast [19]. It is useful to note that the original radiograph
was of relatively low noise.

Figures 4(a) and 4(b) show the results of frequency domain
processing of an inspection radiograph of a reactor fuel element.
Though this particular radiograph was of interest only for qualitative inspection, a similar radiograph from the Oak Ridge National
Laboratory was used for dimensional studies. In that case an
analytic model of x-ray transmission was fitted to the image data.
The fitting parameters were taken as a measure of object dimensions
[20].

Another example of feature extraction arises from the problem
of inspecting laser fusion fuel pellets. This problem may be
typified as a need to examine spherical shells of 50 microns outer
diameter by one micron wall thickness. By means of low-voltage
radiography it is possible to obtain high quality radiographs of
these pellets. However, the numbers of pellets which must be
inspected for wall thickness uniformity and sphericity make an

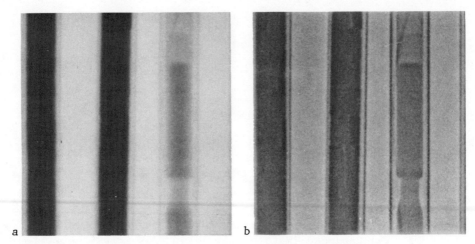

Figure 4.   (a) Standard inspection radiograph of reactor fuel pins.
            (b) Enhanced image of fuel-pin radiograph of 4(a).

automated scheme desirable. Previously developed codes can be used
for locating and sizing simple circular images randomly located in
a microscope field [21]. This code is now being modified so that
it can be used to locate and size both the inner and the outer
surface images in the pellet problem. Though incomplete as of this
writing, the project is mentioned here as a viable application of
image analysis.

Lastly, for image processing seems to offer promise to some
fields of medicine, are Figures 5(a) and 5(b). These form a before-
and-after pair of cephalograms from the Lancaster Cleft Palate
Clinic. With a frequency-domain sharpening followed by a histogram
equalization it is possible to create an image in which the various
cephalometric dimensions [22] can be measured with greater precision.

SUMMARY

Image processing can range from conceptually simple enhancement
through statistical restoration to feature extraction and pattern
recognition. Many of these activities are applicable to radiography,
primarily for making easily missed, but still visible, material
defects more readily detectable. In other applications, image
analysis can be used for mensuration. In this usage the mensuration can be aided by simple sharpening of features, or it may be
automated by statistical analyses of the image data.

Many forms of processing equipment are available. Different

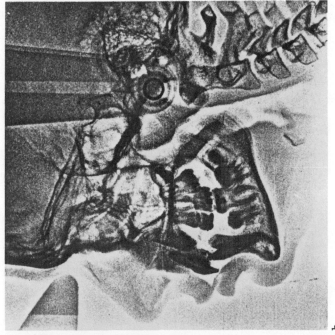

Figure 5. (a) Cephalogram used for measurements of facial detail.
(b) Enhanced cephalogram, obtained by processing 5(a).

types offer different combinations of image quality, throughput, flexibility and cost. A potential user must analyze his individual need prior to installing a processing system.

ACKNOWLEDGEMENT

The author is indebted to several members of the Los Alamos Image Analysis Group for much of the work shown here, specifically R. P. Kruger, H. J. Trussell, R. O'Connor, H. B. Demuth; and to colleagues at other laboratories for examples, specifically E. L. Hall, A. Kumar, A. VanHull, P. Gnadt, and S. W. Oka. This work was primarily funded by the Energy Research and Development Administration under Contract No. 7405-ENG-36.

Much of the background material given in this present paper was supplied by the Picatinny Arsenal as part of its general information for the assistance of laboratories working on this problem.

REFERENCES

1. Berger, H., "Nondestructive Measurements: How Good Are They?", *Mat. Eval.*, 34 (January 1976) 18Aff.

2. Boffey, P.M., "Nuclear Energy: Do States Lack Power to Block Proliferation of Reactors?", *Science*, 191 (January 1976) 360-61.

3. Janney, D.H., B.R. Hunt and R.K. Zeigler, "Concepts of Radiographic Image Enhancement", *Mat. Eval.*, 30 (September 1972) 195 ff.

4. Selzer, R.H., "Digital Computer Processing of X-ray Photographs", Jet Propulsion Laboratory Tech. Report 32-1028, November 15, 1966, p. 12.

5. Pratt, W.K., "Generalized Wiener Filtering Computation Techniques", *IEEE Transactions on Computers*, C-21 (July 1972) 636-41.

6. Wintz, P.A., "Transform Picture Coding", *Proc. IEEE*, 60 (July 1972) 815.

7. Trussell, H.J., "Improved Methods of Maximum a Posteriori Image Restoration", Thesis, University of New Mexico, 1976.

8. Cannon, T.M., H.J. Trussell and B.R. Hunt, "A Comparison of Different Image Restoration Methods", SPSE International Conference on Image Analysis and Evaluation, July 19-23, 1976, Toronto, Canada.

9. Hunt, B.R., "The Application of Constrained Least Squares Estimation to Image Restoration by Digital Computer", *IEEE Transactions on Computers*, C-22 (September 1973) 805-12.

10. Mascarenhas, N.D.A. and W.K. Pratt, "Digital Image Restoration Under a Regression Model", *IEEE Transactions on Circuits and Systems*, CAS-22 (March 1975) 252-66.

11. Hunt, B.R., "Bayesian Methods in Nonlinear Digital Image Restoration", *IEEE Transactions on Computers*, accepted, not yet published.

12. Frieden, B.R., "Restoring With Maximum Likelihood and Maximum Entropy", *J. of Opt. Soc. of Am.*, 62 (April 1972) 511-18.

13. Goodman, J.W., *Introduction to Fourier Optics*, McGraw-Hill Book Co. (1968) 141-97.

14. Dashiell, S.R., "Nonlinear Optical Image Processing with Halftone Screens", Image Processing Institute, U. of Southern California, USCIPI Report 670, May 1976.

15. Kato, H. and J.W. Goodman, "Nonlinear Transformation and Logarithmic Filtering in Coherent Optical Systems", *Optics Communications*, 8 (August 1973) 378-81.

16. Sawchuk, A.A. and S.R. Dashiell, "Nonmonotonic Nonlinearities
     in Optical Processing", IEEE International Optical Computing
     Conference, Washington, D.C., April 23-25, 1975.

17. Casasent, D.P., "A Hybrid Digital/Optical Computer System",
     IEEE Transactions on Computers, C-20 (September 1973) 852-58.

18. Hunt, B.R., "Hybrid Digital/Optical Systems for Nonlinear
     Image Restoration", IEEE International Optical Computing
     Conference, Washington, D.C., April 23-25, 1975.

19. Janney, D.H., "Digital Enhancement of Very Low Contrast
     Radiographs", Mat. Eval., to be published July 1976.

20. Demuth, H.B., "Analysis of Fuel-Bundle Radiographs Using
     Modeling", Los Alamos Scientific Laboratory Report LA-6011-MS,
     July 1975. Also reported at Third International Joint Confer-
     ence on Pattern Recognition, Coronado, Calif., November 8-11,
     1976.

21. Hall, E.L., G. Varsi, W.B. Thompson and R. Gauldin, "Computer
     Measurement of Particle Sizes in Electron Microscope Images",
     IEEE Transactions on Systems, Man and Cybernetics, SMC-6
     (February 1976) 138-45.

22. Enlow, D., Handbook of Facial Growth, W. B. Saunders Co.
     (1975) 251-89.

Chapter 19

AUTOMATED DETECTION OF CAVITIES PRESENT IN THE HIGH EXPLOSIVE FILLER OF ARTILLERY SHELLS

R. P. Kruger, D. H. Janney and J. R. Breedlove, Jr.

University of California, Los Alamos Scientific Laboratory

Los Alamos, New Mexico

ABSTRACT

Initial research has been conducted into the use of digital image analysis techniques for automated detection and characterization of piping cavities present in the high explosive (HE) filler region of 105-mm artillery shells. Experimental work utilizing scene segmentation techniques followed by a sequential similarity detection algorithm for cavitation detection have yielded promising initial results. A description of this work is shown with examples of computer-detected defects.

INTRODUCTION

The U. S. Army's Production Base Modernization Program for ammunition plants includes requirements for high through-put insepction of 105-mm HE projectiles to detect critical cavitation defects and other flaws. Such a projectile is shown in Figure 1. At present this inspection paradigm is shown in Figure 2. With this approach each group of four projectiles are radiographed with a Linatron 4-MeV source and recorded on Kodak AA 35.6- by 42.8-cm sheet film or its equivalent. The film is both automatically loaded and subsequently developed and then presented to one of several film readers for the acceptance or rejection decision. This approach is both costly because of film costs and of doubtful reliability because of reader variability caused in part by fatigue. Reading experience of recent years to these installations has generated considerable skepticism over the validity of qualitative visual analysis and has indicated that the human element is undoubtedly the weakest link in the radiographic inspection process. In

421

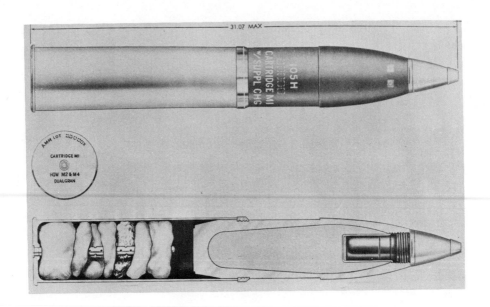

Figure 1.   105-mm HE cartridge.

SCHEMATIC OF

PLANNED SYSTEM

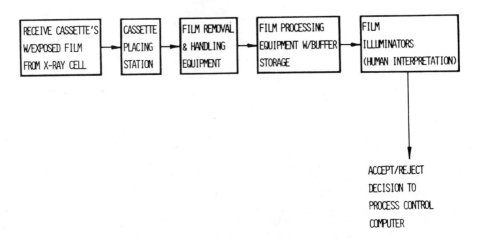

Figure 2.   Present quality control procedure schematic.

view of the fact that defects in the HE filler of artillery shells
can cause premature detonations with the accompanying catastrophic
results, technology to improve the detection of defects is of
utmost importance.

A modern, automated HE melt-our system currently under development will be designed to load at least 800,000 projectiles per month.
Initially 10 percent of these will be radiographed for inspection.
Eventually 100 percent inspection will be required when the process
goes off the controlled phase. Two automation goals should be
considered: 1) elimination of film as a storage medium, 2) elimination of the humanly derived accept/reject decision. This chapter
will deal specifically with defect detection from already existing
film-recorded imagery. However, references to alternative forms
of image acquisition and storage will be referred to when it is
felt appropriate.

CRITICAL HE FILLER DEFECTS

There are four major categories of defects defined as present
in the HE filler. These include a crack, cavity, porosity, and
base separation. A crack is a feature appearing as a dark irregular
line whose length to width ratio equals or exceeds 20 to 1 and whose
width does not exceed .16 cm. Porosity refers to multiple areas
caused by a multitude of small overlapping or interconnected
cavities such that no individual cavity within the area of involvement is clearly discerned. Base separation occurs at filler-
casement boundary and is characterized by a narrow dark band. The
results reported will concentrate on cavity detection. However, it
is felt that the paradigm will be extendable to the other defect
classes as this research continues.

The HE filler has been divided into four discrete segments as
shown in Figure 3. There exist segment-specific criteria related
to maximum permissible defect sizes for each previously defined
defect type. The criteria are expressed in either linear or projected area units. The maximum permissible defect is generally
larger as one proceeds from Segment A to Segment D. These maximums
are bounded by 0 - 1.9 cm for linear measures, and by 0 - 3.25 sq
cm for projected area measures. It should, therefore, be pointed
out that while detection of all defects is always necessary,
there exist segment-specific upper bounds which define acceptability.
Failure to fall below this bound in any segment is sufficient to
reject the projectile. Our current research effort can be divided
into four sequential phases:

1. Detection of shell macrostructures for the purpose of
scene segmentation.

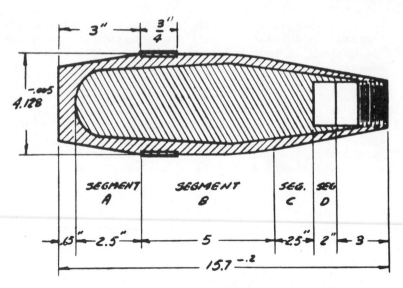

Figure 3.   Delineated segments of HE filler; segmented projectile--
105mm.

     2.   Detection of regions within the segmented shell scene
containing cavitation.

     3.   Measurement of detected anomalies.

     4.   Classification of shells into an accept or reject category.

A brief discussion of the first two phases follows.

Shell Scene Segmentation

     When the human interpreter views a shell to detect and estimate
defects, he first mentally segments the scene into its components.
That is, he notes the regions containing HE filler, casement, and
fuse-well. He may also mentally divide the scene into the four
specified segments A, B, C, and D in order to apply the individual
inspection criteria with respect to each of these segments. Thus,
if one is looking for a pipe cavitation, for example, one must first
isolate the HE filler from the shell's casement. On the other hand,
if one wishes to detect cracks in the shell casement or base region
separation, the detection of the region between the filler and the
shell boundary is necessary.

     With this necessity clearly in mind, an algorithm has been
developed for macrostructure segmentation of the shell into its
component parts on a laboratory computer. Figure 4 shows the

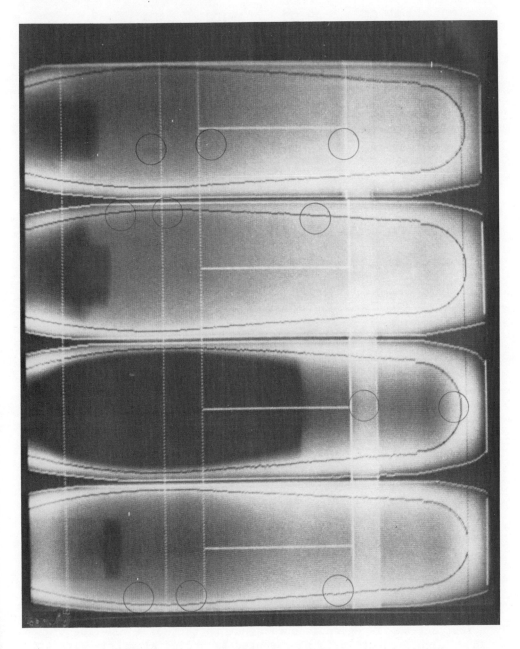

Figure 4. Results of automatic scene segmentation designed to
           isolate shell macrostructure. Encircled regions contain
           cavitations.

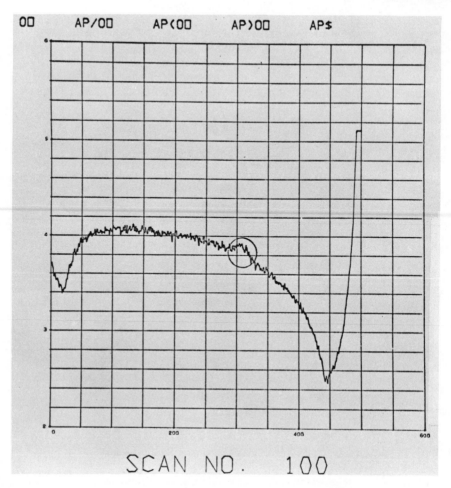

SCAN NO.   100

Figure 5. Representative single scan across a pipe cavitation.
(Note filler-casement boundary appears as points of
maximum relative depression on either side of the shell
midline.)

results of an automatic detection of the HE filler boundary, casement
boundary, rotation band, shell midline, and the four previously
referred-to segments. The algorithm was written and run without
modification on all shells examined. The digital spatial resolution
across a 4-shell region is equivalent to that of a 500-line vidicon.
Thus, each projectile is represented by a digital matrix with 444
rows and 128 columns. The boundary detection techniques involved
rather straightforward threshold detection and boundary following
techniques. Figure 5 shows a typical scan line across one projectile.
The encircled area represents a typical density depression when a
cavitation is encountered. It also appears that the least dense

portion of the scan line (labeled 1 and 2) occurs at the left and
right casement-filler boundaries. A simple peak detector was
therefore sufficient to find these boundaries on each scan line.
Since it is known a priori, that the background-casement boundaries
must lie outside their respective casement filler-boundaries, only
these portions of each scan line need be searched. The boundary
value criteria in this latter case were the detection of the onset
of either a negative (viewer's left), or a positive (viewer's right)
first derivative. These points (labeled 3 and 4) are also shown
for the scan line in Figure 5. Identical detection criteria were
applied to each scan line beginning at the projectile's base and
extending to a point above the fuse-well. Once these boundaries
were successfully detected, the four segment boundaries were determined by the known dimensions of the projectile. All processing at
this level was undertaken in a raster scan (single line at a time)
manner. Thus, no large digital image arrays need be stored in
memory. It appears that this macrostructure scene segmentation
algorithm will be sufficient to isolate the major shell components.

Cavity and Crack Detection

Our method for defect detection involves the use of a class of
algorithms commonly referred to as Sequential Similarity Detection
Algorithms (SSDA) [1,2]. These algorithms have as their major
attribute the potential to examine, sequentially and efficiently,
large regions of a scene and rapidly discard all those regions
except those containing a point of interest, i.e., a crack or cavitation. Thus, they are ideally suited for a detection task where
all but a small percent of the scene is not suspect. We shall
describe SSDA within the context of cavitation detection. It should
be pointed out that, while our initial efforts are confined to a
specific defect, the SSDA concept is easily extended to the other
defect categories. All SSDA are based on the premise that there
exists a finite number of prototypes (templates) which adequately
describe regions of a scene of interest, in this case cavitations.
These templates can be denoted by a set of vectors C defined

$$C \equiv [\underline{t}^1, \ldots, \underline{t}^K] \tag{1}$$

where K = 10 is the number of prototype templates and each $\underline{t}^K$ is a
vector of length m = 16.

In our initial research we are treating cavitations as characterized by profiles of one dependent variable. Several such cavitations are shown in Figure 6. We are assuming that a cavitation
profile will be detectable as a small (.02-.06 density unit)
depression in a raster line scan occurring perpendicular to its
long axis. (Such a depression is shown in Figure 5.) Thus, if one

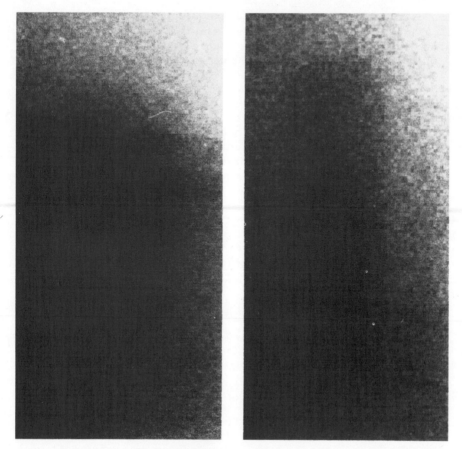

Figure 6. Magnified regions containing cavitations.

detects indications of such a depression on successive raster scans
of the HE filler region of a shell, this defines a suspect region
for this class of defect. In this case each $\underline{t}^K$ represents a vector
of length m = 16 comprising a prototype cavitation profile. Two
such profiles are shown in Figure 7. These one-dimensional profiles
are found by digital integration of representative cavitations in
a direction parallel to their long axes. Let us now segment each
raster line across the shell into 3 parts. Part 1 represents that
portion of the line representing a segment to the viewer's left of
the HE filler. Part 2 is the filler segment itself, and Part 3 is
the line segment to the viewer's right of the HE filler. Segment 2
for each horizontal raster scan across the shell's filler is
detected from the macrostructure analysis phase discussed earlier.
Denote the filler segment on each scan line by a vector $\underline{X}$ of length

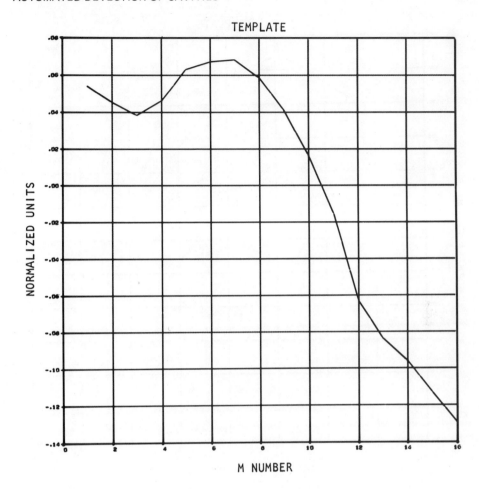

Figure 7(a) and (b) [(b) is on following page]
        Two representative pipe cavity profiles used as templates
        for pipe cavity detection. (These templates constitute
        16-line averages across teh center portion of a
        cavitation.)

n where $n \geq m$, and $n < 128$. The vector $\underline{X}$ is likewise divided into
overlapping subectors $\underline{x}^{\ell}$ of length m. There are $\ell = n - m + 1$ such
subvectors, denoted $\underline{x}^{\ell}$, possible across each $\underline{X}$ when a template
window of length m is slid in unit increments across $\underline{X}$. The detection task is conceptually simple. Given subimage $\underline{x}^{\ell}$, does there
exist a template $\underline{t}^{k}$ such that $\underline{x}^{\ell}$ matches $\underline{t}^{k}$? Since a perfect match
is not expected, an error criterion between each $\underline{x}^{\ell}$ and all members
of C must be established. We will define this normalized error
measure as:

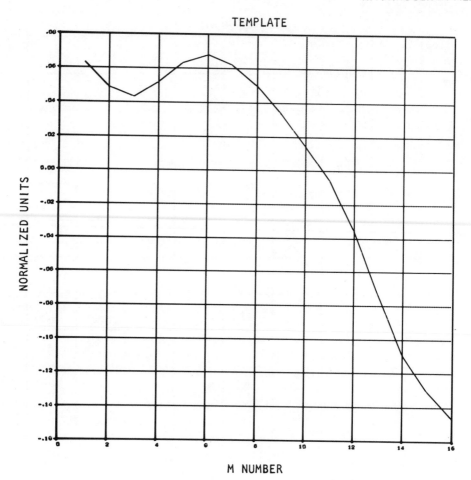

Figure 7(b)

$$e(x_i^\ell t_i^k) \equiv \left| (x_i^\ell - x^{-\ell}) \Big/ \sum_{j=1}^{m} |(x_j - x^{-\ell})| \right.$$

$$\left. - (t_i^k - t^{-k}) \Big/ \sum_{j=1}^{m} |(t_j^k - t^{-k})| \right| \qquad (2)$$

$|\ |$ denotes magnitude

where $x^{-\ell}$ is the mean value of subvector $\underline{x}^\ell$ and t^{-k} is the mean of template k. The cumulative error function E, and threshold function T_k, defined over all elements of $\underline{x}^\ell$ and $\underline{t}^k$ for each template k = 1, ...,K, is defined:

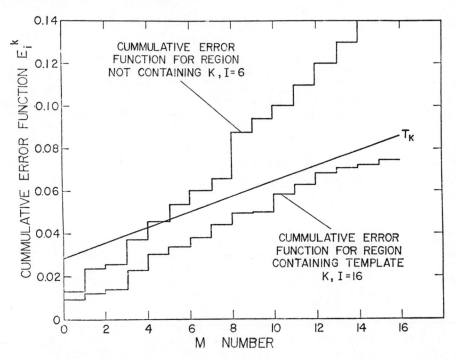

Figure 8.   Cumulative error function and threshold function for
            representative template.

$$E_i^k (\underline{x}^\ell \underline{t}^k) \equiv \sum_{i=1}^{M} e(x_i^\ell, t_i^k) \tag{2}$$

$$\text{for } i = 1,\ldots,M \qquad 1 \leq M \leq m$$

A representative cumulative error funtion E and threshold function
T_k for an $\underline{x}^\ell$ window containing a cavitation is shown in Figure 8.
Also shown are E functions for several $\underline{x}^\ell$ windows which do not
contain cavitations. Note the more rapid error build-up in the
latter case. It should also be noted that the error function E_i
is accumulated with a random index i. This has the effect of
maximizing the amount of new information considered at each step.
Clearly, if an $\underline{x}^\ell$ matches $\underline{t}^k$ perfectly, then $e(x_i^\ell t_i^k)$ is zero for
each i and the monotonic nondecreasing function defined by E_i^k is
a horizontal line of zero slope and intercept. Any degree of
mismatch will increase the slope and intercept of E_i^k. If one
bounds E_i^k from above by threshold function T_k, then the template
$\underline{t}^k$ for which M is maximized (slope of E minimized) for each $\underline{x}^\ell$ is

the best candidate for a match. Expressed mathematically this
maximum is

$$I^k(\underline{x}^\ell) \equiv [i \le m \;/\; \sum_{i=1}^{r} e(x_i^\ell \; t_i^k) \le T_k \; \nabla \; r \le m] \qquad (4)$$

$$k = 1,\ldots,K \qquad\qquad \text{/ denotes "such that"}$$

$$\nabla \text{ denotes "for each"}$$

If one has K such relative maxima for each window $\underline{x}^\ell$, then choose
the I^k for which $i \le m$ is maximized. That is

$$I \equiv \max [I^1,\ldots,I^K] \qquad (5)$$

As each $\underline{x}^\ell$ for each raster line is examined in succession, a
surface of these I maxima defining the filler region will be formed.
Regions of high relative I will define regions of high cavity probability. Only these local regions need be examined in the subsequent measurement and classification phases.

Figure 9 shows the output of the SSDA previously described for
four shells containing a total of ten pipe cavitations. The
lightest region is all the scene area outside the HE filler region.
The medium gray region is the HE filler region which extends from
the base of the HE filler to the end of shell segment C at the base
of the fuse well. The darkest regions define the most probable
regions for the occurrence of pipe caviations and represent less
than one percent of the area over a representative HE filler region.
Also note that over 95 percent of the picture elements are eliminated before I = 3. This represents at least a factor of 10 improvement in computer time over classical correlation measures for K =
10. In this example $1 \ge 12$ was chosen as constituting these most
probably regions. All of the cavitations have been bound along
with several false positive regions which do not contain obvious
voids. For instance, the right-most shell contains a long region
of dark on the right HE filler boundary. This false positive
region is due to the absence of HE filler in the shell. It should
be pointed out, however, that such a shell would have been previously eliminated from the process in a production algorithm
because of inordinately large high densities caused by the missing
HE filler. It was included only because it contained a cavitation.

The goal of the detection phase was to find all cavitations
possible at a cost of several flase positive regions. These
suspected regions would then be examined in the subsequent
measurement phase. It is expected that the absence of a detectable
boundary of the proper shape will eliminate these false positive
regions at that time. The detection phase just described is

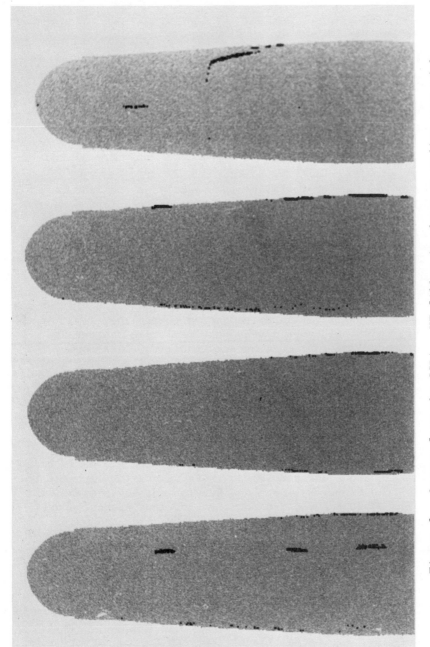

Figure 9. Output from the SSDA. HE filler is shown as medium gray with cavitations shown as black areas. Fully 90 percent of the image area is eliminated from examination before $Ik = 3$. Encircled regions contain detected cavitations.

critical to the success of the total process because it must
isolate all potential "needles in the haystack", i.e., small
suspect regions.

CONCLUSIONS

We have chosen to concentrate initial feasibility efforts on
Phase 2, the detection phase. If reliable detection is possible,
regions with cracks or cavitations can be measured with already
existing texture or boundary tracking techniques. Since the major
defect classification criteria are based on permissible area
involvement as a function of shell segment, the summed area of
boundary enclosed regions in specific segments should offer little,
if any, difficulty.

It should be noted that SSDAs are easily generalized to
characterize two-dimensional space if some defect classes require
this. These SSDAs offer the best hope for a rapid sequential
large region search technique designed to detect the relatively
small subregions containing cracks or cavitations. In this application, K = 10 templates were checked against each subvector $\underline{x}^{\ell}$ in
about the same amount of time necessary to perform on classical
correlation over the same population of subvectors. It also
appears probable that an intensified 500- or 1000-line vidicon may
possess enough sensitivity and resolution to replace film. At the
very worst, this would then involve the real time capture and
storage of an entire four-shell scene in digital mass memory.
Such is completely possible with current technology. The SSDA
algorithms could then either be hardwired or programmed using
existing microprogrammable devices.

ACKNOWLEDGEMENT

The authors wish to thank E. M. Sandoval for his assistance
during the research just completed. Work was performed under the
auspices of the Energy Research and Development Administration,
Contract No. W7405-ENG-36, and for the Picatinny Arsenal under
Memorandum of Understanding AT(29-2)-2138, Modification 2.

REFERENCES

1. Barnea, D.I. and H.F. Silverman, "A Class of Algorithms for
    Fast Digital Image Registration", IEEE Trans. on Computers,
    C-21 (February 1972).

2. Ramapriyan, H.K., "A Multilevel Approach to Sequential Detection
    of Pictorial Features", IEEE Trans. on Computers, C-23, No. 1
    (January 1976).

Chapter 20

ADVANCED QUANTITATIVE MAGNETIC NONDESTRUCTIVE EVALUATION METHODS – THEORY AND EXPERIMENT

J. R. Barton, F. N. Kusenberger, R. E. Beissner, and
G. A. Matzkanin

Southwest Research Institute

San Antonio, Texas

ABSTRACT

The scale of fatigue crack phenomena is reviewed in relation to the size detection capabilities of nondestructive evaluation methods. Several features of the fatigue phenomena which should be considered in developing nondestructive characteriziation methods are tabulated and briefly discussed. A qualitative assessment of such factors in relation to the inspection of ball and roller bearing components suggested that magnetic methods were very promising. The basis of the magnetic methods is magnetic domain phenomena and several aspects are briefly reviewed, including interaction of domains and inclusions and the influence of stress and magnetic field on domains. While magnetic calculations from first principles are extremely complicated when applied to engineering specimens, simplified treatments have been developed and will be reviewed. Experimental results will also be described which indicate that in many instances the simplified calculations can be used to predict many features of the experimental results. A cursory comparison of results predicted by the simple analytic model and other models in which finite element computer analysis predictions have been made do not agree for certain features. Experimental results and analyses obtained on rod-type fatigue specimens which show the experimental magnetic measurements in relation to featues such as crack opening displacement, crack opening volume, crack depth and other features show much promise in providing methods for greatly improved characterization of cracks in relation to fracture mechanics analyses and life pre-diction.

435

INTRODUCTION

Magnetic methods are one of the oldest nondestructive testing approaches and perhaps are more widely used than any of the conventional nondestructive evaluation methods except radiography [1-3]. The most commonly used approach is the magnetic particle method [2,4] and, essentially, consists of magnetizing a part and then flowing fine magnetic particles or flowing magnetic particles suspended in a solution over the region to be examined. Flaws are indicated by the accumulation of particles attracted to the magnetic poles formed by discontinuities; often a fluroescent coating is added to the particles and an ultraviolet light source is used to illuminate the region being examined, thereby enhancing the visual indication. A number of specialized magnetic methods have been used for selected applications and the magnetic approaches offer potential for application to a wide-range of specialized problems.

It is somewhat surprising that magnetic methods have not been treated more thoroughly from a mathematical and theoretical viewpoint, since they were extensively used prior to the widespread use of ultrasonic and eddy current methods. The latter two approaches have received very thorough theoretical and analytical investigations in the NDE context and it is suggested that somewhat similar efforts could yield significant benefits when applied to magnetic methods. Recently a few analytical treatments have emerged in the NDE context [5-7] and, hopefully, such efforts will be continued and expanded. Magnetic methods not only can provide high sensitivity for flaw detection, but potentially can also provide information on thermo-mechanical processing and stress and strain through the magneto-elastic effects. Accordingly, it is somewhat surprising that the methods have not been placed on a firmer scientific basis. Perhaps the reason for this situation is merely the fact that the numerous interactions and nonlinear magnetic parameters, such as permeability, make it extremely difficult to unravel the influence of the individual material conditions which it is desirable to ascertain.

The physical basis for the magnetic properties of ferromagnetic materials (practially all steels, some cast irons, nickel and some nickel alloys) are best described by the magnetic-domain theory [8-17]. This theory, which now has comprehensive experimental confirmation, postulates that the material is comprised of local regions called ferromagnetic domains, each magnetized to saturation, but aligned according to the state of local magnetization. Adjacent domains are separated by a thin, highly localized magnetic transition region called a domain wall or Bloch wall. Even in the demagnetized state, all domains are still magnetized to saturation, but the orientation of the individual domain magnetization vectors is random, which results in the net magnetization

of the specimen being zero. The application of a magnetic field
or a stress [9,16-20] can change the configuration of the domains,
principally by wall movements.

Extensive analytical and experimental investigations have
confirmed that with carefully prepared specimens, usually single
crystals, it is possible to calculate domain wall positions and
configurations based on minimum energy considerations [8,10,21,22].
Magnetic pole distributions around inclusions have been calculated
and experimentally observed and an excellent example is shown in
References 8b and 22. However, with complex shaped polycrystalline
specimens and engineering components, precise calculations based on
domain interactions have not been accomplished. For such cases
classical magnetostatic approaches have been useful and will be
briefly discussed later.

The fact that magnetic domain walls can be forced to move
under the influence of a changing applied magnetic field or a stress,
provides the fundamental basis for the Barkhausen noise method of
nondestructive residual stress measurement [18-20,33,39]. In 1917
Barkhausen [24] discovered that voltages induced in an electrical
coil encircling a ferromagnetic specimen produced a noise when
suitably amplified and applied to a speaker, even though the magnetization applied to the specimen was changed smoothly. From such
experiments he inferred that the magnetization in the specimen does
not increase in a strictly continuous way, but rather by small,
abrupt, discontinuous increments, now called Barkhausen jumps.
Such jumps are caused, principally, by the discontinuous movements
of mobile magnetic boundaries, Bloch walls, between adjacent magnetic domains and occasionally by the initiation of new magnetic
domain walls. Furthermore, the direction and magnitude of the
mechanical stress existing in a macroscopic ferromagnetic specimen
strongly influences the detailed dynamics of the domain wall motion
and correspondingly influences the Barkhausen noise. Figure 1
shows several photographs of magnetic domains (made visible by a
magnetooptic method) on the surface of a carefully prepared single
crystal of silicon-iron and illustrates the manner in which domains
change under the influence of an applied stress [18,20]. The
magnetic field was applied from left to right in the illustration
and this was also the axis of the applied compressive stress.
Arrows have been placed on some domains to indicate the direction
of magnetization within the domain. A careful examination of these
photographs shows marked changes occurring as the compressive stress
is increased; for example, the domains are principally oriented
approximately 45° upward and to the left in the top photograph,
while several domains have grown in a direction oriented upward
and to the right in the lower illustration. As these domains move,
Barkhausen pulses are generated. Figure 2 shows many of these
pulses sensed by an induction coil near the surface of a low carbon
steel bar.

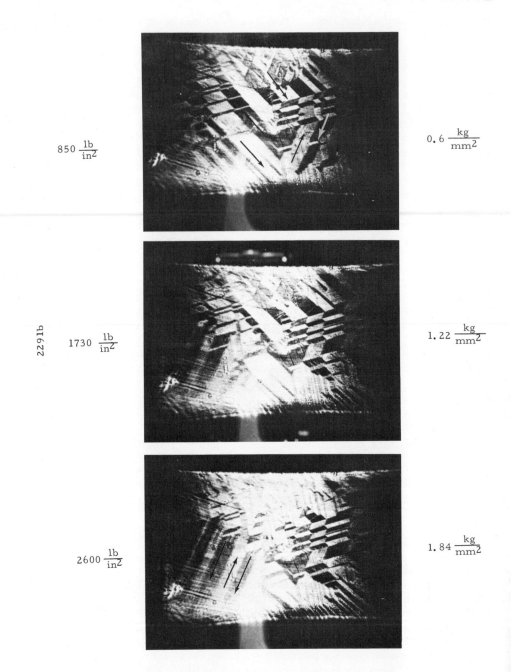

Figure 1.   Influence of compressive stress on ferromagnetic domains
            in single crystal of silicon iron (domains made visible
            by a magnetooptic method).

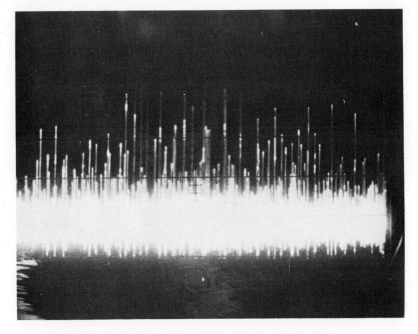

Figure 2. Typical Barkhausen noise pulses.

The foregoing has provided backgroun and a perspective and this chapter will briefly review results obtained over a number of years using advanced magnetic evaluation methods. The work began in the early 1960's when an industrial client contracted with Southwest Research Institute to develop more sensitive methods for subsurface flaw detection in bearing components such as shown in Figure 3. Subsequently, the work has been extended to the early detection of fatigue damage [10,25-38] and more recently to the measurement of residual stress using the Barkhausen noise phenomena; Also several specialized inspection systems have evolved. The research has been concerned primarily with the detection of very small flaws and small fatigue cracks and it is instructive to examine Schijve's scale [39] of fatigue crack phenomena, Figure 4, to provide a dimensional perspective for the problem. On this scale the region in which conventional nondestructive methods have been effective is the range from approximately 0.04 to 0.25 inch and larger; it is pointed out that several recent investigations [40-43] have been conducted to assess the statistical confidence obtainable with the conventional methods and the foregoing numbers were deduced from such investigations. As will be shown, the advanced magnetic methods have been useful in detecting fatigue cracks with a minimum

Figure 3.   Precision rolling element bearing components.

size of approximately 0.003 inch [26-28,30,35] and, under some
conditions, for example in ball bearing components, inclusions as
small as 0.001 inch in diameter have been repeatedly and reliably
detected [18,29,31,37,38,44,45].   An example of such an inclusion
is shown in Figure 5.   As can be observed, the inclusion is approximately 0.001 inch in diameter and is about 0.001 inch beneath the
metal surface.

A number of methods were investigated analytically, using
simple models, to determine their applicability for detecting small
inclusions in bearing steels and magnetic field methods appeared to
offer the most promise.   The analysis for the magnetic problem will
be briefly reviewed and several interesting elements of the analysis
pointed out.

Assume a spheroidal region (inclusion) of constant permeability
μ_0 imbedded in an infinite ferromagnetic matrix of constant permeability μ and with a magnetic field H_0 applied along the X axis as
shown in the upper part of Figure 6.   An exact solution [46,47] for
the magnetostatic potential at any point P (x, Y) outside the inclusion is given by Equation (1):

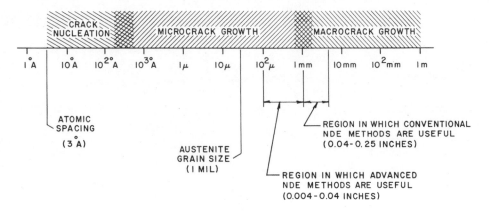

Figure 4.   Schijve's scale of fatigue crack phenomena.

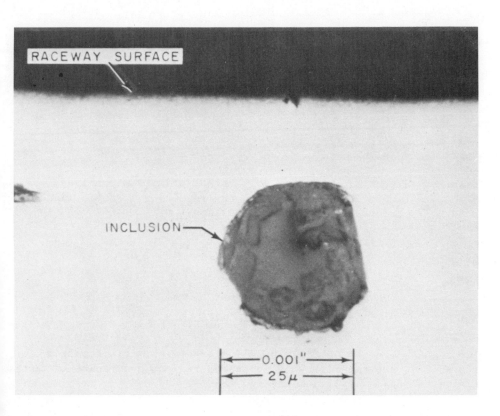

Figure 5.   Subsurface inclusion disclosed by metallurgical
            sectioning.

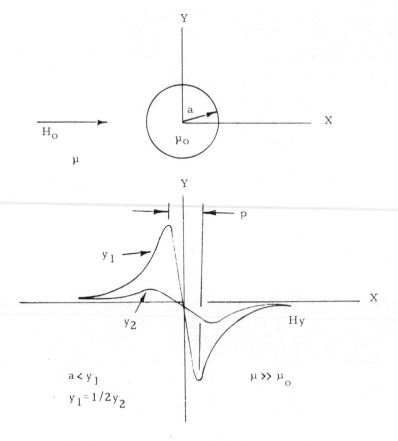

Figure 6.   Sketch of spherical void in an infinite magnetized
            ferromagnetic matrix and the resulting magnetic field
            perturbation in the Y direction.

$$U(x, y) = -[1 - (\frac{\mu_0 - \mu}{\mu_0 + 2\mu}) \frac{a^3}{(x^2 + y^2)^{3/2}}] H_0 x \qquad (1)$$

where: U = magnetostatic potential, μ = permeability of matrix
material, μ_0 = permeability of inclusion region, a = radius of inclusion, and H_0 = applied magnetic field.

        The magnetic field components H_y and H_x can be obtained by
taking the negative partial space derivatives along the X and Y
directions. Equations (2) and (3) are obtained:

$$H_x = \frac{-\partial U(x,\ y)}{\partial x} = H_0 \left[1 - \left(\frac{\mu_0 - \mu}{\mu_0 + 2\mu}\right)\left(\frac{a^3}{(x^2 + y^2)^{3/2}}\right)\left(1 - \frac{3x^2}{x^2 + y^2}\right)\right]$$

(2)

$$H_y = \frac{-\partial U(x,\ y)}{\partial y} = 3H_0 \left[\left(\frac{\mu_0 - \mu}{\mu_0 + 2\mu}\right)\left(\frac{a^3}{(x^2 + y^2)^{5/2}}\right) xy\right]$$ (3)

Attention will be concentrated on Equation (3) which is the magnetic field component, H_y, that would be obtained if a magnetometer probe were scanned parallel to the X axis to detect the magnetic field component emerging out of the surface of a bearing race. (The boundary, assumed to be parallel with the X axis, is neglected as a first approximation and can be partially accounted for by the u* correction [8,48]).

A plot of H_y obtained along a line a distance y from the center of the sphere and extending in the X direction is presented in the lower part of Figure 6; values obtained on a line at two different values of y are shown. Note the characteristic shape of the magnetic field perturbation. If it were possible to scan a point probe along these same lines in the material, then this sketch would represent the shapes of the "signatures" obtained from the "flaw".

Since it is desirable to obtain quantiative NDE data, it is informative to determine the relationship between inclusion size, a, distance, y, beneath the surface and signature features which may provide information on these factors. Accordingly, the locations of the positive and negative peaks are obtained by taking the partial derivative of H_y with respect to x and setting the result equal to zero. For conciseness the details are omitted, but the very simple relation x = ± 1/2y is obtained. These peaks occur at the locations x = ± 1/2y or the peak to peak separation, p in Figure 6, is equal to the distance, y, from the scan path to the centroid of the flaw independent of the flaw size. This is a very important observation since for many practical cases the flaw depth is a primary parameter for assessing criticality. The amplitude of the signature at this location can be obtained by substituting the values x = ± 1/2y into Equation (3). After simplifying, Equation (4) is obtained.

$$H_y\ (x = \pm 1/2y) = K \frac{\mu_0 - \mu}{\mu_0 + 2\mu}\ \frac{a^3}{y^3}$$ (4)

Examining this equation it is seen that the signature amplitude is directly proportional to the volume of the flaw, a^3, and inversely proportional to the cube of the distance, y, from the scan path to the centroid of the flaw. Furthermore, an examination of the

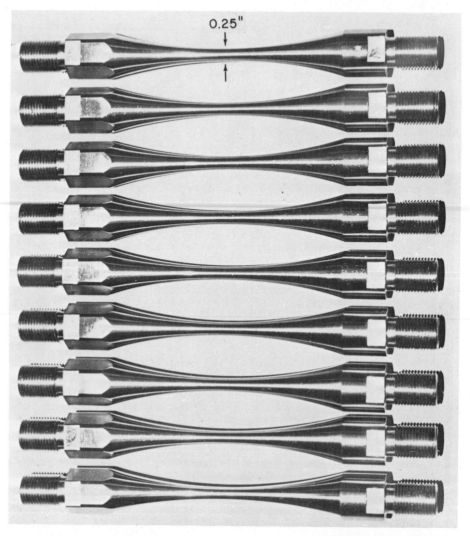

Figure 7.   Photograph of typical AISI 4340 steel rod-type tensile
fatigue specimens.

factor containing the permeability terms shows that when μ is larger
than μ_0, the polarity of the signature would reverse.

Analysis of the case for an infinite cylinder extending along
the Z axis (normal to the page in Figure 6) shows that the signature
amplitude is proportional to a^2/y^2 and again the peak separations are
equal to the depth y.

Although the boundary has been neglected and the permeabilities are constant in these elementary analyses, the results, nevertheless, have been very helpful in guiding the experimental work and in the interpretation of experimental results. Selected results will be reviewed in the next section.

EXPERIMENTAL RESULTS

An extensive continuing experimental investigation [25-28,49-51] of fatigue damage mechanisms and nondestructive fatigue damage detection methods has been conducted on the specimens shown in Figure 7. These are conventional rod type tension specimens, containing no notches, and the surfaces are carefully finished to remove all machining and grind marks. For the magnetic perturbation experiments the specimens have been AISI 4340 steel, heat treated to a yield strength of approximately 180,000 psi and, usually, stressed to 130,000 psi in uniaxial tension (r = 0). The handbook fatigue life under these conditions is approximately 100,000 cycles. Typically, data are acquired at a relatively high and a relatively low magnetic flux density by scanning a small magnetometer along the longitudinal axis of the specimen. The magnetometer is a Hall effect probe with a minute sensing element approximately 0.001 in. X 0.004 in. thereby providing very high resolution. Adjacent scans, 3/4 inch long, are made approximately every 2° to cover the entire center section of the specimen. Data are obtained before cycling and at selected intervals during cycling; the specimen is monitored continuously by means of six ultrasonic surface wave transducers operating at 10 MHz and when the pulse echo responses show an indication of change, the cycling is interrupted and new magnetic perturbation data acquired.

Figure 8 shows typical perturbation signatures obtained under several different magnetic field conditions and at different cycling intervals. In these records the vertical axis is the magnetic perturbation signature amplitude and the horizontal axis is distance along the specimen with the record typically covering approximately 3/4 inch, centered on the mid point of the specimen. The upper left hand record, obtained at low magnetic flux density (near residual) after accumulating 2.62×10^6 stress cycles, shows no distinct features and only a general upward curve of the baseline. By contrast, the record at the upper right, obtained under high magnetic field, shows several sharp vertical excursions or signatures. These signatures are caused by surface and subsurface inclusions and very small surface pits. The lower right hand record was obtained under identical NDE conditions, but after the addition of 40,000 stress cycles during which time a small fatigue crack developed and was propagated until the length was approximately 0.010 inch long. A comparison of the upper right hand and lower right hand records at the signature indicated by the arrow shows essentially no

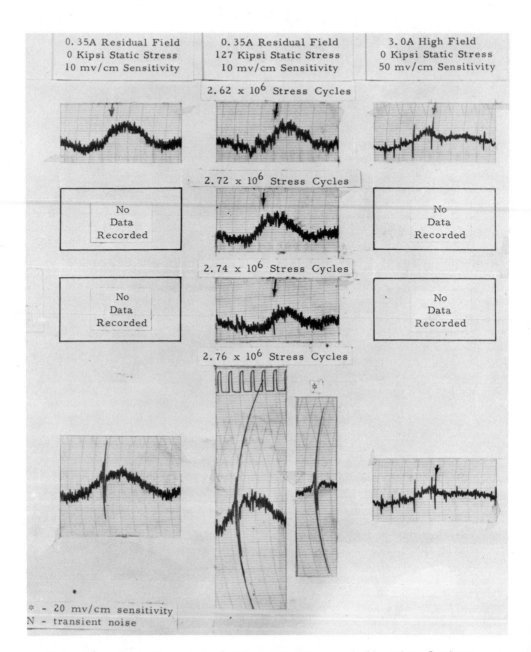

Figure 8.   Magnetic perturbation signatures indicating fatigue
            damage.

difference, even though this signature coincides with the location
of the fatigue crack. Accordingly, it must be concluded that these
conditions are not useful for detecting small fatigue cracks. The
signature amplitude is determined primarily by: the volume of the
inclusion, the distance from the scan path to the centroid of the
inclusion and the applied magnetic field H_0. Since in general the
permeability of the inclusion will be approximately 1, the permeability term of Equation (3) becomes approximately 1/2.

A strikingly different result is obtained under low magnetic
field conditions as is obvious by an examination of the record at
the upper left and the lower left. While no signature at all
which is above electronic system noise was obtained under low field
conditions before the fatigue crack developed, an outstanding signature is obtained after development of the fatigue crack. In this
case the signature is caused primarily by the volume of "fatigue
damaged material" rather than the inclusion volume and the permeability term becomes very complicated since the permeabilities
involved are the permeabilities of the matrix material u and the
permeability associated with the fatigue damaged region of the
matrix material u_d (u_d varies within the damaged volume). The
enhancement of the fatigue crack signature by the application of
a load under low magnetic field conditions is obvious in the lower
row middle column of records; an approximate four-fold increase in
signature amplitude is caused by the application of the load. The
volume of the opened crack along with the complicated interaction
of the applied load and its associated influence on the local
permeability of the material in the vicinity of the fatigue crack
probably accounts for the large signal amplitude. However, at this
point a precise analysis has not been attempted.

The foregoing has provided some insight into the complicated
response of steel to magnetic test conditions and, also, has illustrated the sensitivity and versatility of the magnetic perturbation
method. Furthermore, it has shown the crucial importance of
controlling the magnetic field. Subsequent results will expand on
the quantitative aspects of the method.

QUANTITATIVE ASPECTS OF MAGNETIC NDE

Records have been obtained from a number of ball bearing inner
races and a high degree of correlation between inclusions as confirmed by metallurgical investigation and magnetic perturbation
signatures has been obtained (29,37,38,44,45]. Typical results
are presented in Figure 9. In this illustration the upper record
shows a complete magnetic perturbation scan around the circumfer-
ence of a small bearing. The upper trace is the magnetic perturbation response and the probe covered a strip of the material

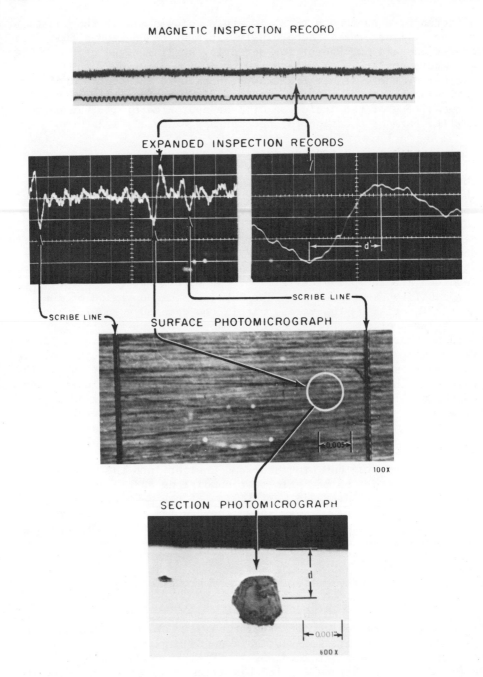

Figure 9.   Inspection record and metallurgical sectioning
correlating inclusion with signature.

approximately 0.025 inch wide. The lower trace is an angular mark
every 5° which can be used to establish the physical location of
the signature region on the surface of the bearing. Attention is
called to the vertical excursion slightly to the right in the upper
record. In preparation for metallurgical sectioning, two fine lines
are scored on the surface of the race using a diamond stylus.
These lines are usually approximately 0.05 inch apart and it is
attempted to bracket the signature location between these two lines.
After making the scribe lines a new magnetic record is obtained at
an expanded sweep and such a record is shown at the left in the
second row; two signatures are generated by the scribe lines and the
separation of the two signatures is, of course, the same as the
physical separation of the two scribe lines on the bearing surface.
The distance on the magnetic record between the scribe lines and
the magnetic perturbation signature from the region to be investigated can be marked on the surface of the race and this region is
indicated by the 0.005 inch diameter circle in the photomicrograph.
A section of the ring is removed by cutting along radial lines and
then the cross-section is metallurgically polished. Examination
is made at each 0.0005 inch increment of material removed until
the location of the inclusion is reached. The inclusion causing
the signature is shown in the lower photomicrograph, and the
distance between the bearing surface and the centroid of the
inclusion is indicated by d. The distance d is proportional to the
distance d' shown in the greatly expanded record of the signature
in the second row, right. Metallurgical investigation of a number
of inclusions and machined flaws has confirmed a proportionality
between peak separation and depth with the peak separation being
approximately 0.5 x depth for near surface flaws and 0.3 x depth
for deep flaws. Figure 10 is a plot of typical results for
several bearings.

To show the relationship between inclusion diameter and signature amplitude, data have been collected from the rod-type tensile
experiments in which the inclusion diameter was measured during
metallurgical investigations. Figure 11 is a plot of the data
and it is observed that the signal amplitude increases directly as
the volume of the inclusion [50]. Another interesting and important
result is shown in Figure 12 in which a plot of signal amplitude
versus the cycles to initiate the propagate a fatigue crack to a
length of 0.003 inch is presented [50]. It is seen that the cycles
to develop the small fatigue cracks is a strong, almost logarithmic
function of signal amplitude and in turn inclusion volume.

The high resolution of the small Hall-effect probe has permitted
detailed investigations of small fatigue cracks as shown by the
signatures obtained on multiple scan paths across the fatigue crack
shown in Figure 13 [49]. A plot of signal amplitude versus crack
opening displacement, COD, measured near the center of the crack

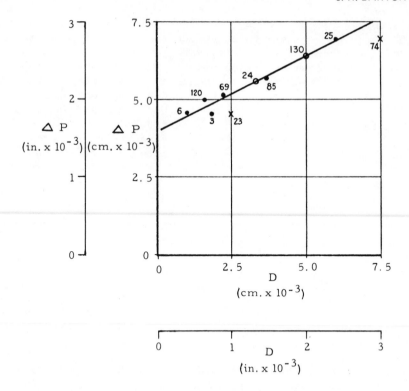

Figure 10.   Experimental results indicating linear relationship
             between peak separation, ΔP and depth of inclusion D.

is presented in Figure 14.   Since the area of the opening between
the crack interfaces should be approximately linearly related to
crack openign displacement, the observed linear relationship
between signal amplitude and COD again shows correspondence with
predictions of Equation (5) modified for the two dimensional case.
Note that at very small CODs less than approximately 50×10^{-6}
inches, there is little of no change of signal amplitude with COD.
A plot of signal amplitude obtained at different locations across
the fatigue crack is presented in Figure 15.   The points are an
excellent fit to a parabola and it should be remarked that a parabolic relationship between crack opening displacements at various
locations along the crack has been predicted by Francis [51] using
a fracture mechanics analysis.   The relationship between peak
separation and crack depth has been estimated on the basis of a
surface entering (half-penny shaped) crack and it is observed that
an approximately linear relationship exists between crack depth
and peak separation at small depths, but departs from this
relationship at greater depths.

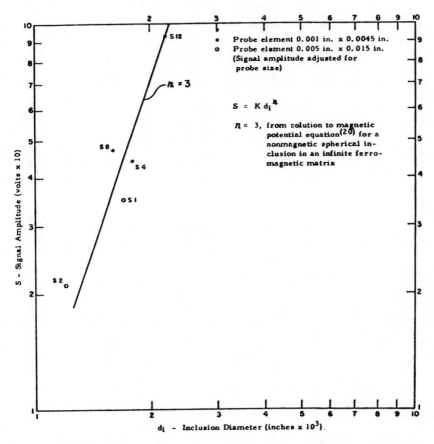

Figure 11.   Functional dependence of magnetic signal amplitude on inclusion size.

Peak-to-peak separation data have been obtained on a crack as it was propagated to various lengths and the data were obtained with zero applied stress and with a peak stress of 130 ksi which was the peak cycling stress.   These data are plotted in Figure 16 and, although there is significant scatter, it is observed that the peak separation changes as the crack gets longer.   Since the cracks are approximately half-penny shaped, the depth would be approximately 1/2 the crack length; furthermore, if it is assumed that the crack opening is wedge shaped, then the centroid of the crack area would be approximately 1/3 the depth, independent of the COD, i.e., stress, and the data are in reasonable agreement with such a prediction.   It is very significant that within the limits of the experimental measurements the peak separation does not change as the load is changed from zero to 131 ksi, although

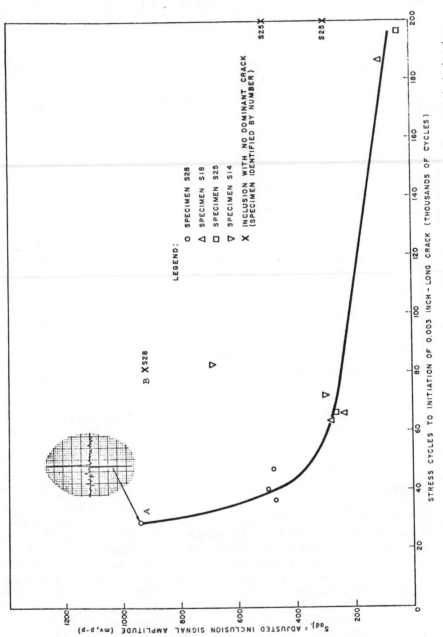

Figure 12. Precycle inclusion signal amplitude versus stress cycles to initiation of 0.003 inch long crack.

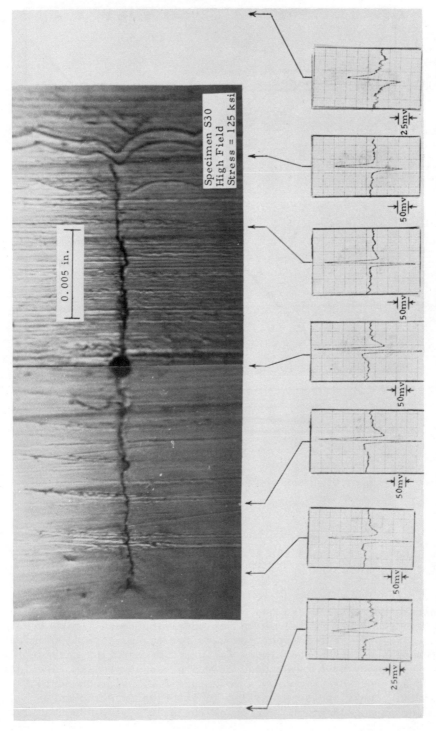

Figure 13. Photomicrograph showing a fatigue crack with associated magnetic records.

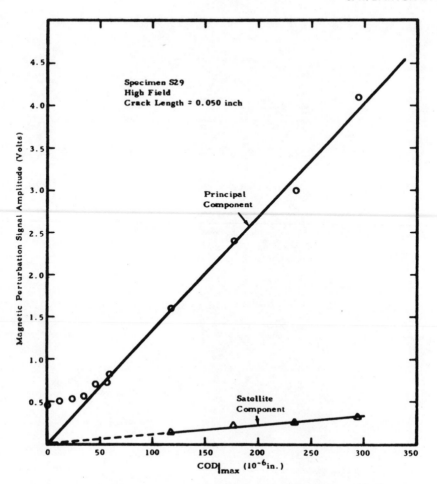

Figure 14.   Magnetic perturbation signal amplitude versus COD_{max}.

this load change causes the peak signal amplitude to increase by a factor of greater than 9 as shown by the data of Figure 14.

A striking example of the influence of permeability on the signature characteristics has been obtained on aircraft engine piston pins and the experimental setup for obtaining the data is shown in Figure 17 [52]. The inspections were for the purpose of eliminating piston pins which had developed fatigue damage during service. In the preliminary investigations, it was necessary to simulate fatigue damage and for such purposes a reference flaw was induced in the specimen. The flaw was a shallow indentation impacted into the exterior surface near the center of the piston pin. Such a flaw

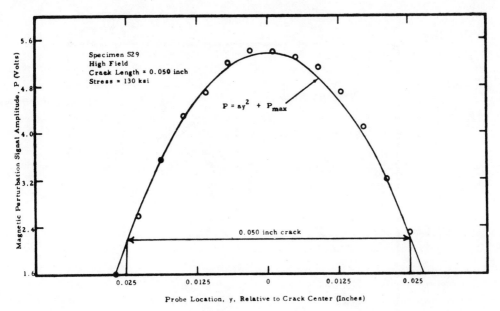

Figure 15.   Variation of principal component amplitude along a
crack interface.

has two features, namely, a void or air volume caused by displacement of the metal and, importantly for simulating fatigue cracks,
a highly stressed and/or strained local volume in the vicinity of
the indent. Several magnetic perturbation inspection records obtained at different applied magnetic fields are shown in Figure 18.
The large vertical excursions at the left and right are end effects
and the region of the indent is near the middle. The first record
shows the signature has an upward and downward going characteristic;
as the magnetic field is gradually increased, this component of the
signature decreases and then the downward and upward going part of
the signature begins to dominate as the applied magnetic field is
increased to near saturation. This change of polarity can be explained as follows. At low magnetic fields the configuration of
the magnetic domains is primarily determined by the local stress
and strain volume surrounding the indent, and not by the magnetic
pole distribution caused by the hole; however, at very high magnetic
fields, the influence of the stress and strain is overcome by the
very high applied magnetic field and the domain configuration then
essentially is determined by the void volume of the indent. An
examination of the signatures also shows that the peak separation
of the upward and then downward going signatures are significantly
greater than are those obtained on the downward and then upward going

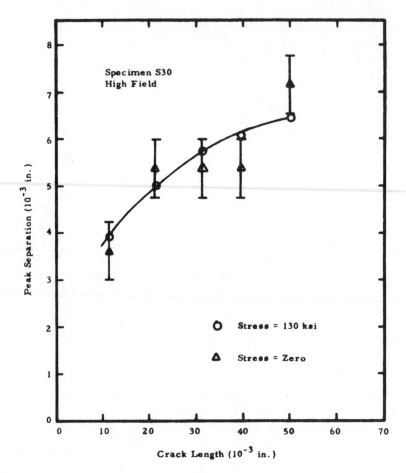

Figure 16.   Peak separation of magnetic perturbation signature
             versus crack length.

or void signatures. The foregoing has shown the very complicated
interactions that are obtained with magnetic methods and also emphasizes the need for control of the magnetic field if quantitative
information is to be obtained.

     The fatigue investigations on the rod-type tensile specimens
have disclosed that with few exceptions the fatigue cracks initiate
at small surface and near-surface inclusions. Other extensive investigations [29,37,38,44,45,53] have also shown that such inclusions
are primary sources of failure in ball bearing races where a complex
three-dimensional stress system exists. A typical example is shown
in Figure 19. In this illustration the upper trace is a magnetic
perturbation record obtained from a scan track approximately 0.025

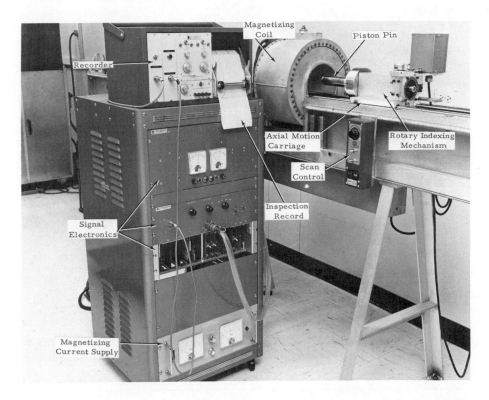

Figure 17. Experimental arrangement for fatigue damage detection
           in piston pins.

inch wide and shows one complete revolution. The outstanding magnetic perturbation signature at approximately 12°, indicated by the
arrow, was generated from a subsurface inclusion and a surface photomicrograph of the region containing the inclusion is shown in the
third row right. It is obvious that the surface is good and there is
no indication of a flaw. The bearing was endurance tested and the
second record was obtained approximately two years later after the
bearing failed; note the correspondence of the signature from the
spall and the original inclusion signature at 12°. A photograph of
the spall is shown in the third row, left illustration, and the
square superimposed on this photomicrograph shows the same surface
area as the photograph at the right before failure. The bottom row
shows respectively, at the left a photograph made through the microscope with bright field vertical illumination, the same area with
polarized light and it was observed that the small dot in the center
of the circle fluoresced when illuminated with polarized light and
is, in fact, the inclusion still remaining in the leading edge of
the spall as shown by the double exposure in the right hand illustration. This small inclusion caused the original magnetic perturbation
signature and was also the primary source of the failure. A large

Field-Amps

Sensitivity mv/cm

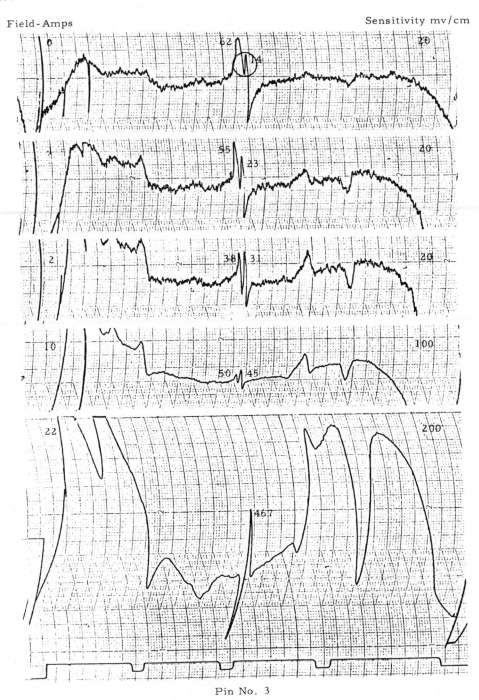

Pin No. 3

Figure 18.   Changes in magnetic perturbation signature as influenced by applied magnetic field.

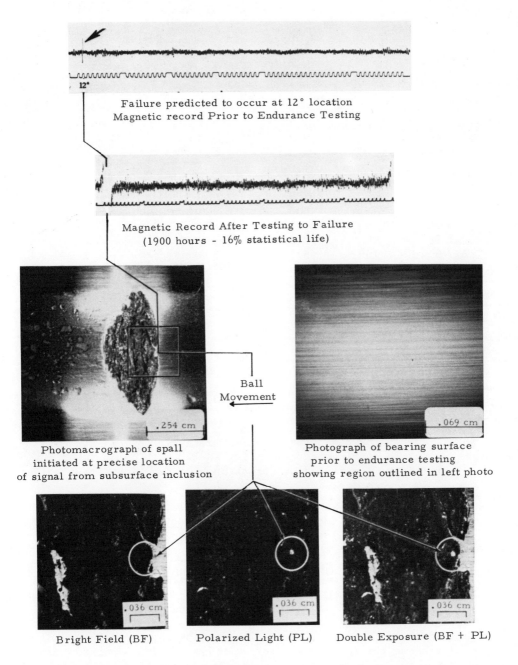

Failure predicted to occur at 12° location
Magnetic record Prior to Endurance Testing

Magnetic Record After Testing to Failure
(1900 hours - 16% statistical life)

Photomacrograph of spall
initiated at precise location
of signal from subsurface inclusion

.254 cm

Ball
Movement

Photograph of bearing surface
prior to endurance testing
showing region outlined in left photo

.069 cm

Bright Field (BF)   Polarized Light (PL)   Double Exposure (BF + PL)

.036 cm   .036 cm   .036 cm

Figure 19.   Life predication-failure initiated at subsurface
inclusion.

TABLE I

Quantitative NDE Results on Ball Bearings

Signal Amplitude	Ring No.	Axial Location	Hours	Sectioning Size cm x 10^{-3} (in x 10^{-3})
67	24	7.7°	1,902	6.8(2.70)F
60	48	18.2	1,826	-
58	130	2.5	1,333	- F
53	6	347.0	P 10,000	2.54(1.00)C
49	84	347	4,000	7.0 (2.75)
47	120	357.5	P 10,000	2.3 (0.90)
39	3	3.5	4,000	3.8 (1.50)C
38	69	23.5	P 10,000	3.05(1.20)
36	82	357.5	P 10,000	-
35	148	18.2	P 9,655	-
30	25	352.2	P 9,940	2.8 (1.10)
29	110	347	3,635	-
25	129	2.5	4,000	-
25	74	2.5	3,126	2.5 x 11.5 x 0.25 (1.0 x 4.5 x 0.1)
22	23	7.7	P 10,000	3.8 x 6.3 x 0.5 (1.5 x 2.5 x 0.2)
22	89	357.5	P 10,000	-
20	96	18.2	-	-
20	7	2.5	3,077	-

Notes: a. 103 bearings inspected; b. 23 inclusion signatures;
c. 54 bearings endurance tested (20 with inclusion signatures; F = failed at inclusion; C = cracking at inclusion;
P = prestressed.

number of bearings, including the bearing of Figure 19, were endurance tested and the results are presented in Table I. The magnetic
perturbation signal amplitudes obtained from individual specimens
is shown in the left column in order of decreasing amplitude of
signature, the next column indicates the bearing serial number,
the next column shows the angular location across the bearing race
and, since the bearings were radially loaded, the position at 0°
(also 360°) was the region of highest load stress and would therefore be the most critical region for inclusions, and the next
column shows the endurance time accumulated. It is pointed out
that the two early failures, Serial No. 24 and No. 130, failed
respectively at 16% and 6% statistical life based on Weibull
statistics. These two components could have been eliminated by
discarding all specimens with a magnetic perturbation signature

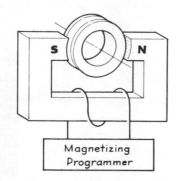

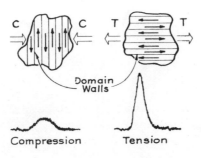

Figure 20.   Schematic illustration of Barkhausen noise stress
              measurement method applied to bearings.

amplitude in excess of 50 mv and located in the region of high
load stresses, that is within ± 10° of the center line or 0°
location.   Other extensive endurance testing results have provided
somewhat similar data, but for conciseness will not be reviewed.

     The Barkhausen Noise Analysis method for residual stress
assessment as applied to bearings is schematically illustrated in
Figure 20.   At the bottom the characteristic signatures obtained
after processing the Barkhausen noise pulses are illustrated and
it is pointed out that low amplitude signatures are obtained from
compression regions and high amplitude signatures are obtained
from tension regions.   This method has been used to obtain measurements on bearings endurance tested in the laboratory and also on
several different types of bearings before and after actual gas
turbine engine service [54].   Typical examples of results obtained
on J79 engine fuel pump bearings are shown in Figure 21.   In this
illustration the solid curve is a plot of the signal amplitude
obtained on the radial bearing in which the load track is located
near 0°.   Data obtained from the mating angular contact bearings
are also shown as the dashed line plot.   Note that the peak of this
plot occurs near 342°, which is the gauging contact angle or region

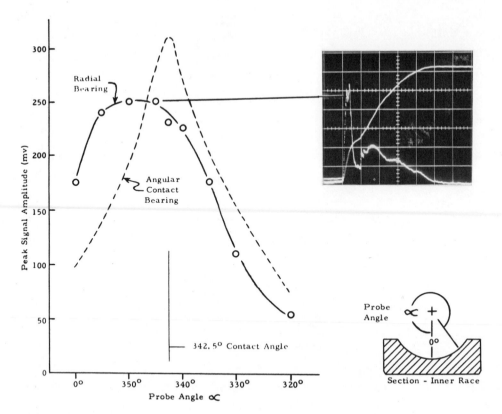

Figure 21.   Major Barkhausen signature changes caused by service on
J79 fuel pump radial and angular contact bearings.

of highest load stress, and is a relatively sharp peak. By contrast
the load track on the radial bearing must vary somewhat to accomodate thermal expansion and other axial motions and, accordingly,
the Barkhausen signature plot is somewhat flat-topped and not as
high amplitude as was the case for the angular contact bearing.
These results are in quantitative agreement with the anticipated
running conditions. The photographic insert shows a typical Barkhausen noise processed signature and it is emphasized that prior to
service the large amplitude spike at the left of the signature did
not exist since the bearings were in residual compression. The
large spike signatures indicate development of tension stresses in
the load track region as a result of service. A recent publication
[55] provides an explanation for the residual stress changes and
attributes these changes to a subsurface material transformation
caused by the high repeated load stresses. The residual stresses
associated with such transformations have been measured using x-ray
diffraction at successive surfaces exposed by removal of thin layers
of the material. Results obtained on bearings that have been

endurance tested and also on bearings that have not been tested
are compared in the plots of Figure 22. Note that on the endurance tested bearing a tension stress peak of approximately 30,000
psi occurs 0.005 inch beneath the surface; an inclusion located
at this region or slightly above would have a high probability of
developing cracks and causing a failure. Accordingly, the measurement of such residual stresses are important in forcasting bearing
life, particularly for bearings that are to be reused and/or
restored for further service as contemplated by the recent Army-
NASA investigation [56].

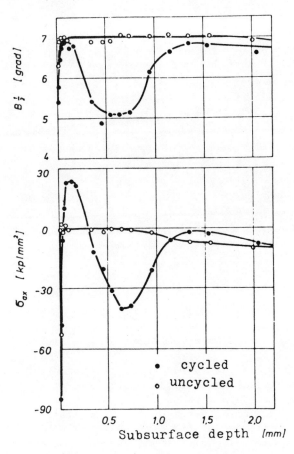

Figure 22.   Line width and residual stress below the contact area
              of a cycled and an uncycled rolling element [55].

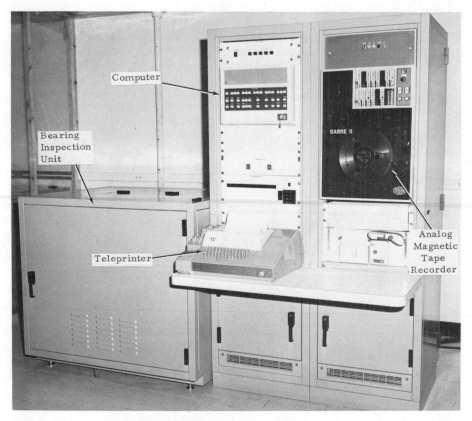

Figure 23.   Overall view of Critical Inspection of Bearings for
             Life Extension (CIBLE) equipment.

TRANSLATION OF METHODOLOGY INTO EQUIPMENT

     Success of the magnetic methods in providing NDE measurements
of critical importance in determining the material performance of
bearings has provided the technical basis for significantly improved
nondestructive evaluation procedures. Gradually, during many discussions, with personnel of the Army, Navy, Air Force and commerical
airlines, concepts evolved for incorporating the improved methodology
into an integrated program with the acronym CIBLE – Critical
Inspection of Bearings for Life Extension. For conciseness,
details of the program have been omitted and are described elsewhere [57]. However, a brief description of the NDE equipment which
has evolved and preliminary but important results which have already
been obtained will be reviewed.

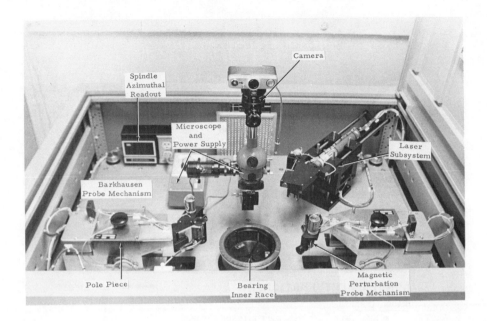

Figure 24. Closeup showing bearing component inspection deck.

An overall view of the CIBLE equipment is shown in Figure 23. The cabinet at the left encloses the electromechanical assembly on which bearings are actually inspected. The two equipment racks contain a minicomputer, a magnetic recorder, power supplies, signal conditioners, etc. A close-up view of the inspection deck is shown in Figure 24. The mechanical assembly essentially consists of: (1) a precision spindle, driven by a DC motor with precise speed control, for rotating the inner and outer bearing races, a 5000 count/rev. digital shaft encoder is directly attached to the spindle shaft for indicating angular shaft location; (2) a magnetic circuit assembly with a range of pole pieces which can accomodate a wide range of inner and outer bearing race sizes, probes and preamplifiers attached to pole pieces which match the contour of the race surfaces; (3) a low-power laser integrated into a power driven mechanical assembly to position the laser for scattered radiation inspection, a power-driven mirror scans the beam across the bearing race; (4) a light-beam system for detecting a reference mark on each bearing thereby permitting precise correlation of all data to the specific location generating the signature, etc. Supervisory control of all functions is accomplished by an HP 9600 computer data acquisition system integrated with the mechanical, electrical and electronic subsystems of the bearing inspection system.

TABLE II

Description of Equipment Features and Inspection Capabilities

CIBLE

(Critical Inspection on Bearings for Life Extension)

AUTOMATED BEARING INSPECTION SYSTEM

FEATURES

COMPUTER SUPERVISED AND CONTROLLED INSPECTION
RAPID FIXTURING CHANGEOVER FOR DIFFERENT BEARINGS
COMPUTER SETUP OF PARAMETERS FOR DIFFERENT BEARINGS
COMPUTER PRINTOUT OF SIGNAL LOCATIONS
PERMANENT RECORD ON MAGNETIC TAPE
DIAGNOSTIC PRINTOUTS AND SAFETY INTERLOCKS

SPECIFICATIONS

INSPECTION METHODS	CONDITIONS DETECTABLE	SCAN PATTERN
MAGNETIC PERTURBATION		
RADIAL FLUX		
● HIGH FIELD	SURFACE PITS, INCLUSIONS,	
● LOW FIELD	SPALLS AND INDENTATIONS	0.025-INCH WIDE CIRCUMFERENTIAL STRIPS WITH 20% OVERLAP
CIRCUMFERENTIAL FLUX		
● HIGH FIELD	SUBSURFACE INCLUSIONS, AND SPALLS AND DEEPER SURFACE ANOMALIES	
● LOW FIELD	FATIGUE DAMAGED REGIONS AND INDENTATIONS	12 TO 60 SCANS PER INSPECTION METHOD
LASER-SCATTERED LIGHT		SYNCHRONIZED SCANS
SURFACE ANOMALY	SURFACE SCRATCHES, PITS, SPALLS, AND INDENTATIONS	
SURFACE FINISH	RELATIVE SURFACE FINISH	
BARKHAUSEN NOISE		
	RELATIVE SURFACE AND NEAR-SURFACE RESIDUAL STRESS CONDITIONS	PROGRAMMED SAMPLING 0.050 × 0.050-INCH REGIONS 9 TO 15 LOCATIONS
	SERVICE MODIFICATION OF RESIDUAL STRESS	

In addition to system supervision and parameter input, the computer will be used in data analyses. Software also includes a number of diagnostic routines which aid in identifying specific malfunctions of the equipment.

The equipment automatically performs seven different nondestructive evaluations as shown in Table II.

At the present time a system has been completed and is being used for further research, and also, to provide specialized bearing inspection services at SWRI; a system has been delivered to the Army at the Corpus Christi Overhaul Depot and is being integrated into the bearing inspection processing, and a third system is being completed and will be delivered soon to the Air Force.

As part of the overall Air Force program a number of bearings are to be endurance tested in the laboratory to assist in developing an understanding of the detailed mechanisms involved in the spalling phenomena. Candidates for such experiments are selected on the basis of specific signature characteristics occurring in and near the operating load track of the bearings. Before endurance testing, comprehensive data will be permanently recorded using the CIBLE equipment. The endurance runs will be interrupted at selected intervals, additional inspection data obtained, and then the endurance testing will be continued. Two bearings can be tested simultaneously and at the present time, four bearings have been partially tested. The first two bearings completed 100 hours endurance testing, were removed from the equipment, and two other bearings installed and testing initiated. After only 16 hours of testing one of these latter two bearings failed and a photograph of the failed or spalled region is shown in Figure 25. The computer generated printout obtained during routine inspection of the bearing component is presented in Figure 26; only one magnetic perturbation signature and numerous laser signatures were obtained. In the printout the column labeled:

TY indicates the inspection parameters of the particular inspection being conducted,

ST is the probe step location across the raceway surface,

BR is the location of the signature from a permanent reference line on the bearing race, and

SR is the location of the signature with respect to the 0 location of the digital shaft encoder.

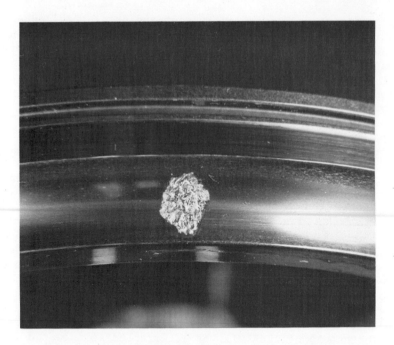

Figure 25. View of spalled region on inner race of J57-#4 bearing
            (S/N 10161) which developed after 16 hours of endurance
            testing.

In the lower group of data the symbol LA is for laser scattered
light associated with surface pits and other surface anomalies.
Concentrating attention on the first row of data the signature was
obtained under circumferential flux orientation high magnetic flux
density, CH, and such signatures are caused primarily by inclusions
and voids. In the next column the number 0007 is a probe step
location coinciding with the load track of the inner bearing race.
The fact that no radial flux orientation, RH, signature was obtained
and also the fact that no signature was obtained with the laser
scattered radiation coincident with the location of the circumferential high field signature are all indicative of the fact that the
signature is from a subsurface source. A record of the signature
is shown in Figure 26 and the polarity corresponds with that obtained
from inclusions. Furthermore, an expanded horizontal axis record
of the signature is presented in Figure 27 and the peak-to-peak
separation indicates that the source is approximately 0.002 to
0.003 inch beneath the surface being inspected. Detailed analyses
of the inspection data have established conclusively that the
precise location of the signature was the extreme leading edge of
the spall and a magnified view illustrating the results is presented
in Figure 27.

```
                        ┌ Bearing Serial No. , J57-#4, Inner Race, Loaded Half
ENTER BEARING│PARAMETERS SEPARATED BY A COMMA OR SPACE
11 S10161@│6599 056 0 2 0 1 0 M B L
ENTER THE FOLLOWING 21 DIGITS ON THE TAPE ENCODER
1 1 8 3 6 4 1 0 1 6 1 6 5 9 9 0 0 5 6 2 2 0
TYPE YES WHEN DONE YE

LOAD P-TAPE NO. 110
TYPE YES WHEN DONE YE
LOAD BEARING, CLOSE DOOR, THEN TYPE YES  YE

ENCODER READING AT BEARING REF. MARK DETECTION 2967
BEARING REF. MARK FROM SHAFT ZERO 4725

FLAWS
TY  ST  BR  SR           Legend: CH-Circumferential Flux High
CH 0007 4850 4575                  Flux Density
                              ST-Probe Position Across Race
                              BR-Circumferential Position From
                                  Ref. Scribe Line
                         Note:  Bearing was selected for endurance
                                testing on basis of signal @ 0007 and 4850
DECREASE TAPE SPEED TO 3. 75 IPS, THEN TYPE YES YE
INCREASE TAPE SPEED TO 30. 0 IPS, THEN TYPE YES YE

ENCODER READING AT BEARING REF. MARK DETECTION 2967
BEARING REF. MARK FROM SHAFT ZERO 4725

ENCODER READING AT BEARING REF. MARK DETECTION 2967
BEARING REF. MARK FROM SHAFT ZERO 4725

FLAWS
TY  ST  BR  SR
LA 0000 3182 2907
LA 0001 3182 2907
LA 0002 3157 2882
LA 0002 3205 2930
LA 0003 3170 2895
LA 0004 3170 2895
LA 0010 2633 2358
LA 0012 1028 0753
LA 0012 4161 3886
```

Figure 26.   Computer printout and associated magnetic perturbation
             signature (CH) indicating subsurface inclusion in
             J57-#4 bearing S/N 493-1.

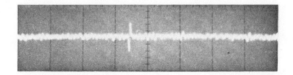

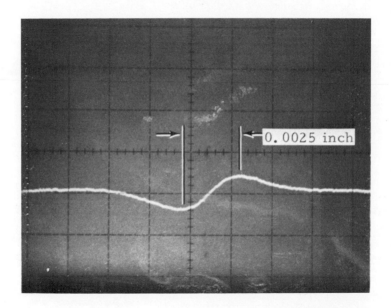

Figure 27.   Signature from subsurface inclusion at BR 4850 before
             endurance test.

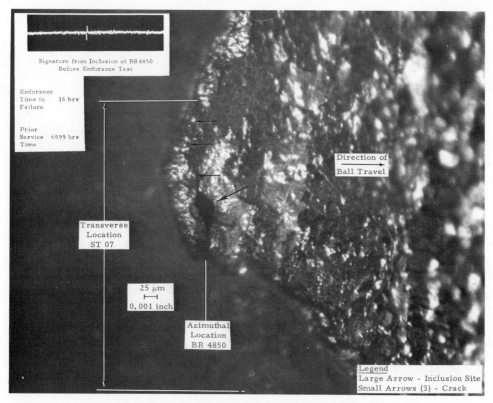

Figure 28. Photograph of endurance tested bearing showing precise correlation between signature.

The inclusion which generated the magnetic perturbation signature and which also caused the failure is visible and is indicated by the arrow; in addition, what appears at this magnification to be cracks extending from the inclusion are also indicated by several arrows. It is emphasized that this bearing which failed after only 16 hours of endurance testing was selected for the endurance test on the basis of the circumferential high field, CH, printout of Figure 26 and the signature associated with this printout location shown in Figure 27. Accordingly, this is considered a striking example of the effectiveness of the methodology and hardware which has been developed. It is also important to point out that prior to the endurance test this J57 #4 position engine bearing had accumulated a total of 6600 hours service time and apparently was near failure from the subsurface inclusion. If this bearing had been returned to service the probability of an early engine failure and associated damage would have been extremely high.

CLOSING REMARKS

The preceding has presented an overview intended to be of sufficient detail to provide a convincing basis illustrating the sensitivity, versatility, and effectiveness of advanced magnetic nondestructive evaluation methods. It should be apparent that further analytical investigations which incorporate the boundary conditions at the surface of components being inspected, an extension to a variety of flaw configurations, and the refinements to incorporate the effects of varying permeability, etc., are desirable. Results from such analyses could perhaps provide guidance for inspection parameters which could more precisely define the flaw parameters indicated by fracture mechanic analyses to be of importance in assessing flaw criticality.

The research and evolution of the bearing inspection equipment has spanned more than 15 years and has required significant long-term dedication of resources; however, prospects for an early and substantial savings based on use of the equipment and methodology are highly promising. It is hoped that success of the overall program can provide a perspective to aid in justifying commitment on a long-term basis of resources and effort to attack other difficult NDE problems which demand solutions if we are to achieve high reliability; acceptable life cycle costs; high component and equipment performance; efficient use of scarce materials; and component utilization to a significant fraction of individual lives without incurring unacceptable risks of failure. Ever increasing energy costs will accentuate requirements and could become a cardinal factor for justifying commitments.

ACKNOWLEDGEMENT

The information presented has been abstracted from research covering nearly two decades and the technical contributions of the numerous personnel involved is gratefully acknowledged. In addition, funding and encouragement of many government agencies including the Air Force, Army, Navy, and NASA is gratefully acknowledged. The long-term continuous research funding from the Air Force Office of Scientific Research is cited as being of particular significance with the emphasis on basic understanding of phenomena and associated analytic methods.

REFERENCES

1. McMasters, R.C., ed., Nondestructive Testing Handbook, Vols. I and II, Soc. for Nondestructive Testing, Roland Press, 1963.

2. Betz, C.E., Principles of Magnetic Particle Testing, Magnaflux Corp., 1963.

3. Gardner, C.G., Nondestructive Testing - Asurvey, NASA SP-5115, 1973.

4. Doane, F.B. and C.E. Betz., Principles of Magnaflux, Magnaflux Corp., 1941.

5. Zatsepin, N.N. and V.E. Shcherbinin, "Calculation of the Magneto-Static Field of Surface Defects, I. Field Topography of Defect Models", Defektoskopiya, 5 (Sept.-Oct. 1966) 50-59; "II. Experimental Verification of the Principal Theoretical Relationships", Defektoskopiya, 5 (Sept.-Octo. 1966) 59-65.

6. Hwang, J.H. and W. Lord, "Finite Element Modeling of Magnetic Field/Defect Interactions", J. of Test. and Eval., 3, No. 1 (Jan. 1975) 21-25.

7. Hwang, J.H. and W. Lord, "Magnetic Leakage Field Signatures of Material Discontinuities", Proc. of 10th Symposium on Nondestructive Evaluation, San Antonio, Texas, April 1975.

8a. Kittel, C., "Physical Theory of Ferromagnetic Domains", Reviews of Mod. Phys., 21, No. 4 (October 1949).

8b. Kittel, C. and J.K. Galt, "Ferromagnetic Domain Theory", Solid State Phys., 3 (1956).

9. Bozarth, R.M., Ferromagnetism, New York: D. Van Nostrand Co., Inc., 1951.

10. Williams, H.J. and W. Shockley, "A Simple Domain Structure in an Iron Crystal Showing a Direct Correlation with the Magnetization", Phys. Rev., 75, No. 1 (January 1949).

11. Goodenough, J.R., "A Theory of Domain Creation and Coercive Force in Polycrystalline Ferromagnetics", Phys. Rev., 95, No. 4 (August 1954).

12. Kittel, C., "Theory of the Formation of Power Patterns on Ferromagnetic Crystals", Letters to the Editor, Phys. Rev., 76, No. 10 (November 1949).

13. Bates, L.F. and F.E. Neale, "A Quantitative Study of the Domain Structure of Single Crystals of Silicon-Iron by the Powder Pattern Technique", Proc. of Phys. Soc., Vol. 63.

14. Graham, C.D., Jr. and P.W. Neurath, "Domain Wall Orientations in Silicon-Iron Crystals", J. Appl. Phys., 38 (July 1957).

15. Stewart, K.H., "Domain Wall Movement in a Single Crystal",
     Proc. of Phys. Soc., LXIII, 7-A.

16. Corner, W.D. and J.J. Mason, "The Effect of Stress on the
     Domain Structure of Cube-Textured Silicon Iron", Proc. of
     Phys. Soc., 81 (1963).

17. Dijkstra, L.J. and U.M. Martius, "Domain Pattern in Silicon-
     Iron Under Stress", Rev. of Mod. Phys., 25 (January 1953).

18. Gardner, C.G. and J.R. Barton, "Recent Advances in Magentic
     Field Methods of Nondestructive Evaluation for Aerospace
     Applications", Propulstion and Energetics Panel Advisory Group
     for Aerospace Research and Dev., London, England, April 1970.

19. Pasley, R.L., "Barkhausen Effect - An Indication of Stress",
     Mat. Eval., 28 (1970) 157.

20. Gardner, C.G., G.A. Matzkanin and D.L. Davidson, "The Influence
     of Mechanical Stress on Magnetization Processes and Barkhausen
     Jumps in Ferromagnetic Materials", Int. J. Nondestructive
     Testing, i.

21. Neel, L., "Effet des cavités et des inclusions sur le champ
     coercitif", Cahiers Phys., 25 (1944b) 21-44.

22. Williams, H.J., "Direction of Domain Magnetization in Powder
     Patterns", Phys. Rev., 71 (1947).

23. Barton, J.R., "Residual Stresses in Gas Turbine Engine Compon-
     ents from Barkhausen Noise Analysis", Gas Turbine Conference
     ASME, Zurich, Switzerland, March-April 1974. Published in
     J. of Engr. for Power, October 1974.

24. Barkhausen, H., "Zwei mit Hilfe der neuen Verstarker entdeckte
     Erscheinungen", Physik Z., 20 (1919).

25. Barton, J.R., "Early Fatigue Damage Detection in 4140 Steel
     Tubes", Proc. Fifth Annual Symposium on NDE of Aerospace and
     Weapons Systems Components and Materials, South Texas Chapter
     ASNT, San Antonio, Texas, April 1965.

26. Kusenberger, F.N., B.E. Leonard, J.R. Barton and W.L. Donaldson,
     "Nondestructive Evaluation of Metal Fatigue", Air Force Office
     of Scientific Research, Scientific Report AFOSE 67-1288, April
     1967.

27. Barton, J.R., W.L. Donaldson and F.N. Kusenberger, "Nondestruc-
     tive Evaluation of Metal Fatigue", Proc. of the Office of
     Aerospace Research, Research and Applications Conf., Arlington,
     Virginia, March 1969.

28. Kusenberger, F.N., P.H. Francis, B.E. Leonard and J.R. Barton,
     "Nondestructive Evaluation of Metal Fatigue", AFOSR Scientific
     Report, 69-1329TR, April 1969.

29. Barton, J.R. and F.N. Kusenberger, "Magnetic Perturbation Inspection to Improve Reliability of High Strength Steel Components", Paper 69-DE-58, Design Engineering Conf. of ASME, New York, New York, May 1969.

30. Kusenberger, F.N., J. Lankford, Jr., P.H. Francis and J.R. Barton, "Nondestructive Evaluation of Metal Fatigue", Air Force Office of Scientific Research, AFOSR 70-1206 TR, March 1970.

31. Barton, J.R. and F.N. Kusenberger, "Fatigue Damage in Gas Turbine Engine Materials", Chapter on Fatigue Damage Detection, ASTM STP 495 (1971) 193-208.

32. Kusenberger, F.N., J. Lankford, Jr., P.H. Francis and J.R. Barton, "Nondestructive Evaluation of Metal Fatigue", Scientific Report, AFOSR-TR-71-1965, April 1971.

33. Kusenberger, F.N., J. Lankford, Jr., P.H. Francis and J.R. Barton, "Nondestructive Evaluation of Metal Fatigue", Scientific Report AFOSR-TR-72-1167, April 1972.

34. Kusenberger, F.N., W.L. Ko, J. Lankford, Jr., P.H. Francis and J.R. Barton, "Nondestructive Evaluation of Metal Fatigue", Scientific Report AFOSR-TR-73-1070, April 1973.

35. Lankford, J., Jr. and F.N. Kusenberger, "Initiation of Fatigue Cracks in 4340 Steel", Met. Trans.,4 (1973) 553.

36. Barton, J.R., F.N. Kusenberger, P.H. Francis, W.L. Ko, and J. Lankford, Jr., "Nondestructive Evaluation of Metal Fatigue", Scientific Report AFOSR-TR-74-0820, February 1974.

37. Barton, J.R., "Quantitative Correlation Between Magnetic Perturbation Signatures and Inclusions", ASTM Intl. Symposium on Rating of Nonmetallic Inclusions in Bearing Steels, Boston, Mass., May 1974.

38. Barton, J.R., "Advanced Magnetic Methods of Flaw Detection", Workshop for NDE Sponsored Jointly by ARPA and AFML, Thousand Oaks, Calif., June 1974. Technical Report AFML-TR-74-238, November 1974.

39a. Bush, H.D. and R.S. Tebble, "The Barkhausen Effect", Proc. Phys. Soc., 60 (1948) 370.

39b. Schijve, J., "Significance of Fatigue Cracks in Micro-Range and Macro-Range", Fatigue Crack Propagation, ASTM STP 415, Am. Soc. for Testing and Materials, 1967.

40. Packman, P.F., H.S. Pearson, J.S. Owens and G.B. Marchese, "The Applicability of a Fracture Mechanics-Nondestructive Testing Design Criterion", Technical Report, AFML-TR-68-32, Air Force Materials Laboratory, May 1968.

41. Sattler, F.J., "Nondestructive Flaw Definition Techniques for Critical Defect Determination", Contract NAS 3-11221, Jan. 1970.

42. Rummel, W.D., P.H. Todd, S.P. Frecska and R.A. Rathke, "The Detection of Fatigue Cracks by Nondestructive Testing Methods", NASA CR 2369, Feb. 1974.

43. Gulley, L.R., Jr., "An Investigation of the Effectiveness of Magnetic Particle Testing", Tech. Memorandum AFML/MX 73-5, Oct. 1973.

44. Barton, J.R., J. Lankford, Jr., and P.L. Hampton, "Advanced Nondestructive Testing Methods for Bearing Inspection", SAE Paper No. 720172, Automotive Engineering Congress, Detroit, Mich., Jan. 1972.

45. Barton, J.R. and J. Lankford, Jr., "Magnetic Perturbation Inspection of Inner Bearing Races", NASA CR-2055, May 1972.

46. Harnwell, G.P., Principles of Electricity and Electromagnetism, McGraw-Hill, New York, pp. 406-07.

47. Kraus, J.D., Electromagnetics, McGraw-Hill, New York, 1953, pp. 538-41.

48. Craik, D.J. and R.S. Tebble, Ferromagnetism and Ferromagnetic Domains, John Wiley & Sons, Inc., 1965.

49. Kusenberger, F.N., G.A. Matzkanin, J.R. Barton and P.H. Francis, "Nondestructive Evaluation of Metal Fatigue", Scientific Report AFOSR-TR-76-0384, March 1976.

50. Kusenberger, F.N., W.L. Ko, J. Lankford, P.H. Francis and J.R. Barton, "Nondestructive Evaluation of Metal Fatigue", Scientific Report AFOSR-TR-73-1070, Air Force Office of Scientif Research, April 1973.

51. Kusenberger, F.N., G.A. Matzkanin, J.R. Barton and P.H. Francis, "Nondestructive Evaluation of Metal Fatigue", Scientific Report AFOSR-TR-75-0937, Air Force Office of Scientific Research, March 1975.

52. Barton, J.R. and F.N. Kusenberger, "Nondestructive Detection of Early Fatigue Damage in Piston Pins", Final Report, Contract N 204(64)38999, Naval Air System Command, 1 Dec 1966.

53. Parker, R.J., "Correlation of Magnetic Perturbation Inspection Data with Rolling-Element Bearing Fatigue Results", ASME J. of Lubrication Technology (April 1975).

54. Kusenberger, F.N. and J.R. Barton, "Barkhausen Noise Stress Measurements on Bearing Races Before and After Service", Final Report, SWRI Project 15-2888, AF Contract No. F09603-70-C-5547, June 1974.

55. Bohm, K., H. Schlicht, O. Zwirlein and R. Eberhard, "Nonmetallic Inclusions and Rolling Contact Fatigue", Bearing Steels: The Rating of Nonmetallic Inclusion, ASTM STP 575, Am. Soc. for Testing and Materials (1975) 96-113.

56. Proc. of Joint Army-NASA Seminar, Bearing Restoration by Grinding, May 20-21, 1976, St. Louis, Mo. Sponsored by Army Aviation Systems Command and NASA Lewis Research Center.

57. Barton, J.R., F.N. Kusenberger, P.L. Hampton and H. Bull, "Critical Inspection of Bearings for Life Extention - CIBLE", Proc. of 10th Symposium on Nondestructive Evaluation, April 23-25, 1975, San Antonio, Texas.

Chapter 21

POSITRONS AS A NONDESTRUCTIVE PROBE OF DAMAGE IN STRUCTURAL
MATERIALS

J. Wilkenfeld and J. John

IRT Corporation

San Diego, California

ABSTRACT

Laboratory studies have shown that one can relate features of
the two photon annihilation of positrons with electrons in a material
to the degree of damage created by plastic deformation, thermal
cycling, irradiation, or implanation of impurities. We have examined
the feasibility of translating the growing body of experimental data
and theoretical modeling on the annihilation of positrons in matter
into an NDI technique for the detection of structure damage in
metals and the applicability of this technique for predicting
component failure. As a test case, the mean life of positrons
was measured in an aluminum structural alloy ($A\ell$-clad 7075-T6) in
which varying degrees of plastic strain had been introduced. A
correlation useful for nondestructive inspection has been found
between the positron mean life and the degree of strain. The effect
of the T6 tempering process was also reflected in the range of lifetimes observed. These data suggest that it may be possible to
create positron annihilation life history curves for structural
components. The advantages of the positron annihilation technique
are (1) its sensitivity to small amounts of damage, i.e., small
strains, (2) that it is nondestructive, and (3) that the surface
or the interior of a sample may be probed by proper choice of positron source. We outline a developmental program to utilize positron
annihilation as a nondestructive inspection tool for detection of
material damage which may lead to component failure.

INTRODUCTION

When positrons are injected into a sample with energies of
several hundred keV, they are arpidly slowed down through ionizing
collisions with sample atoms and reach energies of 0.1 eV or less.
A positron will annihilate with an electron in the sample yielding
packets of electromagnetic energy in the form of gamma quanta. The
energy released in the annihilation is comprised mainly of the rest
mass of the electron-positron pair,

$$2E_0 = 2m_0c^2 = 2 \times 0.511 \text{ MeV} \tag{1}$$

where m_0 is the rest mass of the electron = $E_0/c^2 = 9.1 \times 10^{-28}$ g,
and c is the velocity of light in vacuum = 3×10^{10} cm/sec. There
is also an additional small but significant contribution to the
total energy of the pair which is largely a measure of the energy
of the electron with which the positron annihilates. In metals,
all positrons annihilate with the emission of two gamma quanta
which are detectable by conventional nuclear techniques. The rate
of annihilation, inversely proportional to the mean lifetime of
positrons in a material, depends on the overlap between the positron
wave packet and the spatial distribution of target electrons
sampled by the positrons. The range of energies of the annihilation quanta, as marked by their departure from 0.511 MeV, is a
measure of the energy and momentum distributions of the material's
electrons probed by the positrons. Thus, the annihilation behavior
of positrons in a material, as revealed by the annihilation quanta
emitted, is a manifestation of the electronic structure of the
material. This fact forms the basis for the use of positron annihilation as a nondestructive probe for the study of structural
properties of matter. This use has been well documented in the
proceedings of three international conferences [1-3] and several
review articles [4-7].

During the past ten years, it has been increasingly realized
that positrons can be used to probe defects in materials created
by such diverse processes as radiation damage, fatigue, or thermal
cycling. In general, the defects can be vacancies, dislocation
networks, grain boundaries, foreign precipitates, microvoids, and
other irregular regions in the material, such as might be created
by surface corrosion. In these regions, the annihilation behavior
of the positron is different from that in the body of the material.
This depends on the fact that the electron environment seen by the
positron before annihilation in regions containing defects is
different from that in an ordered crystalline domain. In many
cases this leads to experimentally detectable effects which can
be used to probe the nature of disordered regions in a material.
For some problems, such as the determination of the concentration

and energy of formation of vacancies in metals, positron annihilation has become a valuable and sensitive tool. In this regard, a
substantial body of experimental data has been accumulated, and
can be accounted for by a relatively simple model which is in
accord with the results from defect studies by other means, such
as thermal expansion, x-ray diffraction, and calorimetric measurements [8]. In regard to the investigation of other kinds of
structural defects, an increasing amount of data is available
which correlates characteristics of positron annihilation to such
phenomena as stress-induced fatigue, thermal creation of dislocations through quenching and annealing cycles, and the structure of
alloys [9-11]. While the theory for the annihilation of positrons
in disordered regions of metals and alloys is not as well established
as that for point defects, patterns which correlate annihilation
behavior with material damage are discernible. These experimental
observations suggest that it is possible to employ the body of techniques and conceptual framework developed over the past 25 years in
the laboratory as a foundation to establish positron annihilation
as an NDI technique to characterize the structure of materials and
to study damage in them.

As a test case, we have performed a feasibility evaluation of
the positron lifetime technique to detect simulated life-service
fatigue in a standard aircraft structural alloy, Al-clad 7075-T6,
QQ-A250/13. A definite correlation has been found between the
mean life of the positrons and the degree to which the alloy samples
have been strained up to fracture. In addition, the T6 tempering
process is also seen to change the mean life of positrons in the
material if compared to that in crystalline well-annealed aluminum.

In the second section of this chapter we summarize relevant
aspects of the annihilation behavior of positrons in condensed
matter. The significant point is that positrons which annihilate in
disordered regions of a material display experimentally distinguishable behavior if compared to those which annihilate in the ordered,
crystalline domains.

The third section of this chapter describes the experimental
technique employed, i.e., a measurement of the mean lifetime until
annihilation. We then discuss our data. It can be seen that a
monotonic correlation can be made between the degree of strain
created in the aluminum samples and features of the annihilation
process. Finally, in the last section of the chapter, we outline
the steps needed to promote the development of the positron annihilation technique as a nondestructive inspection tool for detecting
life-service damage in structural materials.

BEHAVIOR OF POSITRONS IN MATTER

Positrons injected into condensed matter rapidly lose energy
by electron ionization and excitation until they reach energies of
the order of 0.1 eV or less in a time estimated to be a few picoseconds for metals. This time is short compared to the lifetime
of positrons in metals which are typically 100 to 200 psec [12,13].
The energy released by the annihilating electron-positron pair is
composed of two parts, $E \simeq 2m_0c^2 + E_{kin}$, where $m_0c^2 = 0.511$ MeV is
the rest mass of an electron or positron, and E_{kin} is the kinetic
energy of the electron-positron pair. As the kinetic energy of
the positron is ≈ 0, E_{kin} is essentially that of the sample electron
with which the positron annihilates, typically ~ 10 eV for valence
electrons in metals.

Conservation of momentum and energy requires that at least two
particles other than the electron-positron pair participate in the
annihilation process. It can be shown that the most common mode of
annihilation in condensed matter is that in which two gamma quanta
are emitted subsequent to annihilation [7,14]. If the positron-
electron pair were at rest, each gamma ray would have an energy =
m_0c^2 and a momentum of m_0c, and would be emitted in opposite
directions as shown in Figure 1(a). Thus, the presence of gamma
rays of energy ≈ 0.511 MeV is the signature of the annihilation
process. If we assume that the positron is essentially at rest,
it can be shown that the emerging gamma rays will come off at a
relative angle which differs from 180° by an amount θ, which is
proportional to p_t, the component of momentum perpendicular to the
direction of the emitted gammas where:

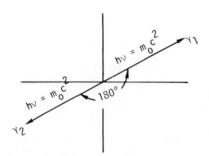

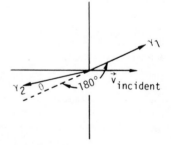

(a) Electron-positron pair   (b) Electron moving toward
     at rest   positron with a relative
                                        velocity $\vec{v}$

Figure 1.   Two-gamma annihilation of an electron-positron pair.

$$\theta \simeq \frac{P_t}{cm_0} \tag{2}$$

to order $v/c < 1$, where v is the electron velocity, as shown in Figure 1(b). The two photons will have slightly different energies

$$E_{1,2} \simeq m_0 c^2 \pm \frac{cp_\ell}{2} \tag{3}$$

to order v/c, whose spread about 0.511 MeV is proportional to p_ℓ, the component of momentum parallel to that direction. By measuring the angular dependence of the emerging quanta or the broadening in energy about the 0.511 MeV annihilation line, one can infer the momentum and energy of the electron distribution with which the positrons annihilate. This requires the measurement of angles of order $\theta \simeq 0.01$ radian, and energy differences of order $\Delta E \simeq cp_\ell/2$, or about 1.5 keV for 10 eV electrons.

The fundamental parameter characterizing the interaction between a positron and electron is the cross section for the annihilation process. This quantity, which has the dimensions of areas, is essentially the size of an electron as seen by a positron before annihilation. The magnitude of this quantity depends on the mode of annihilation, i.e., the number of quanta emitted, the relative orientation of the electron-positron spins or magnetic moments, and the relative velocity of the pair. For the most common case, which applies to the annihilation of positrons in metals in which the spins of the electron and positron are randomly oriented, and the relative velocity of the pair v, largely due to that of the electron, is much less than the velocity of light (i.e., $v/c < 1$), then the annihilation cross section for 2-gamma annihilation is

$$\sigma_{2\gamma} = \pi r_e^2 \frac{c}{v} \tag{4}$$

where $r_e = 2.82 \times 10^{-13}$ cm, is the classical radius of the electron. Thus, for 10 eV electrons, where $v = 1.9 \times 10^8$ cm/sec, $\sigma = 3.9 \times 10^{-23}$ cm^2. The rate of annihilation in a material where there are N_e electrons per unit volumve which can interact with the positron, possessing an average velocity, v, is then

$$\Gamma_{2\gamma} = N_e \sigma_{2\gamma} v = \pi r_e^2 c N_e \tag{5}$$

Equation (5) is the crucial relationship between positron annihilation characteristics and the structure of matter. It states that the mean annihilation rate for an ensembel of slow positrons in a material is proportional to the average electron density sampled by the positrons before annihilation. This rate is equivalent to

$1/\tau$, where τ is the mean life of the positron, an experimentally measurable quantity.

Because of the Coulomb interaction between the positrons and electrons in a metal, N_e is not equivalent to the mean density of conduction electrons. The positron perturbs its surroundings. This fact complicates the interpretation of positron annihilation data for metals. One can estimate to an order of magnitude the mean life of positrons in a metal. The effective electron density seen by a positron is

$$N_e = \frac{N_o d Z_{eff}}{A} \tag{6}$$

for a metal of density d, atomic number A, and where Z_{eff} is the effective number of electrons per atom probed by the positron. Typically, $Z_{eff} \geq$ the number of valence electrons in a metal. For aluminum, d = 2.7 g/cm^3, A = 27, and the number of valence electrons per atom is 3, yielding $\tau \leq$ 750 psec. In fact, the mean life of positrons in annealed aluminum single crystals is $\sim$165 psec, while in a defect or dislocation in aluminum this lifetime is about 242 psec.

The manner in which Equation (5) has been written conceals all of the elegant physical modeling which has been developed to describe the complicated interaction of a positron with electrons in the environment of a solid. For nondestructive inspection purposes, however, the essential point is that laboratory studies have demonstrated that regions of different structural order in a material, such as the ordered, crystalline, undamaged domains in a metals, or the disordered domains related to the presence of defects, dislocations, etc., have detectably different electron environments. These differences are sensed by positrons and lead to observably different annihilation behavior, as evidenced by differing lifetimes, energy shifts of the annihilating quanta from $m_0 c^2$ = 511 keV, or the departure from 180° of the angle between the two emerging gammas.

There are then three basic measurement techniques commonly employed which yield complementary evidence about the annihilation behavior of positrons in matter. These are based on sampling the three significant features of the two emergent gamma quanta which carry away all information about the annihilation of a given electron-positron pair. The three techniques are:

1. Lifetime measurements in which the time between the creation of the positron and its annihilation is marked by the detection of two gamma quanta, the first emitted simultaneously with the creation of the positron by a source radionuclide, and

the second which is one of the two 0.511 MeV annihilation quanta. This technique measures the electron density distribution of a sample seen by the positrons according to Equation (5).

2. Angular correlation measurements in which θ, the departure from 180° of the angle between the directions of the two emergent annihilation quanta, is determined. This technique measures the distribution of electron momenta samples by the annihilating positrons according to Equation (2).

3. Doppler broadening measurements in which the energy distribution of the annihilation gammas are determined. The annihilation line shape is broadened because of the increment to the photon energies about m_0c^2, as given by Equation (3).

A more detailed discussion of these techniques can be found in Reference 11.

To provide a background for an understanding of our data, some highly condensed remarks will be made about the annihilation of positrons in metals.

An idealized metal can be viewed as a regular array of atomic ion cores embedded in a sea of conduction electrons. These electrons, which come from the valence shells of individual atoms, are not attached to any single atom, but in many respects behave like an ideal gas. Because the ion cores have a net positive charge, the positrons tend to be found in the regions between ion cores, being kept away from them by the Coulomb repulsion between like charges. Thus, the positron samples primarily those electrons outside the ion cores, i.e., conduction electrons. If one assumes that the positron samples only conduction electrons whose unperturbed density, N_e, can be calculated for various models, then the annihilation rate for a particular metal can be calculated from Equation (5). It has been found that to correctly match experimental and theoretical annihilation rates, one must take into account the fact that the electron environment seen by the positron is screened by the conduction electrons moving with it through the solid as a consequence of the Coulomb attraction between unlike charges. The local electron density around the positron is enhanced above it unperturbed value at a given location. As our simple calculation indicated, the observed annihilation rates are always much larger than those predicted from models of the electronic structure of metals which neglect the polarization effect of the positron on electronic structure. In addition, the annihilation rate is further enhanced because of a small additional contribution from annihilation of positrons with electrons attached to the positive ion core. While the positron annihilation lifetime will reflect, to some degree, the electronic structure of the crystal, it also alters that environment. To be successful, models based on annihilation data which attempt to describe morphological

features of metallic lattices must factor out the perturbing effect
of the positron.

The electron environment in a defect is different than in a
crystalline domain. In simple cases it is clear why this is so.
For example, when vacancies are created in a metal, ion cores are
freed from their equilibrium sites in the lattice and transferred
to the surface of the material. The absence of the positively
charged ion core leaves regions of net negative charge. To re-
establish charge neutrality in the interior of the metal, conduction
electrons expand away from the region of the defect. This leaves
a site with a deficiency of conduction electrons, i.e., a lower
electron density, as well as space in which the positron can be
preferentially trapped as it diffuses through the lattice. In
fact, one does observe the trapping of positrons in point defects.
The experimental consequences of such trapping are:

1. an increase in the mean lifetime of the positrons trapped
in such defects, as the mean electron density is lower;

2. a narrowing of the angular correlation spectrum because
of a diminished contribution from annihilations with high momentum-
energy core electrons to the total annihilation rate; and

3. a narrowing of the Doppler-broadened annihilation line
for the same reason.

It is relatively easy to correlate features of the annihilation of positrons in point defects to simple structural models,
such as the trapping model of Brandt, Bergerson, Stott, Connors,
and West [5]. Positron annihilation has become an important tool
in the study of point defects, not only in metals, but also in
insulators [8]. Because the positron is preferentially trapped
in such defects, it is more sensitive in determining their presence
than classical methods such as dilatometric or x-ray diffraction
techniques.

Positron annihilation can be used to examine other kinds of
disorder in materials. These include the dislocation loops created
in metals subsequent to plastic stress, microvoids induced by the
irradiation of metals, and the presence of impurities with different
chemical properties which might be introduced as a consequence of
corrosion. The physical nature of the trapping process in these
cases cannot usually be described in terms of a simple model, as
in the case of trapping of positrons by point defects, if indeed
it can be definitely identified at all from features of the annihilation. The technique is often not sensitive enough to distinguish between the annihilation of positrons in the different kinds
of defects which may be simultaneously present in a material.
However, positrons may still be utilized as a probe for these
defects, provided that: (a) the electron environment seen by the

positron in the disordered region of the material be different
from that in crystalline domains; (b) the positrons be preferentially
attracted to the disordered region, whether because of the polarization force set up in a dislocation, because of an attraction
between the positron and an impurity atom, or because of the
absence of an ion core which leaves space for a positron; and
(c) a significant fraction of the positrons remain in the vicinity
of the defect long enough, i.e., be trapped, for annihilation to
occur.

A great deal of experimental work has been carried out to
elucidate features of the annihilation behavior of positrons in
metals and alloys containing defects, and particularly in aluminum.
In general, it has been found that in metals:

1. defect sites represent regions where positrons can be
preferentially trapped;

2. annihilation of positrons from these defect sites is
experimentally observable;

3. as the relative concentration of defects increases, the
observed positron mean life increases, the width of the angular
correlation curve is narrowed, and the width of the Doppler-
broadened annihilation line also narrows (the results for the
three kinds of measurement are mutually consistent); and

4. for a sufficiently large concentration of defects, all
positrons annihilate in defects.

It is important that characteristics of the annihilation of
positrons be related to specific morphological features of a material
to advance our fundamental understanding of what happens to a
material when damaged. Such a detailed analysis can be difficult,
and not usually necessary for developing positron life history
curves in a material for NDI purposes. A single parameter characterization often yields sufficient information. In this case, one
assumes that (1) each distinguishable annihilation state, i.e.,
for the undamaged material, or the defect region, is manifest in
a weighted, linear fashion in some detectable annihilation characteristic, and (2) that the kinds of such states to reasonably
describe positron annihilation in a material are limited. If F is
such a characteristic, then one can measure the average value of
this parameter

$$\bar{F} = \sum_{j=1}^{N} F_j I_j \tag{7}$$

where F_j is the value for annihilation from the jth state, and I_j
is the relative fraction of annihilations in that state. Measurable
parameters that are presumed good and linear in the above sense

include:

1.   mean lifetime, $\bar{\tau}$;

2.   ratio of peak height to area of an angular correlation
distribution, h; and

3.   the peak-height-to-area ratio of the Doppler-broadened
annihilation line, S.

The basis for use of Equation (7) can be found in Reference 5.
Its strongest support lies in the fact that conclusions about the
annihilation behavior of positrons in a given material derived by
any of the three methods agree.

A growing body of evidence has been found which demonstrates
that one can correlate in a qualitative manner features of the
annihilation behavior of positrons, in the spirit of Equation (7),
to the life history of materials. For example, MacKenzie, Eady,
and Gingerich [15] have shown that the Doppler-broadened lineshape
parameter, S, increases monatomically with deformation induced by
the cold rolling of pure Cu, and in 6061 Aℓ as shown in Figure 2.
The percentage of positrons trapped in defects, which is related
to the increase in S, also increases, reaching 100% for ∿12%
deformation. These data are typical of positron life history
curves. The annihilation behavior is sensitive to small degrees
of damage which reflect the preferential trapping of positrons in
defects produced by the deformation. It can be seen that 50% of

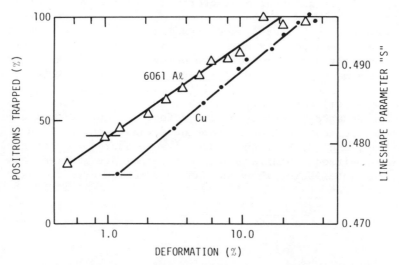

Figure 2.   The lineshape change with the percentage reduction in
            sample thickness resulting from unidirectional cold
            rolling [15].

the positrons are trapped in defects for ∿1% flattening. Even in
the undeformed material, a significant fraction of positrons
annihilate in defects present in all polycrystalline materials.
The effect saturates when the defect concentration becomes large
enough so that all positrons are trapped before annihilation.
Such life history curves correlate well with other global measurements of deformation, such as surface hardness, as shown in
Figure 3. Unlike hardness measurements, the positron annihilation
technique is completely nondestructive.

EXPERIMENTAL

Measurements of the positron lifetime spectrum was the technique employed to monitor changes in the structure of the aluminum
alloy samples incrementally strained. The positron lifetime
spectrum comprises the number distribution versus lifetime until
annihilation of a large number (10^5 to 10^6) of positrons which
annihilate in the sample. The number of annihilations recorded is
usually determined by the requirement that the statistical uncertainty of the results obtained should be less than that due to
instrumental error. In positron physics, one nearly always looks
at the characteristics of the annihilation process to determine
the behavior of positrons in a material, rather than observing the
particles themselves. Thus, the lifetime of a positron is determined by measuring the time interval between the detection of a
gamma ray emitted coincident with the birth of the positron, and
the detection of one of the 0.511 MeV annihilation quanta. The
observed range of positron lifetimes in condensed matter runs
from about 100 psec up to a few nsec.

The experimental setup for the measurement of positron lifetimes is shown in Figure 4. The positron source was the isotope
^{22}Na. This radioisotope emits a 1.275 MeV gamma quantum within a
few picoseconds of the positron emission, and is used to mark the
injection of the positron into the sample. This is about the same
time that it takes the positron to thermalize in a metal at room
temperature. The detectors employed in lifetime measurements are
typically plastic scintillators coupled to photomultipliers. Such
a combination has the high gain and speed required to provide the
sharply defined pulses (with nsec rise times) necessary for sub-
nanosecond timing. The main component of a lifetime spectrometer
is the time-to-amplitude converter (TAC). This device is essentially
a very stable integrator which is turned on when the start photomultiplier (PMA) receives a 1.275 MeV gamma, and turned off when
the stop photomultiplier (PMB) intercepts a 0.511 MeV annihilation
quantum. The TAC produces a voltage proportional to the time
interval between these two events. The output of the TAC is input
to a multichannel pulse-height analyzer (MCA) which stores the

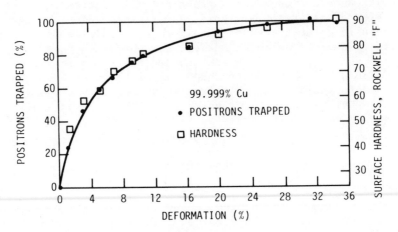

Figure 3.   The correlation between hardness and positron trapping
in copper of high purity [15].

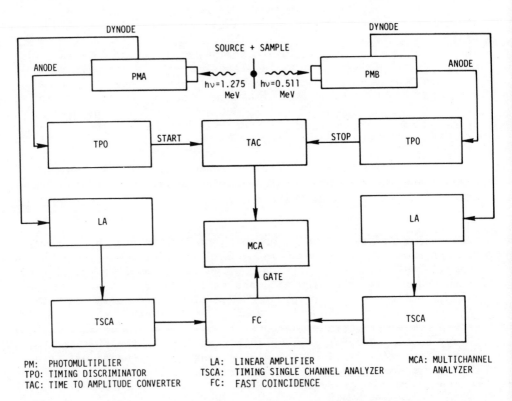

Figure 4.   Positron lifetime spectrometer.

positron lifetime spectrum as the number of events occurring in
a time interval corresponding to the width of one channel versus
time. The remaining modules are employed to provide standardized
pulses for timing (TPO), and to provide a coincidence signal (LA,
TSCA, FC) which tells the MCA that a valid event has occurred,
so that it is gated to accept a pulse from the TAC. For this
experiment, a valid event is the receipt of a 1.275 MeV gamma in
the START photomultiplier, followed by a 0.511 MeV gamma detected
by the STOP photomultiplier. Standard NIM electronic modules are
available from which positron lifetime spectrometers can be
constructed.

The resolution of a positron lifetime system is given by its
prompt spectrum, which is the response of the lifetime spectrometer
to a gamma cascade in which two photons are emitted nearly simultaneously. Typically, ^{60}Co is used as a calibrating source. The
shpae of the prompt spectrum can be well approximated by a gaussian,
and characertized by its full-width-at-half-maximum (FWHM) and the
slope of the wings, expressed as an equivalent half-life. This
slope sets a lower limit on the smallest lifetime of a multicompon-
ent lifetime spectrum which can be resolved. For our system, the
FWHM was typically 285 psec, and the slope at 1/5 maximum corresponded to a half-life of 70 psec.

The lifetime spectrometer was calibrated by inserting precision,
1 nsec air dielectric coaxial delay lines (GR type 874) in series
with cables going to the start and stop inputs to the TAC. The
shift in the centroid of the prompt spectrum as a function of
inserted delay yields the time calibration. Our system, calibrated
in this manner, showed a differential linearity of better than 2%.

It was the aim of this study to provide a realistic test of
the applicability of the positron annihilation technique to detecting
fatigue in aluminum structural alloys in the field. To this end,
actual samples of an aircraft structural alloy were studied. The
samples were sheets of an aluminum-clad, aluminum alloy, type
7057-T6, QQ-A250/13. The sample thickness was 0.050 inch. The
cladding material, which was about 0.003 inch to 0.004 inch thick,
was type 7072 aluminum alloy. No cleaning or annealing of the
samples was carried out before investigation. It was felt advis-
able to examine samples which had both the normal components of
defects and dislocations and of surface corrosion and oxide formation. Five sample pairs were made by cutting 1-inch-square pieces
from a sheet of the alloy. Sample pairs 1 and 2 were used as cut.
Sample pair 3 was strained by bending each piece approximately 22°
in a sheet metal brake. Sample pair 4 was strained by bending each
piece 45° in the brake without fracture. Sample pair 5 was strained
by bending each piece 45° in the brake, with a resultant fracture
of both pieces.

The positron source was a solution of ^{22}NaCℓ. About 10 μCi of activity was deposited at the strained area in one of the two pieces of metal identically stressed. This activity was chosen as a compromise between the deisre for reasonable count rates proportional to activity, and the need to minimize the background, which is proportional to the square of activity. Experiment times were several hours. For NDI purposes in which rapid data accumulation is desirable, a stronger source may be used and less stringent requirements imposed on gating the MCA. The source was fixed by evaporation forming a spot ∿3 mm in diameter. The second piece of the sample pair was laid over the first, with strained areas aligned forming a sandwich with the activity in the middle. Such a procedure localizes the positron activity so that the majority of positrons are stopped in the stressed volume. One can estimate the sample volume probed by the positrons, given the fact that the energy distribution of the emitted positrons is that of a β^+ emitting radionuclide whose endpoint energy is 0.545 MeV. From the range energy relationship for electrons and positrons [16], the endpoint range in aluminum is 0.74 mm, which is less than the thickness of one sheet (1.27 mm). In fact, one can show that the intensity of the positron beam drops to 1/e of its initial intensity in about 0.11 mm in aluminum. Thus, the majority of positrons from this external source were stopped in a depth slightly greater than the cladding thickness. One can vary the sample depth probed with such external sources by choosing other radioisotopes whose β^+ endpoint energies are different, for example, ^{44}Ti/^{44}Sc, whose positrons have an endpoint energy of 1.47 MeV corresponding to a range of 2.4 mm, and an e fold attenuation length of 0.34 mm in aluminum. To probe greater thicknesses, one may use neutron activation or photonuclear (γ,n) reactions [11].

The measured positron lifetime spectrum is distorted by the instrumental resolution, as characterized by its prompt spectrum. To account for the finite time resolution of the spectrometer, the lifetime data is usually fit to the convolution of a gaussian prompt spectrum and a sum of exponentially decaying components, one for each resolvable mode of annihilation. When there is more than one component present, this fit is done numerically by computer. The product of such a fit is a set of lifetimes and relative annihilation intensities. In practice, two components which have lifetimes which lie close together (less than a factor of two or so) are difficult to resolve. The contribution to the total annihilation spectrum is typically fit by a single component whose mean life is a weighted average of the two. A detailed discussion of how lifetime data is reduced can be found in Reference 11, and a discussion of the relationship between this data and models which describe the annihilation of positrons in metals can be found in References 5, 6, 8 and 9.

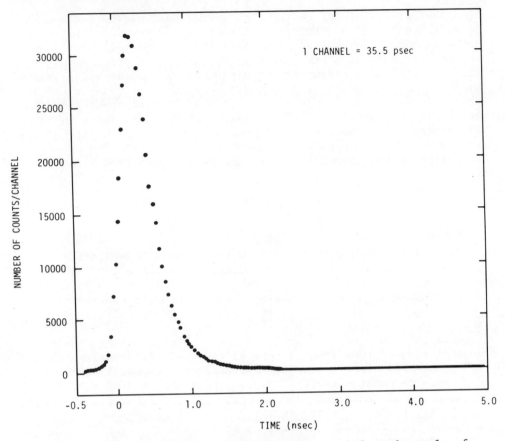

Figure 5. Positron lifetime spectrum in an undeformed sample of
aluminum alloy, aluminum-clad 7075-T6, QQ-A250/13.

DISCUSSION

Figure 5 shows the data for a typical lifetime measurement.
Its form is that of the number of counts per channel versus channel
number. This is equivalent to obtaining the decay rate, i.e., the
number of annihilations per unit time versus time. The peak of
the counting rate is shifted by about the mean lifetime of positrons
in damaged aluminum, or about 230 psec from the location of the
centroid of the prompt spectrum. A significant number of annihilations appear to occur at negative times. This is an artifice of
the measuring system, for what is observed is not the actual decay
rate, but the convolution of the decay rate with the prompt or
instrumental function of the system.

The lifetime spectrum for each sample pair was determined at least three times to obtain an estimate of the reproducibility of our measurements. The annihilation lifetimes and relative inter-action were extracted from the data by a computer program which fit the lifetime spectrum to the minimum number of decaying exponen-tial components, which give a good fit as measured by a chi-squared test. In each case it was found that the lifetime spectra could be fit to two exponentially decaying lifetimes components and a background. The shorter lifetime component had a mean life of 218 to 245 psec, and comprised 95 to 99% of all annihilations. The mean life of the longer lifetime component ranged from about 0.5 to 1.3 nsec.

Based on many studies carried out in well-characterized aluminum samples, the origin of the two resolved components may be assigned [17-20].

1. The component with the shorter mean lifetime is that of positrons which annihilate in the interior of the alloy. This component represents a weighted average of at least two pendants unresolved in these measurements. One of these is contributed by positrons which annihilate in the ordered, undamaged regions of the alloy, primarily with electrons of aluminum atoms. The life-time of this component is probably identical to that for the annihilation of positrons in pure aluminum single crystals, $\tau_{cr} = 165$ psec. The second component arises from the annihilation of positrons in damaged regions, i.e., in defects or dislocations created by the tempering process or by strain, or at precipitation clusters. Positrons which annihilate in any of these sites in aluminum have comparable lifetimes which cannot be resolved and are of the order of $\tau_d \simeq 242$ psec [18].

2. The longer lived component is commonly attributed to the annihilation of positrons in the surface of the material where they sense impurities, such as oxide layers, or mechanical damage resulting from specimen preparation [19,21].

One can minimize or eliminate entirely this latter component in the laboratory, at least, by using internal sources and/or carefully cleaning sample surfaces and carrying out measurements in an inert atmosphere, or under vacuum. Our samples were examined as received, with no treatment.

For this study, we are primarily interested in the behavior of the shorter lived component. The relevant data are summarized in Table 1. The lifetimes tabulated are the maximum likelihood estimate of the mean value for each lifetime weighting each datum by the square of its reciprocal standard deviation derived on the basis of counting statistics.

TABLE 1

Positron Mean Lifetimes in Aluminum–Clad 7075–T6

Sample	Deformation	$\bar{\tau}$ (psec)	$\sigma_{\bar{\tau}}$ (psec)	Range of Lifetimes Observed (psec)
1	None	228.5	0.5	218 to 238
2	None	230.5	0.6	226 to 239
3	22° Bend	237.0	0.4	235 to 238
4	45° Bend	242.5	0.5	233 to 247
5	45° Bend and fracture	242.0	0.6	230 to 245

Our findings may be summarized as follows.

1. The mean life of positrons in unstrained samples (1 and 2) is 229.5 +_ 1 psec.

2. The mean life of positrons in the most heavily strained samples (4 and 5) is 242 psec.

3. There is a correlation between the amount of additional strain introduced in the sample and the positron mean life, as indicated by the mean life observed in sample 3, 237 psec, with a lesser degree of strain than samples 4 and 5.

In assessing the degree of disorder or damage in the samples studied, it is convenient to compare our results to those obtained for deformed pure aluminum. In Figure 6 we have plotted the angular correlation data of Connors [22] and the lifetime data of Hautojarvi et al. [20], along with the Doppler-shift measurements of MacKenzie et al. [15] in terms of the fraction of positrons trapped versus the percentage of plastic deformation. As we have discussed, this can be obtained from the measurement of an annihilation characteristic, $\bar{F}$, which can be the mean lifetime, $\bar{\tau}$, the ratio of the peak height to area in an angular distribution, h, or the peak-height-to-area ratio of the Doppler-broadened annihilated line, S. In the case where N = 2, i.e., where positrons annihilate either in the bulk material or in a defect site, then Equation (7) can be written as

$$f = \frac{\bar{F} - F_c}{F_d - F_c} \tag{8}$$

where f is the fraction of positrons annihilating from defect or
impurity sites, F_c is the value of the parameter for bulk annihilations, and F_d is the value of the parameter for annihilations from
damaged or impurity sites. Where $\bar{F} = \bar{\tau}$, the mean life on aluminum,
then

$$f = \frac{\tau - \tau_{cr}}{\tau_d - \tau_{cr}} = \frac{\bar{\tau} - 165}{242 - 165} \tag{9}$$

It can be seen that the data which were obtained by different measurement techniques, primarily on deformed pure aluminum, form a
single curve as a function of deformation, which reaches a saturation value for plastic strains in excess of 20% where essentially
all positron annihilate from defect sites. The solid line has
been drawn to summarize the trend in the data. This curve also
shows: (1) the rapid rise in the fraction of positrons trapped
with initial deformation; (2) that a significant fraction of
positrons annihilate in defects even in unstrained samples; and
(3) that for large deformations, all positrons are trapped in
defects.

The lifetime data from this study have been placed on the
curve for comparison. It is to be noted that the strain in our
samples was not produced in the same manner, nor did we start
with annealed crystalline aluminum. In any case, the correlation
between $\bar{\tau}$ and the fraction of positrons trapped should still hold.
It is evident that even for the unstressed sample pairs (1 and 2),
approximately 80 to 85% of all the positrons annihilate near
defects or impurities. For the samples strained by a 45° bend,
essentially all of the positrons annihilate at such sites. The
change between the unstressed and fractured samples is relatively
small, but nevertheless statistically significant.

To understand these observations, some remarks must be made
about the thermal history of the aluminum alloy sheets [23]. In
the manufacture of an aluminum-clad aluminum alloy, the core and
cladding are first hot-rolled to produce sheets of approximately
nominal thickness. The sheets are then subject to a tempering
process to strengthen or harden the alloy. The T6 tempering
treatment involves the following procedure. The clad sheets are
solution heat treated at 488 to 499°C for approximately 30 minutes
to form a solid solution of the aluminum and alloying agents,
after which they are hard quenched by rapidly cooling to room
temperature in water. Any final rolling to produce sheets of
the desired degree of flatness and thickness is then carried out.
Finally, the sheets are artificially aged, i.e., tempered by
precipitation hardening via heating at 121°C for 24 hours.

The defect of the tempering process is to create a large

number of defect sites, in this case primarily loop dislocations,
as a consequence of the solution heat treatment and hard quenching,
and insoluble precipitates in the alloy solid solution as a result
of the precipitate hardening [24]. It has been shown that the
defects and dislocations created as a result of plastic deformation
or thermal cycling in metals serve as preferential trapping sites.
In the case of aluminum-zinc alloys, it has also been demonstrated
[25] that the precipitate clusters formed in the precipitate hardening of such alloys also serve as trapping sites for positrons.
As the lifetime of positrons in all three kinds of trapping sites
are apparently similar, it is difficult to resolve the contributions for each kind without additional information correlating
defect type to thermal history.

From our lifetime data, it is clear that the tempering process
creates enough trapping sites (i.e., vacancies, dislocations,
precipitate clusters, etc.) so that the majority of positrons are
trapped in, and annihilate from, these sites, even before stress
is applied to the alloy. While a determination of the relative
concentrations of the various trapping sites was not made, we can
estimate the fraction of positrons trapped in defects, dislocations,
and precipitates. Cotterill et al. [18] measured $\bar{\tau}$ in aluminum,
resolving the contribution from trapping in vacancies and from
dislocations by a thermal cycling process which first created
vacancies and then dislocations by annealing of the point defects.
To accomplish this, they first heated their samples to an initial
high temperature and then rapidly quenched. This produced primarily
vacancy defects which were frozen in by keeping the sample at
liquid nitrogen temperature. The specimens were then annealed at
80°C to condense the vacancies into dislocation loops. This process,
in essence, is similar to the T6 tempering treatment undergone by
the alloy. Instead of annealing at 80°C, the alloy is aged at
121°C for 24 hours. If we assume that the maximum defect concentration is determined by the initial temperature to which the alloy is
heated before quenching, and that this concentration is not affected
by the presence of the minority alloy constituents, then the lifetimes obtained by Cotterill et al. after the partial anneal to
produce dislocation loops allow us to obtain an estimate of the
fraction of positrons trapped in dislocations. The balance of
∿80% of positrons annihilating in traps must come from annihilation
of positrons bound to precipitate clusters. For a heat treatment
temperature of 490°C, followed by a partial anneal at 80°C, they
found that the mean positron lifetime, $\bar{\tau}$, is 180 psec. According
to Equation (9), this implies that f (dislocations) = 0.2. The
balance of annihilations are comprised of ∿60% from positrons
bound to precipitation clusters, vacancies, and at given boundaries,
and 20% from positrons annihilating in the undamaged alloy. Plastic
straining is not likely to increase the concentration of precipitation clusters, while it does increase the concentration of

dislocations. Thus, the observed increase of $\bar{\tau}$ with increasing strain probably results from an increase in the number of positrons annihilating in dislocation sites as their concentration increases as a consequence of strain applied to the alloy.

It is evident from the data plotted in Figure 6 that the positron annihilation technique is most sensitive to detecting relatively low concentrations of disorder, i.e., point vacancies, dislocations, precipitation clusters, as the positrons are preferentially trapped in and annihilate within these disordered sites. Normal manufacturing processes, such as tempering, which create large concentrations

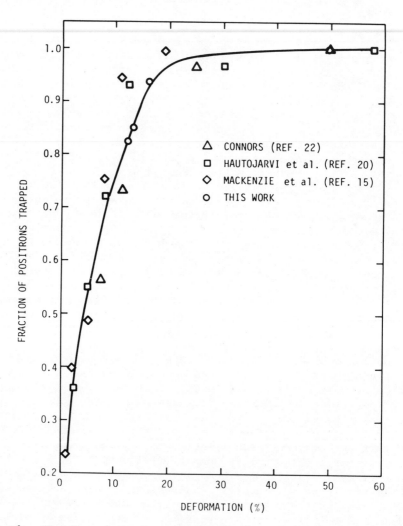

Figure 6.   The fraction of positrons trapped in defects in aluminum versus percent deformation.

of defects or impurity sites, ensures that most positrons will
annihilate in these sites, even before the structural component
suffers any further degree of damage as a consequence of its life
service. This is borne out by a study of the aluminum–clad alloy.
However, the degree of sensitivity of the technique is such that
even in this disordered alloy one can still observe the progression
of strain until rupture.

CONCLUSIONS

In this study we have measured the change in the positron mean
life with life service damage as stimulated through the creation
of plastic strain introduced by bending in an aluminum aircraft
structural alloy, aluminum clad 7075–T6, QQA250/13. We found that
there is a correlation between the positron life and the degree of
strain. The effect of the T6 tempering process is reflected in
the range of lifetimes observed. Thus, we have demonstrated that
it is possible to monitor the effect of fatigue, as simulated by
plastic strain, on this alloy by the change in the measured positron lifetime. It is to be pointed out that this change was
superimposed on a much larger concentration of annihilations in
defects during the tempering process. If these were not present,
the observed change would have been even more pronounced.

To evaluate the applicability of positron annihilation techniques to the development of an NDI probe for the field detection
of life service damage, some comments must be made. Positron
annihilation is a microscopic NDI probe. It is microscopic in
the sense that individual psoitrons annihilate with electrons,
which are part of the atomic constituents of damaged areas. These
can range from single atom sites such as point vacancies, to
groups of atoms forming precipitation clusters, loop dislocations,
or to the regions along grain boundaries, or in microvoids.
Because positrons annihilate preferentially in the disordered
regions of materials where there is a relatively low electron
density, and with electrons of lower average momenta, they are
sensitive to the presence of small concentrations of defects. If
expressed in terms of point defects, the range of relative
concentrations detectable through positron annihilation lies between
10^{-7} to 10^{-4} [8]. In terms of plastic strain, the technique is most
sensitive to the initial strain applied to a material, as indicated
in Figure 6. These finding suggest that one application of positron annihilation as an NDI probe lies in monitoring the change
in materials during their life before failure, i.e., in a preventive
maintenance role. These changes would be manifest by some rearrange-
ment on the molecular level, which produces a greater degree of disorder in the material. This is in distinction to NDI techniques
such as radiography, in which macroscopic damage is monitored.

To develop the positron annihilation technique as a nondestructive tool, several steps must be taken. First, additional fundamental research must be carried out which relates features of a material's microstructure, especially the atomic environment typical of a specific mode of damage or disorder, to annihilation characteristics. To date this work has been carried out primarily by solid-state physicists. It is encouraging to see that more materials science groups are now using positrons to study material properties. It is important to correlate findings based on positrons as a probe with other means of material analysis such as electron microscopy, radiography, etc. Second, one must develop positron annihilation life history curves for technologically important metals and alloys subject to well-defined processes which simulate fabriaction and damaging processes such as annealing, hot and cold working, stress cycling, and irradiation, which relate the positron behavior to damage created. Third, laboratory measurements should be performed on real components subject to simulated life cycling to establish benchmarks for field tests. Finally, a practical NDI positron probe which can be used by relatively unskilled personnel in the field must be developed.

The positron lifetime technique was used to monitor the stress-induced changes in the aluminum-clad alloy, because it appeared to offer the optimum sensitivity for detecting these changes. From the point of view of developing a nondestructive inspection apparatus which can be used in a field environment, it is probably not the optimum technique. It is our feeling that it would be better to use the change in the shape of the Doppler-broadened annihilation line width to monitor the life history of a material. Its prime advantage is that it is a one-detector method. This means that data can be taken rapidly in about in minutes per point, rather than the hour or more required for angular correlation or lifetime measurements. The fact that only one detector is needed means that such a system can be fielded in more difficult environments. Data analysis is relatively simple if compared to the lifetime method. Its principal drawback is that the resolving power for observing the anticipated changes is not as sensitive as the lifetime or angular correlation techniques. However, it is clear from the studies of MacKenzie and coworkers [21] that commerically available equipment can be put together in systems of adequate sensitivity to study the changes in positron annihilation characteristics concommitant with the damage introduced into metals as a consequence of life service.

SUMMARY

The relationship between the annihilation behavior of positrons and the structure of the material in which they annihilate has been

discussed. It was shown that the positron annihilation is a useful
probe for the nondestructive study of materials in which damage is
created, because one can experimentally distinguish between positrons which annihilate in ordered, crystalline regions of a material
and those which annihilate in defects. A feasibility study was presented in which a correlation was found between the positron mean
life and the degree of strain in a commercial aluminum alloy. An
assessment is made of the steps required to transform this technique
into a tool for the nondestructive inspection of materials in which
life service induced failure is problematic.

ACKNOWLEDGEMENT

Portions of this work were supported by the Naval Air
Development Center, Warminster, Pennsylvania, under Contract
N62269-75-M-6111. The valuable assistance of Vesa Junkkarinen
in performing the feasibility study is gratefully acknowledged.

REFERENCES

1. Stewart, A.T. and L.O. Roellig (eds.), Positron Annihilation, New York: Academic Press, 1970.

2. Hogg, B.G. and A.T. Stewart (eds.), Proceedings of the Second International Conference on Positron Annihilation, Kingston unpublished (1970).

3. Proceedings of the Third International Conference on Positron Annihilation, Otanemi, Finaldn (1973); published in Appl. Phys., 4 and 5 (1974).

4. Goldanskii, V.I., "The Physical Chemistry of the Positron and Postronium", At. Energy Rev., 6 (1968) 3-148.

5. West, R.N., "Positron Studies of Condensed Matter", Advances in Phys., 22 (1973) 263-383.

6. Dekhtyar, I.Ya., "The Use of Positrons for the Study of Solids", Phys. Letters, Pt. C, 9 (1974) 245-353.

7. Wallace, P.R., "Positron Annihilation in Solids and Liquids", Solid State Physics, ed. by F. Seitz and D. Turnbull. New York: Academic Press (1960), Vol. 10, 1-69.

8. Seeger, A., "Investigation of Point Defects in Equilibrium Concentrations with Particular Reference to Positron Annihilation Techniques", J. Phys. F., Metal Phys., 3 (1972) 278-94.

9. Goland, A.N., "Positron Annihilation and Its Application to Defect Studies in Metals", Brookhaven National Laboratory Report 16517 (1972).

10. Tao, S.J., "Detection of Vacancy Concentration and Defects In Aluminum, Titanium, and Their Alloys, and Nickel Base Super-alloys by Positron Annihilation", New England Institute, Ridgefield, Ct., Naval Air Systems Command, Contract Report 000019-74-C-0170, March 1975.

11. Wilkenfeld, J.M., J. John and V. Junkkarinen, "Feasibility Evaluation of Positron Annihilation Techniques to Detect Void Defects and Residual Stress in Aircraft Structural Components", IRT Corporation, San Diego, Calif., Naval Aird Development Center Contract Report N62269-75-M-6111, July 1975.

12. Lee-Whiting, G.E., "Thermalization of Positrons in Metals", Phys. Rev., 97 (1955) 1557-58.

13. Carbotte, J.P. and H.L. Aurora, "Thermalization Time of Positrons in Metals", Can. J. Phys., 45 (1967) 387-492.

14. deBenedetti, S. and H.C. Corben, "Positron Annihilation", Annual Rev. Nucl. Sci., 4 (1954) 191-214.

15. MacKenzie, I.K., J.A. Eady and R.R. Gingerich, "The Interaction Between Positrons and Dislocations in Copper and in an Aluminum Alloy", Phys. Let. A, 33 (1970) 279-80.

16. Evans, R.D., The Atomic Nucleus, New York: McGraw-Hill (1955) 621-29.

17. Snead, C.L., T.M. Hall and A.L. Goland, "Vacancy-Impurity Binding Energy in Aluminum - 1.7% Zn Using Positron Annihilation Lifetimes", Phys. Rev. Let., 29 (1972) 62-65.

18. Cotterill, R.M.J., K. Petersen, G. Trumpy and J. Traff, "Positron Lifetimes and Trapping Probabilities Observed Separately for Vacancies and Dislocations in Aluminum", J. Phys. F, Metal Phys., 2 (1972) 459-67.

19. Berko, S. and H. Weissberg, "Positron Lifetimes in Metals", Phys. Rev., 154 (1967) 249-57.

20. Hautojarvi, P., A. Taminin and P. Jauho, "Trapping of Positrons in Dislocations in Aluminum", Phys. Rev. Let., 24 (1970) 459-61.

21. Campbell, J.L., T.E. Jackman, I.K. MacKenzie, C.W. Schulte and C.G. White, "Radionuclide Emitters for Positron-Annihilation Studies of Condensed Matter", Nucl Instr. Methods, 116 (1974) 369-80.

22. Connors, unpublished (1971), as reported in Ref. 5.

23. We wish to thank F. Jackson of Kaiser Aluminum for details of the manufacture and tempering process involved in making aluminum-clad 7075-T6.

24. Friedel, J., Dislocations, London: Pergamon Press (1964) 368-84.

25. Sedov, V.L., V.A. Teimourazova and K. Berndt, "Interaction of Positrons with Precipitations in a Diluted Aℓ-Zn Alloy", Phys. Let. A, 33 (1970) 319-20.

Chapter 22

QUANTITATIVE METHODS IN PENETRANT INSPECTION

P. F. Packman and J. K. Malpani of Vanderbilt
University, Nashville, Tennessee

G. Hardy of AFML/MX WPAFB, Ohio

ABSTRACT

This chapter deals with the quantitative measurement of detectability of surface defects by penetrant inspection. The chapter reviews studies concerned with the quantitative evaluation of penetrant inspection systems and presents statistical analysis techniques which can be used for rating penetrants and for optimization of inspection process variables, namely, analysis of various ANOVA and multiple comparison tests. These analysis techniques should ultimately help in establishing the penetrant inspection method as quantitative relative to detection capability rather than qualitative or based on chemical and procedural standards.

INTRODUCTION

Liquid penetrant inspection is a deceptively simple and efficient method of determining the presence and severity of surface discontinuities, or flaws on materials. The ability of the penetrant process to detect small crack-like surface defects depends primarily on the capillary ability of the liquid penetrant system to penetrate into the surface discontinuity, and remain in place during the subsequent action of the water wash, emulsifier or solvent, and to be made visible by the subsequent developer action.

Present day penetrant inspection procedures involve a series of steps designed to enhance the signal of the indication and yet not produce too large a background indication and to minimize the

number of false indications. Significant increases in penetrant
sensitivity are produced when going from colored dye tracers to
fluorescent tracers to post emulsifier fluorescent systems. This
in turn necessitates the need to improve the methods of cleaning
and the initial wash procedures so that false indications of defects
due to residues of entrapped penetrant in nonsignificant areas
would not obscure the actual defect indications. Thus, while
measurements of fluorescent threshold and those factors influencing
the flaw entrapment efficiency are of immediate concern in quantify-
ing penetrant systems, they are not dealt with within the scope of
this chapter.

This chapter is concerned with the general problems of the
detectability of defects by penetrant systems and particularly with
quantitative measure of the detectability of small crack-like defects
by penetrant inspection. Such a measure would ultimately develop
into a ranking system or scale against which a wide range of penetrants could be evaluated. The results could then be applicable
to evaluation and acceptance of penetrant systems only as far as
the quantitative ranking applies to the particular type of inspection standard being used. One would not necessarily expect that any
single specimen type would provide a ranking of penetrants under
other operating conditions or for use on all types of materials
and defects.

QUANTITATIVE PENETRANT EVALUATION METHODS

Although a large number have been proposed there is no universal standardized method or test specimen that has been developed
and accepted to evaluate penetrants. This is due primarily to
(1) the large number of penetrants available, (2) the differences
in the surface conditions of the parts to be inspected, (3) differences in the type of defect to be inspected, (4) the differences
in levels of sophistication required by the user. Table I presents
a summary of the type of tests and the specimens used in recent
evaluative processes for penetrants.

An early procedure for evaluating penetrant materials was
presented by Miller [1] using a set of precision ground sleaves
fitted onto a threaded bolt. When the bolt was tightened to 30
ft.-lbs., all the dye penetrants inspected gave distinct high-
contrast indications of the sleave interface. If, however, the
torque applied to the system was 300 ft.-lbs., the dye penetrant
indications of the sleave interface became much more indistinct
for one of the penetrants while the other dye penetrants' radiations
remained relatively distinct. This showed that one of the penetrant
systems had, at this time (1958), a lower degree of sensitivity to
this type of defect. No tests were presented on fluorescent penetrants.

TABLE I

SUMMARY OF RECENT PENETRANT STUDIES LEADING TO PENETRANT STANDARDS

Specimen Type	Reference	Study	Measure
Chrome Cracked Brass	Sherwin	Effect of Dwell Mode	Qual.
Chrome Cracked Brass	McCalley & Van Winkle [3,4]	Crack Detection Eff.	Quant.
Chrome Cracked Brass	Monsanto [3,4]	Crack Detection Eff.	Qual.
Ti6Al-4v SCC	Lord & Holloway [10]	Penetrant Eff.	Qual.
Ti6Al-4v Fat	Lord & Holloway	Penetrant Eff.	Qual.
Ti6-6-25n Forge Porosity	Lord & Holloway	Penetrant Eff.	Qual.
Chrome Cracked Steel	Fricker [7]	Developer	Qual. SBS
Titanium NaCl Crack			
Fatigue Cracked Panel	Borucki [10]	Effect of Wash Times on Brightness	Quant.
Fatigue Cracked Cylinders 7075-T6 Aluminum 4335V mod Steel	Packman [6]	Effect of Crack Size Accur. of Crack Length	Quant.
Torque Loaded Sleave	Miller [1]	Effect of Crack	Qual.
Turbine Blade Cracks	Lomerson [14]	Penetrant Rating	Quant.
Chrome Cracked Brass	Canadian AF [15]	Penetrant Rating	Quant.
Cracked CF-104 Ldg. Gear	Canadian AF [15]	Penetrant Rating	Quant.
Quenched Cracked Aluminum	Jones [8]	Penetrant Rating	Quant.
Screw Tightening Steel Plates	Swamy [2]	Penetrant Development	Qual.
Brinnelled Anodized Plate	P & W [14]	Penetrant Rating	Qual.
Cracked Aluminum (Anodized)	Alburger [12]	Penetrant Rating	Qual.

S. Swamy [2] developed a test block consisting of two ground steel plates with tightening screws that changed the degree of one crack-like defect. He used this specimen to develop a dye penetrant system for use in India.

McCauley and Van Winkle [3,4] conducted extensive studies into the relationships between the properties of penetrants and the observable detection behavior of penetrants. These results are shown in Table II. They developed a standard chrome-plated brass plate which can be cracked to produce a pattern of surface defects. The panels are divided into three categories, depending upon the width of the defects: coarse (500 microinches), medium (90-130 microinches), and fine (19 microinches).

They define a crack detection efficiency (CDE) to be the ratio of the number of cracks/inch detected by the penetrant, to the number of cracks/inch counted under a 100 x magnification. A plot of CDE versus the fluorescent absorption coefficient, a measure of the energy absorbed by the fluorescent dye, is shown in Figure 1. Two roughly parallel lines can be seen corresponding to different amount of detergent in the penetrant system. It can be seen that the higher detection efficiencies are found in the penetrants that contain the most fluorescent material.

An evaluation of solvent remover penetrants was performed by the Canadian Aircraft Maintenance Development Unit using chrome cracked panels as well as cracked CF-104 aircraft main landing gear links [5]. The landing gear links had extremely tight cracks as narrow as 10 microinches. The relative sensitivity was evaluated by measuring the cumulative length of the defect indications. The standard use was the penetrant system which eventually proved the most sensitive. Cumulative lengths obtained using the other penetrant systems were compared to this standard. Comparison of the ranking of penetrant using the cracked chrome test panel and the cumulative length procedure showed that the first two penetrant system rankings did not change, and only one penetrant system was interchanted of the five examined [5]. Their conclusion was that the chrome cracked panels are satisfactory for the evaluation of penetrant sensitivity.

An analysis of the sensitivity of penetrants using the measurement of actual crack length was conducted by Packman et al. [6] for aluminum and steel cylinders containing fatigue cracks. In this experiment, an assessment was made of the accuracy of the length of the crack indication by penetrants. The length of the defect indication was compared to the length of the actual indication obtained by fracture of the specimen at the completion of the program. There appeared to be no influence of the length of the defect on the length of the defect indication.

TABLE II

Penetrant Parameters [2]

CDE*	SPP**	Kc***	Q****	KcQ
35	31.5	629	33.6	21100
51	31.6	320	33.6	10750
54	29.4	55.9	35.3	1970
63	33.4	700	30.1	21100
66	28.5	120	35.9	4300
75	30.4	542	32.7	17700
76	34.0	3060	30.3	92600
80	37.0	7540	32.9	248000
89	39.6	11060	22.4	248000
91	33.2	1730	31.9	55260

* CDE – Crack density efficiency (%)
** SPP – Static penetrability parameter dyne cm^{-1}
*** Kc – Absorption coefficient
**** Q – Fluorescent efficiency

T. R. Fricker [7] of the Naval Rework Facility reports a series of tests conducted using cold-rolled steel chrome plated and cracked specimens bent and subsequently straightened to produce lateral cracks ranging from 0.001 to 0.030 inches in width on the center section of the specimen. A machined groove was placed perpendicular to the crack direction to enable side-by-side comparison. Tests were also conducted on titanium strips cracked by a salt water stress corrosion cracking fixture followed by a 800°F soak that produced a wide variety of crack widths. Tests were made to examine the sensitivity of dry, water soluable, and nonaqueous developers. The conclusion was reached that water washable penetrants provide a more reliable inspection process than the post emulsifier types for the new requirements. They indicated that both deep and shallow broad discontinuities are made more visible with greater consistency. One observation reported that the effect of over-washing was more pronounced on the titanium specimens than on the steel strips.

Jones [8] at NASA Marshall evaluated several penetrant materials using quench cracked 2219-T86 blocks for use on the Saturn V.

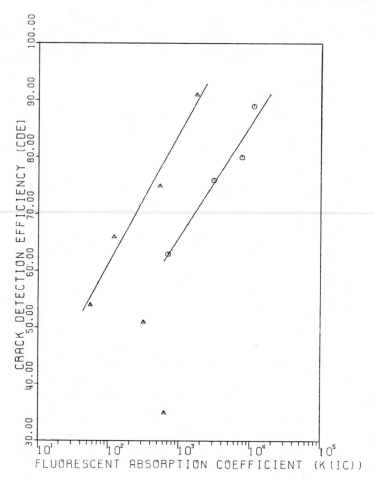

Figure 1.   Relation between crack detection efficiency and
            fluorescent absorption coefficient.

Measurements were made by counting the number of visual indications,
and then by sectioning the specimen and performing a metallurgical
count. An interesting observation was that the most sensitive
penetrant was not acceptable because of the high number of false
indications, i.e., indicated consistantly higher crack counts than
actually present.

    Most work reported concerning the influence of dwell time on
sensitivity does not differentiate between the dwell mode types,
immersion and drain. Sherwin [9] reviewed the influence of the

mode of dwell on the number of indications found in the chrome plated cracked specimens and concludes that a drain dwell is preferable to an immersion dwell, given the same performance characteristics on the same specimen. Flaw indications were easier to see, brighter, and more complete, on specimens examined both with and without developer.

Titanium 6Al-4V stress corrosion cracks produced by stressing with an anhydrous-methanol-NaCl solution were examined by post emulsifiable and water washable penetrants in a study reported by Lord and Holloway [10]. Their study also reported results obtained by inspection of single fatigue cracks in Ti6Al-4V, as well as porosity in hand forged billets of Ti641-6V-25n. The stress corrosion cracks were about 0.0002 inches wide, compared to the fatigue cracks which were about 0.001 inches wide. Visual examination and photographs of the panels exposed to the different penetrants were made to evaluate the effect of dwell time penetrant effectiveness. The results are summarized in Table III. The results for gross cracks (0.001 inches) agrees with Alburger [11] in that all penetrant types were equally as efficient in detection while noticeable differences for the tighter cracks were found for different penetrants and developers.

Alburger [12] develops a Performance Merit Factor which relates the diameter of the spot on the meniscus lens test to a measure of the "smallest flaw dimension that can be detected". The actual values of the smallest flaw that can be detected is almost always larger than this value because (1) 100% entrapment efficiency is rarely obtained; (2) residual dirt, surface contamination and cleaning residue significantly change the surface of the defect from that of the glass meniscus spot test; and (3) the sensitivity of the penetrant system is not necessarily dominated by any single property.

Borucki [13] reported on the influence of wash times on the decrease in percentage indication brightness as a function of wash times for water washable and post emulsifier, hydrophylic emulsifier spray, lipophilic elulsifier (slow and fast action) penetrant systems. In all cases the brightness indication decreases as a function of increasing wash or emulsification times. The brightness indications were measured on a part through fatigue cracked flat test panel similar to those developed for the fracture control demonstration programs.

Lomerson [14] developed a test to rate penetrant performance on a quantiative basis which he called the "Two Fold Congruency" test. The method consists of observing the number of congruent observations obtained on a set of turbine blades processed through three separate runs on the same penetrant. The congruent indication

TABLE III

Effect of Crack Type on Developer Effectiveness

Crack Type	Developer Type	
	Post Emulsifiable Penetrant	Water Washable Penetrant
Gross Cracks	1) Nonaqueous wet, dry & aqueous equally effective	1) Nonaqueous wet, dry & aqueous equally effective
	2) No developer	2) No developer
Porosity	1) Dry 2) Nonaqueous wet 3) No developer 4) Aqueous	All types equally effective
Tighter Cracks	1) Nonaqueous & wet 2) No developer 3) Aqueous	1) Nonaqueous wet 2) No developer 3) Aqueous

is an indication obtained at the same position on any two runs. The mean number of congruent indications is calculated by averaging the number of congruent indications for each penetrant under investigation. If an indication was found by one penetrant inspection and not by the subsequent inspections, it is not included in the analysis.

The relative sensitivity of the penetrant is calculated by comparing the results obtained on the same set of blades by using the standard penetrant to those obtained using the unknown penetrant. The use of congruencies rather than a total number of indications tends to eliminate occasional random indications from the assessment, and thus incorporate some measure of the reproducibility into the evaluation process.

Lomerson ranked 16 penetrant systems, both water washable and post emulsifier systems, against a penetrant standard. It was pointed out that some penetrants exhibited an extremely large spread in the results, while others showed comparatively small variations. There was no correlation between the sensitivity of the penetrant (compared to the standard penetrant) and the spread in results. All penetrants that ranked above 100 (implying greater sensitivity than the chosen standard) showed scatter values greater than the standard penetrant.

Using the two fold congruency procedure Hyam [15] was able to show the influence of some factors affecting the sensitivity of a penetrant system, including:

1. The effect of rinse time on water washable penetrants, increasing the rinse time from 2 to 12 minutes decreased the sensitivity for two water washable systems examined.

2. The effect of contact time with a hydrophilic remover system on the sensitivity of two penetrants was shown to be relatively small. The concentration of the hydrophilic remover had a significant influence on the relative sensitivity, in general the higher the concentration of the hydrophilic remover, the lower the sensitivity rating.

3. The influence of remover contact time using lipophilic removers was shown to be significant. Increasing the time of contact also decreased sensitivity.

4. Developer type on the sensitivity index was shown to be significant, in all cases the results with no developer were lower than those using a developer. The sensitivity was shown to increase using wet suspension, dry powder and solvent suspension developers, in ascending order.

ANOVA AND CRACK LINE INTERSECTION COUNTING

The work of McCauley and Van Winkle [3,4] made use of a measure of the crack line density as a measure of the efficiency of the penetrant. In the following section some work extending that concept is presented.

If a panel containing surface cracks of a wide variety of crack widths is inspected, one would normally expect that a high resolution penetrant system would find a greater number of cracks than a lower resolution system. The low resolution system would be expected to miss some of the finer, tighter cracks, while the higher resolution system would not be expected to miss as much. Hence, given the same specimens, and assuming that there were no changes in the specimen due to the multiple inspections, and that the subsequent inspection results were not modified by the prior penetrant system, one would expect to see a lower crack density for the lower resolution penetrant system.

The problem which arises when many different specimens are used is how to separate out the influence of the true number of cracks in the specimens so that the influence of only the penetrant can be considered. The purpose of analysis of variance (ANOVA) or multiple analysis of variance (MANOVA) is to separate out the different levels of the various variables. This enables one to determine if the results are due to different specimens, different penetrants, or simply a random error [17].

   In this analysis four specimens were used with differing crack distributions. A typical panel pair is shown in Figure 2. Six different inspection procedures were used to inspect each panel, these procedures including changes in penetrant system, changes in dwell time, changes in emulsifier, etc. The specimens were photographed under identical conditions. Four random lines were drawn on each photograph corresponding to each treatment. Thus there were 6 x 4 = 24 photographs and each photograph had 4 random lines. For each specimen panel photograph for any treatment line locations and geometries were kept constant. Number of lines intersection/ lines were counted, and are tabulated in Table IV as replicates 1 through 4 for each treatment–specimen panel combination. Table V also shows row sums (total number of line intersections for all four lines for each 'treatment' – specimen panel combination), column sums (replicate totals) and grand total.

   The analysis of variance model assumes that any observation x_{ijk} in Table IV out of a total of 96 can be written as a sum of population mean u, replicate effect α_i, treatment effect β_j, specimen effect γ_k, treatment–specimen interaction effect $\beta\gamma_{jk}$, and chance error ε_{ijk}.

   in other words:

$$x_{ijk} = u + \alpha_i + \beta_j + \gamma_k + \beta\gamma_{jk} + \varepsilon_{ijk}$$

The analysis of variance table is computed as follows:

$$\text{Correction term} = \frac{(\text{Grand Total})^2}{\text{Total No. of Observations}}$$

$$= \frac{(4579)^2}{96} = 218408.76$$

$$\text{Total Sum of Squares S.S.T.} = \Sigma\Sigma\Sigma \; x_{ijk}^2 - (C)$$

$$\text{S.S. (Replicates)} = \frac{(\Sigma_j \Sigma_k \; x_{ijk})^2}{n_j n_k} - (C)$$

$$\text{S.S. (Treatments)} = \frac{(\Sigma_i \Sigma_k \; x_{ijk})^2}{n_i n_k} - (C)$$

$$\text{S.S. (Specimens)} = \frac{(\Sigma_i \Sigma_j \; x_{ijk})^2}{n_i n_j} - (C)$$

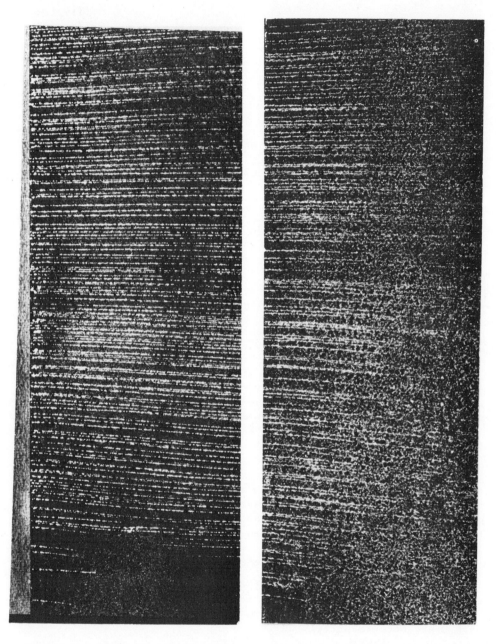

Figure 2. Penetrant indications on chrome plated cracked steel plates. With of plate is 4 inches.

$$\text{S.S. (Treatments - Specimen)} = \frac{(\Sigma_i x_{ijk})^2}{n_j} - (C) {- \text{S.S.(Treatments)} \atop - \text{S.S.(Specimen)}}$$

Where

n_i	= number of replicates
n_j	= number of treatments (inspection procedures)
n_k	= number of specimen panels
$n_i n_j n_k - 1$	= total degrees of freedom
$n_i - 1$	= replicates degrees of freedom
$n_j - 1$	= treatment degrees of freedom
$n_k - 1$	= specimen degrees of freedom
$(n_j - 1)(n_k - 1)$	= treatment-specimen interaction degree of freedom
$(n_j - 1)(n_j n_k - 1)$	= error degrees of freedom

Mean sum of squares (M.S.S.) is obtained by dividing S.S. by corresponding degrees of freedom = sum of squares/corresponding d.f.

$$\text{F ratio} = \frac{\text{M.S.S.}}{\text{Error M.S.S.}}$$

If F ratio is greater than $F_{\lambda_1, \lambda_2, \alpha}$ we conclude there is significant effect due to that factor where λ_1 and λ_2 are degrees of freedom for the numerator and denominator, respectively, and α is level of significance. The value $(1-\alpha)$ is the confidence. All the calculated values are shown in Table V along with F values from calculated statistical tables for 95% and 99% confidence levels.

From Table V the following can be concluded:

1. There are no significant replication effects at both 95% and 99% confidence levels. This means that replicates were not very much different. Hence the number of crack distributions in each plate was essentially uniform.

2. The type of penetrant inspection procedure (treatments) had significant effect on detection of cracks at both 95% and 99% confidence levels.

3. Type of specimen panel also had significant effect on

TABLE IV

Penetrant Data in a Two Way Factor with Four Replicates

Factor Treatments	Specimen No.	Rep. 1 (Line 1)	Rep. 2 (Line 2)	Rep. 3 (Line 3)	Rep. 4 (Line 4)	Total
A	18	43	43	44	50	190
A	19	47	38	120	37	242
A	20	78	57	43	97	275
A	21	67	43	52	51	213
B	18	35	38	24	37	134
B	19	26	30	93	34	183
B	20	61	47	32	68	208
B	21	55	25	27	22	129
C	18	37	47	40	42	166
C	19	39	37	136	48	260
C	20	82	65	60	107	314
C	21	62	46	56	50	214
D	18	24	29	26	21	100
D	19	21	39	96	33	189
D	20	76	63	52	100	291
D	21	54	29	47	43	173
E	18	24	32	29	21	106
E	19	22	31	7	1	61
E	20	65	53	39	62	219
E	21	62	31	32	32	157
F	18	35	40	37	41	153
F	19	38	34	14	75	161
F	20	63	48	46	87	244
F	21	67	41	47	42	197
		1183	996	1199	1201	7579

detection of cracks at both 95% and 99% confidence levels. This means that different specimens were different as to the number of surface flaw distributions.

4, There is no significant interaction between type of specimen panel and type of inspection procedure effect at both 95% and 99% confidence levels. This is important since the specimen panels had significantly different number of flaw distributions and still the type of penetrant inspection procedure did not react differently for different types of specimen panels, flaw distribution, i.e., a good

TABLE V

Analysis of Variance Table

(Including Analysis of Interaction Effects)

Source of Variation	Degrees of Freedom	Sum of Squares	Mean Sum of Squares	Calculated F Ratio	Theoretical at Confidence	
					95%	99%
Replication	3	1237.365	412.544	.99	2.75	4.09
Main Effects: 1. Type of Treatment- A, B, C, D, etc.	5	7598.4275	1519.6855	3.6494	2.36	3.30
2. Type of Specimen No. 18, 19, 20, or 21	3	10779.0317	3593.010567	8.6282	2.75	4.09
Interaction Effect	15	3746.0308	249.73	.5997	1.82	2.32
ERROR	69	28733.385	416.4259			
TOTAL	95	52094.29				

inspection procedure was good for any type of flaw distribution.

STATISTICAL MULTIPLE COMPARISON TESTS

In this section we are interested in determining methods whereby the significant factors affecting the experimental results can be sorted out of the test data. From the ANOVA tests it was concluded that the inspection procedures showed significant differences. However, we are interested in knowing whether one inspection procedure is better than other procedure, at a given confidence level. If, for example, one procedure were better than another at a 95% confidence level, we would conclude that the difference in crack detection capability was significantly improved in going from one penetrant system to the other. If, however, the confidence that we could place in the statement that one penetrant was better than the other were only 50%, one would not obtain much of an improvement in crack detection by changing to the new penetrant. ANOVA tests cannot give such information.

In a statistical analysis, the most important use of the ANOVA or multiple analysis of variance MANOVA is to test whether the mean values of two of more populations are equal, against the alternate that they are not equal. For only two populations the student 't' or similar tests can be used, but for multiple populations the ANOVA or MANOVA must be used. The immediate aim is to sort out the factors that significantly change these mean values. As used in our penetrant analysis, one additional use it to test the hypothesis that other parameters have joint or interactive effects. Here we ask the question, "Was a parameter, such as crack type or specimen type significantly differing in the test program so as to influence the ability of the penetrant to detect the defects?" Statistically we asked, "If the changes in the mean values (crack intersections) attributable to one factor or parameter (penetrant type) are the same at different levels of other factors (specimen type, crack type, etc.)."

Even if the hypothesis of equal mean values is rejected at the desired confidence levels (i.e., 90 or 95%) and it is concluded that other factors do not interact with the main effect, one still has the problem of deciding which of the population means are significantly different.

Multiple comparison tests, MCT, can determine if there is truly a difference between the individual test means. These tests examine a number of test statistics to determine which comparisons among means are significant. In this analysis we have used Dunn's new multiple comparison procedure, also called Dunn's New Multiple Range Test, as the most simple and widely used procedure [16,17].

　　　A multiple comparison test was carried out using the data obtained from the ANOVA test in Table IV. From Table V, the Mean Sum of Squares for Error (MSE) = 416.4259, the number of observations penetrant procedure was n = 96/6 = 16.

　　　Given below are ordered mean values of flaw indications per inspection procedure, decreasing in value from left to right. These values are obtained by summing row totals in Table I for each inspection procedure and dividing by 16 (number of observations per inspection procedures).

Inspection Procedures	E	B	D	F	A	C
Mean Values of Flaw Indications	33.93	40.87	47.06	47.18	57.43	59.62

We set overall confidence for the test at 95%. Therefore, level of significance $\alpha = 0.05$. From Duncan's tables in CRC Handbook [16], we get the following values of statistic $q\lambda, \alpha$ were λ = number of degrees of freedom for error = 69.

$$q_2 = 2.83$$
$$q_3 = 2.98$$
$$q_4 = 3.045$$
$$q_5 = 3.116$$
$$q_6 = 3.172$$

From these critical values for q statistic we can obtain the critical range values (W) as follows:

Estimated value of standard deviation $= \sqrt{MSE}$

$$= \sqrt{416.4259}/16$$

$$= 5.102$$

Therefore, critical ranges are as follows:

$$W_2 = 2.83 \times 5.102$$
$$= 14.44$$

$$W_3 = 2.98 \times 5.102$$
$$= 15.20$$

$$W_4 = 3.045 \times 5.102$$
$$= 15.53$$

$$W_5 = 3.116 \times 5.102$$
$$= 15.89$$

$$W_6 = 3.172 \times 5.102$$
$$= 16.18$$

The range of all six means = 59.62 - 33.93 = 25.69. The range of all six means exceeds W_6 which is obvious because we concluded in analysis of variance section that inspection procedure has significant effects.

The difference of mean values E and A (23.50), B and C (18.75) both exceed W_5 (15.89) and hence we conclude that inspection procedure A and C are better than E and B, respectively, at 95% confidence.

The differences of mean values of E and F (13.25) and D and C (12.56) and less than W_4 (15.53) and hence it cannot be concluded that treatments F and C are better than E and D, respectively. But the difference of B and A (16.56) is greater than W_4 and hence A is superior to B at 95% confidence.

None of the values W_2 and W_3 are exceeded by any corresponding pairs of means and hence these pairs (like E-D, B-F, D-A and E-B, B-D, D-F, etc.) are not significantly different at 95% confidence.

CONCLUSIONS

Of the numerous techniques proposed for the quantitative evaluation of penetrants no one procedure can be classified as acceptable for the wide range of user requirements. Most tests involve side by side comparisons of two different penetrants, and except for the twofold congruency technique, the ANOVA, and multiple comparison tests procedures, do not measure directly the penetrant performance on a quantitative basis. These last procedures, too, while quantitative in ranking performance do not place the process on an absolute scale. Tests such as the spot miniscus test, while measuring properties relating to performance do not measure performance directly. While the twofold congruency test relates one penetrant against another, the ANOVA procedures developed here rate several penetrants against one another, and also indicate whether there is a significant difference between specimens. The ANOVA procedure tells the investigation that there is a difference in the penetrant performance, but one requires the multiple comparison test procedures to be able to rate the different penetrants.

Using this data one can rate the penetrants as follows: penetrant A is superior to B, D, E, F, and penetrant C is superior

to penetrants B, D, E with B, D, E, and F approximately the same
at a 95% confidence level. It appears that penetrant A is superior
to C with the rest all inferior to A and C.

It should be noted that ranking, while able to order the penetrants, cannot as yet produce a quantitative ranking process.
Hence one cannot simply say that penetrant A rates 90 out of a
possible 100 points while C ranks 80 out of 100. Thus while a
statistical procedure enables pairs of penetrants to be rated
against each other at levels of significance, absolute ranking
procedures are not yet available.

ACKNOWLEDGEMENT

The authors wish to acknowledge the support of the Air Force
Materials Laboratory under Air Force Contract No. F 33615-76-R-
5166.

REFERENCES

1. Miller, R.W., "A Method for Evaluating Materials Used in Penetrant Flaw Detection", Weld. J. (Jan. 1958) 30.

2. Swamy, S.G.N., "Indigenous Dye – Penetrants", CMERI #A8, Central Mechanical Engineering Research Institute Durgofus – g (India).

3. McCauley, R.B. and Q. Van Winkle, "Research to Develop Methods for Measuring the Properties of Penetrant Inspection Materials", WADD TR, Part 1 (June 1960).

4. McCauley, R.B. and Q. Van Winkle, WADD TR, Part III (Feb. 1963).

5. Nielson, D.C. and J.G.H. Thomson, "Evaluation of Liquid Penetrant Systems", Mat. Eval. (Dec. 1975) 284.

6. Packman, P.F., H.S. Pearson and G.B. Marchese, "The Applicability of a Fracture Mechanics/NDT Design Criterion for Aerospace Structures", AFML TR 68-32 (1968).

7. Fricker, R.T., "Evaluation of High Sensitivity Water-Washable Fluorescent Penetrants", Mat. Eval., 31 (Sept. 1972) 200.

8. Jones, G.H., "Sensitivity and Comparison Evaluation of Saturn V Liquid Penetrants", NASA TMX – 64769, Marshall Space Center, July 27, 1973.

9. Sherwin, A.G., "Establishing Liquid Penetrant Dwell Modes", Mat. Eval., 33 (March 1974) 63.

10. Lord, R.J. and Holloway, J.A., "Choice of Penetrant Parameters for Inspecting Titanium", Mat. Eval., 34 (Oct. 1975) 249.

11. Alburger, J.R., "Signal to Noise Ratio in the Inspection Penetrant Process", Mat. Eval., 33 (Sept. 1974) 193.

12. Alburger, J.R. and K.A. Skeie, "Super Sensitive Inspection Penetrant Processes", presented to ASNT Convention, Chicago, Ill., November 1966. SEC. 0-02200 Bulletin No. 551007, Vresco Inc., Downey, Calif.

13. Borucke, J.S., "Recent Adances in Fluorescent Penetrant Inspection Trhough the Use of Hydrophilic Penetrant Removal System", presented at Air Transport Assoc., NDT Meeting, Miami, Fla., Oct. 7-9, 1975.

14. Lomerson, E.O., Jr., "Statistical Method for Evaluating Penetrant Sensitivity and Reproducability", Mat. Eval. (March 1969) 67.

15. Hyam, N.H., "Quantiative Evaluation of Factors Affecting the Sensitivity of Penetrant Systems", Mat. Eval., 31 (Feb. 1972) 31.

16. Beyer, W.H., CRC, Handbook of Tables for Probability and Statistics, 2nd Ed., The Chemical Rubber Co., Cleveland, Ohio, 1968.

17. Namboodini, N.K, L.F. Carter and H.M. Blalock, Jr., Applied Multivariate Analysis and Experimental Designs, McGraw-Hill Book Co., New York, 1975.

A-7 nose landing gear, stress corrosion, 173
Acoustic emission waves, optical probing of, 347ff
Adhesive bonded structure, inspection of, 162ff
Aircraft engine components, inspection of, 167ff
Aircraft structures, detection of corrosion, 170ff
Aluminum, recrystallization of, 77
Analysis
  residual stress, 1, 12ff
  x-ray line profile, 92
  x-ray microbeam Debye, 40
Angle-scanning diffractometry, 12
Anisotrophy, elastic, 200ff
Anomalous transmission, 80
Applications of, x-ray diffraction, 1
Artillery shell filler, detection of cavities, 421ff
Atomic scattering factors, 6
Axial residual stress, 51

Back reflection
  camera, 57
  cone, 61
Barkhausen noise
  pulses, 439
  stress, 461
Barkhausen signature change, 462
Berg-Barrett Method, 70

Borrmann effect, 80
Bragg reflection, 60, 70

C-130 aircraft, corrosion, 172
Camera
  back-reflection, 57
  , neutron radiography, 155ff
  x-ray micro, 39
Cavity detection, 427ff
Chemical valences, 3
Circuit, correction, 366
Cold working, 108ff
Composite materials, 161ff
Compositional analysis, nondestructive, 183ff
Cone, back reflection, 61
Correction circuit, 366
Corrosion in aircraft structure, detection of, 170ff
Corrosion
  A-7 nose landing gear, 170
  C-130 aircraft wing, 172
  wing tank, 171ff
Counter system, 24
Crack
  detection, 427ff
  tip of, 205
Critical inspection of bearings, for life extension, 464
Crystalline materials, 22
Crystals, micro defects in, 82
Curve, S-N, 30
Cycling, fatigue stress, 30

DC-9 Aircraft, wing tank corrosion in, 171

Defect structure, x-ray
        transmission in, 81
Deformation substructure, 77
Delaminations, ultrasonic
        detection of, 312
Detection
  cavity, 427ff
  crack, 427ff
  x-ray, 11ff
Differential interferometer,
        361
Diffraction technique, 355
Diffraction, flash x-ray,
        136ff
Diffractometry
  angle-scanning, 11
  double crystal, 83
  energy-scanning, 11
Dipole moments, 3
Direct image, 81
Dispersion, spectral, 186ff
Distributed image information,
        analysis of, 379ff
Distribution, electron
        density, 3
Divergence,
  horizontal, 10
  vertical, 10
Double-crystal diffractometer,
        72
Dynamical theory, 1, 4ff

Effect
  Borrmann, 80
  focus, 365
  Pendellösung, 79
  tilt, 365
Elastic anisotropy, 200ff
Elastic constants,
  X-Ray, 203ff
Electromagnetic transducer,
        323ff
Electron density distribution,
        3
Electron microscopy,
  transmission, 83
Electronics,
  solid-state, 3
Energy dispersion geometry,
        187

Energy-scanning diffractometer,
        11
Exploding gold foil,
  transmission powder, 146
Explosives, metal jacketed,
        158ff
External interference, 2
Extrinsic stacking faults, 6

Factor
  scattering factors, 6
  structure, 6
Fatigue crack
  phenomena, 491
  propagation, 44ff, 52
Fatigue damage, 55ff, 447
  by x-rays, 21ff
  detection, 34
Fatigue fracture surface,
  x-ray line broadening, 249ff
Fatigue in engineering, 28ff
Fatigue
  life, 30
  of steel, 30
  process, 38ff
  size effect on, 30
  stress cycling, 30
  stressing, 35
Faults
  extrinsic stacking, 6
  intrinsic stacking, 6
Film-densitometer system, 24
Flash radiograph, 132
Flash x-ray, 117ff
  diffraction, 136ff
  energy storage, 119ff
  pulse shaping, 119ff
  radiography, 130ff
Flaw definition, quantitative,
        258ff
Focusing geometry, 106ff
Fringe method, 6
Fringes, Pendellösung, 5

Gas adsorption method, 10
Generator, rotating-anode, 11
Geometry,
  constant, 105
  energy dispersion, 187
  focusing, 106ff

Geometry (cont'd)
  single exposure, 104
  wavelength dispersion, 186
Germanium, Li-drifted, 11
Grain size, 109ff

hkl reflection, 12

Image processing in NDT, 409ff
Image processor, digital
          computer, 414
Image system, neutron
          radiography, 154ff
Images
  direct, 81
  dynamical, 81
  intermediary, 81
Imaging and scattering approach,
          272ff
Impact strength test, 63
Interference
  external, 2
  internal, 2
Interferometer
  comparison of, 358ff
  differential, 361
  optical path, 364
  quadrature dual, 362
  system, 361ff
  swept path, 363
  x-ray, 7
Intermediary image, 81
Internal interference, 2
Interplanar spacing, 102
Intrinsic stacking faults, 6

Kinematical theory, 1, 2ff

Laminated structures, 313
Lang Method, 78ff
Laue-Transmission setting, 78ff
Li-drifted germanium, 11
Life prediction failure, 459
Loading stresses, 205ff
Location, diffracted peak, 106ff

Macromolecules, 11
  study of, 10
Macroscopic stresses, 323ff

Magnetic NDE, 447ff
  methods, 435ff
Marx generator, 121
Materials
  Crystalline, 22
  mechanical behavior of, 22,
          23ff
Metal jacketed explosives, 158ff
Method
  Berg-Barrett, 70
  double-crystal, 72
  fringe, 6
  gas adsorption, 10
  Lang, 78
  parafocusing, 25
  parallel beam, 24
  topographic, 70ff
  x-ray diffraction, 3
  x-ray divergent beam, 90
Micro defects, 82
Microbeam techniques, 13
Microlattice strain, 24
Microparameters, 43, 46
  changes in, 47
Microscopic
  internal stress, 338ff
  parameters, 38ff
  structure, 22
Microstrain, 24
Microstructure, 109ff
Moire fringe, 7, 392
Moire method, 385ff
  application to NDT, 385ff
Monochromatic x-ray, 234
Monochromator,
  multiple reflection, 8
  x-ray, 8

NDE, probe of damage, 479ff
NDT, image processing in, 409ff
Neutron radiography, 151ff
  camera, 155ff
  image system, 151ff
  system, 154ff
Nickel, deformation substructure,
          77
Non-Destructive XRD Testing,
          European Advances in,
          195ff

Nondestructive composition,
    analysis of, 183ff
Nondestructive distribution,
    measurements of, 230ff
Nondestructive testing, 306ff
  neutron radiography, 151ff

Optical arrangement, 365
Optical heterodyning, 356
Optical interferometry, 358
Optical methods, knife-edge
    technique, 352
Optical path interferometer,
    364
Optical processors, 412ff
Optical system, microphotometer,
    399
Optical techniques,
    comparison of, 351ff
Oscillating beam principle, 48

Parafocusing method, 25
Parallel beam method, 25
Particle size, 24
Path stabilization electronics,
    block diagram of, 366
Pendellösung
  effect, 77
  fringes, 5
Penetrant inspection,
        quantitative methods in,
    505ff
Piezoelectric strains, 13
Plastic zone size, 228ff
Polychromatic x-ray stress,
    205ff
Polyethylene, 11
Polymers, study of, 10
Portable instrumentation,
        development of, 110ff
Positrons, 479ff
  behavior of, 482ff
Projectile breakup, flash
        radiographs of, 132
Propagation, fatigue crack,
    44ff, 52
Proteins, 11
Pulse-echo intensity, angular
        dependence,
    267

Quadrature dual interferometer,
    362
Quantum chemistry, 3

Radiography, flash x-ray, 130ff
Reflection,
  Bragg, 60, 70
  hkl, 12
Regression equations, 188ff
Research, ultrasonic NDE, 257ff
Residual stress, 36, 209ff
  analysis, 1, 12ff
  and fatigue, 211ff
  distribution, 236ff, 240ff
  due to welding, 215ff
  measurement, 101ff
  techniques for measuring, 102
Rotating-anode generators, 11

S-N curve, 30
Scattering, small-angle, 1, 8ff,
    10
Setting, Laue-transmission, 78ff
Shock wave experiments, 141
Silicon crystal, 5
Size effect on fatigue life, 31
Small-angle scattering, 1, 8ff,
    10
Solid-state electronics, 3
Spacing
  interplanar, 102
  stress, 102
Spark gap, 123ff
  triggerable, 124
Spectral dispersion, 186ff
Spectroscopy
  transducers for, 302ff
  ultrasonic, 299ff
Stainless steel,
  machined type, 236ff
Steel, fatigue of, 30
Strains, piezoelectric, 13
Stress analysers, Wagon type
        x-ray, 25
Stress constant, 105
Stress distribution, 205
  residual, 240
Stress spacing, 102
Stress
  axial residual, 51

Stress (cont'd)
  Barkhausen noise, 461
  fading of residual, 36
  fatigue, 36
  loading, 205ff
  macroscopic, 323ff
  microscopic internal, 338ff
  polychromatic x-ray, 211ff
  residual, 209ff
  residual and fatigue, 211ff
  surface residual, 36
Stresses in metals, acoustic
        interaction, 321ff
Structural materials, 479ff
Structure factor, 6
Structure, microscopic, 23
Sub-grain size, 24
Submacroscopic parameter, 25ff
Surface
  condition, 109ff
  residual stress, 36
  study of, 10
  work hardened layer, 239ff
Swept path interferometer, 363
Synchrotrons, 11
System, film-densitometer, 24

Techniques
  microbeam, 13
  x-ray diffraction, 23ff
Television-analog, 412ff
Test
  impact strength, 63
  ultrasonic angle beam, 312
  ultrasonic harmonic, 338ff
Texture, 108ff
Theory
  dynamical, 1, 4ff
  kinematical, 1, 2ff
Thin crystals, 79
Topographic methods, 70
Topography, 69
  high-speed x-ray, 93
  reflection, 73
  transmission, 79
  x-ray, 69ff, 80, 81, 90
Transducer
  design and application, 287ff
  electromagnetic,
        323

Transmission electron microscopy,
        83
Transmission pattern, 92
Transmission powder pattern, 146
Transmission topography, 79, 81
True strain, 48

Ultrasonic angle beam testing,
        312
Ultrasonic detection, of
        delaminations, 312
Ultrasonic harmonic generation,
        338ff
Ultrasonic NDE research, 257ff
Ultrasonic spectroscopy, 299ff,
        313
Ultrasonic transducers,
        electromagnetic, 283ff

Vertical slit height, 9

Wagon type x-ray, 25
Wavelength dispersion, geometry,
        186
Wing tank corrosion, 171ff
  in DC-9 aircraft, 171ff

X-ray
  beam
    fivefold-reflected, 9
    singly-reflected, 9
  detectors, 11ff
  diffraction
    applications of, 1
    methods, 3
    techniques, 23ff
  diodes, 128
  divergent beam methods, 90
  elastic constants, 203ff
  fatigue damage by, 21ff
  flash, 117ff
  fractography, 241ff
  interferometer, 7
  line broadening, 249ff
  line profile analysis, 92
  microcamera, 39
  microbeam Debye analysis, 40
  microbeam technique, 48
  monochromatic, 234
  Pendellösung fringes, 83

X-ray (cont'd)
  residual stress, 101ff
  spectra, 128
  Stress Analysis, Basic
        Principles, 197ff
  study, 44
  topography, 1, 69ff, 80, 92
    high-speed, 93